Opportunistic Networking

Vehicular, D2D, and Cognitive Radio Networks

Opportunistic Networking

Vehicular, D2D, and Cognitive Radio Networks

Edited by

Nazmul Siddique
Syed Faraz Hasan
Salahuddin Muhammad Salim Zabir

CRC Press is an imprint of the
Taylor & Francis Group, an **informa** business

CRC Press
Taylor & Francis Group
6000 Broken Sound Parkway NW, Suite 300
Boca Raton, FL 33487-2742

Printed on acid-free paper

International Standard Book Number-13: 978-1-4665-9696-2 (Hardback)

Visit the Taylor & Francis Web site at
http://www.taylorandfrancis.com

and the CRC Press Web site at
http://www.crcpress.com

Contents

Editors

Nazmul Siddique is a lecturer in the School of Computing and Intelligent Systems, University of Ulster, Coleraine, UK. He obtained his Dipl.-Ing. degree in cybernetics from the Dresden University of Technology, Germany; his MSc in computer science from the Bangladesh University of Engineering and Technology, Dhaka; and his PhD in intelligent control from the Department of Automatic Control and Systems Engineering, University of Sheffield, England. His research interests include cybernetics, computational intelligence, bio-inspired computing, stochastic systems, vehicular communication, and opportunistic networking. He has published over 150 research papers, including four books published by John Wiley, Springer, and Taylor & Francis (in press). He guest edited eight special issues of reputed journals and co-edited seven conference proceedings on Cybernetic Intelligence, Computational Intelligence, Neural Networks, and Robotics. He is a fellow of the Higher Education Academy, a senior member of IEEE, and a member of different committees of the IEEE—the SMC Society and the UK-RI Chapter. He is on the editorial board of seven international journals.

Syed Faraz Hasan is with the School of Engineering and Advanced Technology at Massey University, Palmerston North, New Zealand, where he leads the Telecommunication and Network Engineering research group. He also sits on the Prime Minister's Chief Science Advisor's panel on Science Policy Exchange. He holds a PhD from the University of Ulster, Coleraine, UK, and a bachelor's degree in engineering (with distinction) from the NED University of Engineering and Technology, Karachi, Pakistan. He frequently reviews book proposals and journal and conference articles for reputed publishers. His main expertise is in wireless networks with more specific interests in 5G networks and device-to-device communication. His ongoing research has appeared in several newspapers including those in New Zealand, India, and the United States.

Salahuddin Muhammad Salim Zabir is a professor at the National Institute of Technology, Tsuruoka College, Japan. He has been involved in research and development of computer networks, IoT, ICT for assisting seniors, machine to machine (M2M), smart cities, e-health, wellness, and disabilities. He earned his PhD and

MS degrees in information science, both from Tohoku University, Sendai, Japan. Before that, he obtained his MSc Engineering and BSc Engineering degrees in computer science and engineering from Bangladesh University of Engineering and Technology, Dhaka. Prior to his current appointment, he served at Tohoku University; Kyushu University, Fukuoka, Japan; Kyung Hee University, Seoul, Korea; and Bangladesh University of Engineering and Technology. He also worked for the French telecom operator Orange at their R&D Labs in Japan and Panasonic R&D headquarters in Osaka, Japan. His research interests include computer networks, network protocols, disaster networks, performance evaluations, IoT technology and applications, ubiquitous computing, smart cities, e-health, and applications of ICT in emerging countries. Dr. Zabir has served on the program/technical committees of various international conferences and is guest editing special issues of scholarly journals. He is a senior member of the IEEE and BCS.

Contributors

Abdullah Alghamdi
Department of Software Engineering
College of Computer and Information Sciences
King Saud University
Riyadh, KSA

Stuart M. Allen
School of Computer Science &
Informatics Cardiff University
Cardiff, UK

M. Anwar Hossain
Department of Software Engineering
College of Computer and Information Sciences
King Saud University
Riyadh, KSA

Jean-Sébastien Bilodeau
Laboratoire d'Intelligence Ambiante pour la Reconnaissance d'Activités (LIARA)
Université du Québec à Chicoutimi (UQAC)
Chicoutimi, Quebec, Canada

Ali Bohlooli
Faculty of Computer Engineering
University of Isfahan
Isfahan, Iran

Bruno Bouchard
Laboratoire d'Intelligence Ambiante pour la Reconnaissance d'Activités (LIARA)
Université du Québec à Chicoutimi (UQAC)
Chicoutimi, Quebec, Canada

Kevin Bouchard
Laboratoire d'Intelligence Ambiante pour la Reconnaissance d'Activités (LIARA)
Université du Québec à Chicoutimi (UQAC)
Chicoutimi, Quebec, Canada

Abdenour Bouzouane
Laboratoire d'Intelligence Ambiante pour la Reconnaissance d'Activités (LIARA)
Université du Québec à Chicoutimi (UQAC)
Chicoutimi, Quebec, Canada

Niaz Chowdhury
Knowledge Media Institute
The Open University
England, United Kingdom

Dany Fortin-Simard
Laboratoire d'Intelligence Ambiante pour la Reconnaissance d'Activités (LIARA)
Université du Québec à Chicoutimi (UQAC)
Chicoutimi, Quebec, Canada

Sebastien Gaboury
Laboratoire d'Intelligence Ambiante pour la Reconnaissance d'Activités (LIARA)
Université du Québec à Chicoutimi (UQAC)
Chicoutimi, Quebec, Canada

Abdolbast Greede
University of Zawia
Zawiya, Libya

Syed Faraz Hasan
School of Engineering and Advanced Technology
Massey University
Palmerston North, New Zealand

S. M. Kamruzzaman
Department of Electrical and Computer Engineering
Ryerson University
Toronto, Canada

Sarang Karim
Institute of Information & Communication Technologies
Mehran University of Engineering and Technology
Jamshoro, Pakistan

K. K. Pattanaik
Department of Information and Communication Technology
ABV-Indian Institute of Information Technology and Management
Gwalior, India

Faisal Karim Shaikh
Institute of Information & Communication Technologies
Mehran University of Engineering and Technology
Jamshoro, Pakistan

and

Science and Technology Unit
Umm Al-Qura University, Makkah, KSA

Yoshitaka Shibata
Faculty of Software and Information Science
Iwate Prefectural University
Takizawa, Iwate, Japan

Norio Shiratori
Global Information and Telecommunication Institute
Waseda University
Tokyo, Japan

Nazmul Siddique
School of Computing and Intelligent Systems
Ulster University
Coleraine, UK

Anshul Verma
Department of Computer Science and Engineering
National Institute of Technology Jamshedpur
Jharkhand, India

Stefan Weber
School of Computer Science and Statistics
Trinity College Dublin
Dublin, Republic of Ireland

Salahuddin Muhammad Salim Zabir
National Institute of Technology
Tsuruoka College
Tsuruoka, Japan

Chapter 1

Introduction

Syed Faraz Hasan, Nazmul Siddique, and
Salahuddin Muhammad Salim Zabir

Contents

1.1 Introduction

The concept of interconnecting computer systems began in the 1970s when a few academic institutes in the United States wanted to share data. This culminated in the development of Advanced Research Projects Agency Network (ARPANET), which connected four academic institutions together. The successful deployment and operation of ARPANET gave rise to limitless discussion on how information can be shared between devices. Topics pertinent to the architecture of a network, its operation, the number and type of interconnected devices, data type, and so on were brought into the discussion. Issues related to the type of architecture were particularly interesting. Two of the earliest architectural designs were infrastructure-based and ad hoc networks [1]. Infrastructure-based networks have a central entity, generally the base station, which governs communication between all participating devices. In contrast, in ad hoc networks, devices communicate with each other directly. Because there is no central authority in ad hoc networks, a variety of research issues emerged that required careful consideration. The number and nature of these issues progressed with the introduction of wireless communication services. On top of everything else, devices can now move from one place to another while still staying connected to the network. An ad hoc network that gives its nodes freedom from wires and supports mobile nodes is known as a *mobile ad hoc network* (MANET) [2].

1.2 Mobile Ad Hoc Networks

By definition, a MANET is a group of devices that relays data from one device to another over several hops until the data reaches its intended destination. Figure 1.1 shows a typical scenario that uses mobile nodes to convey data over a multihop link. MANETs continue to play an important role in present-day communication systems. They are particularly popular in applications that use sensors. Freedom from wires and from a central managing entity makes MANETs an ideal choice for distributed environments. Several communication technologies can be used on such a distributed architecture. For instance, IEEE 802.11 Wi-Fi and IEEE 802.15.4 ZigBee have been extensively used in ad hoc network deployments. Another useful feature of MANETs is that the nodes can commence communication very quickly [3]. Due to this advantage, MANETs work well in emergency communication where access to the network infrastructure has been compromised.

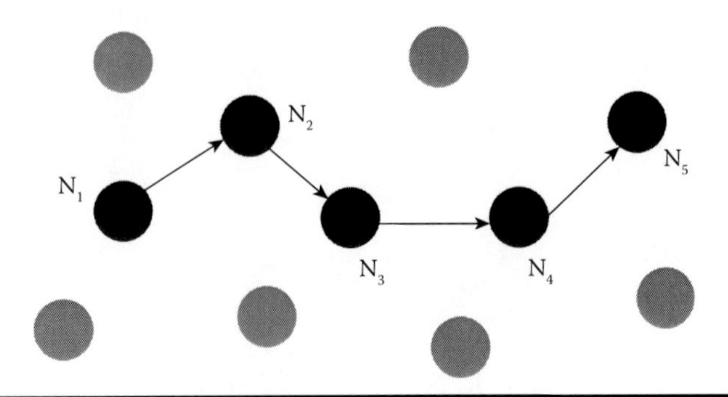

Figure 1.1 A MANET comprising 11 nodes. Node 1 uses Nodes 2–4 as relays to convey information to Node 5.

Several disaster recovery programs make use of MANETs to provide communication services in emergency environments.

Because the devices forming a MANET are mobile, the network architecture (or topology) is dynamic. The architecture changes with the mobility of the participating devices. It is intuitive that the greater the device mobility, the quicker the change in network architecture. This rapidly changing architecture poses considerable challenges in terms of protocol design. Legacy communication protocols, for example Transmission Control Protocol (TCP) and User Datagram Protocol (UDP), were developed for a network with fixed topology. It is therefore imperative for the success of MANETs to develop new protocols that can withstand a sudden change in network architecture.

The main discussion in this book starts with a detailed overview of MANETs in Chapter 2. Several important aspects of MANETs, their operation, associated challenges, and application scenarios are discussed in detail in Chapter 2. Giving due attention to the issue of protocol design for a rapidly changing topology, a new protocol known as *Beatha* is also proposed in Chapter 3. The internal and external delays caused by Beatha and its features related to flow control and retransmission are highlighted. After thoroughly defining the principles and operation of Beatha, its performance is evaluated using Optimized Network Engineering Tool (OPNET) for a variety of static and mobile scenarios. It is shown that Beatha outperforms TCP and its variants in terms of throughput, retransmission count, and time delay. Although Beatha is a routing protocol that is meant for mobile scenarios, we discuss the routing issues in highly mobile scenarios later in Chapters 4 and 5 in this book.

Nevertheless, special-purpose routing protocols such as Beatha keep communication services running even when there is a change in network topology. In certain situations, however, a change in network architecture is so sudden and frequent that communication services run the risk of getting suspended anyway. Such time-critical MANETs are known as *opportunistic networks* [4]. It must be noted that the notion of opportunistic communication is not limited only to MANETs, as highlighted in the following.

1.3 Opportunistic Networking

In opportunistic networks, devices communicate with each other without having a dedicated end-to-end communication path. It is not necessary to know the communication path between the source and its intended destination in a typical opportunistic network. The concept is very similar to the delay-tolerant networking (DTN) paradigm, which also does not require a dedicated path between the source and destination [5]. In an ideal opportunistic network, the network topology may vary drastically, having little or no impact on the quality of service received by the users. The end user receives a high-level impression of being always connected to the network services. In essence, an opportunistic network offers a series of connections one after another. These opportunistic connections can be enabled by various wireless communication technologies. Dedicated wireless communication technologies that could work well in opportunistic networks are not well known in the literature. Most of the works consider the use of existing wireless technologies in opportunistic environments. These include infrastructure-based as well as ad hoc technologies.

The opportunistic scenarios often emerge because of high node mobility. When a mobile node is moving at high speeds, it often faces periods of connectivity and disruption. An interesting example of an opportunistic communication scenario is fast-moving vehicles that try to access sparsely deployed roadside infrastructure [7]. These vehicles give rise to vehicular ad hoc networks (VANETs), which are very similar to MANETs except that they have higher node mobility. This book gives a detailed account of VANETs in the context of opportunistic networking in Chapter 4. Chapter 5 provides a detailed account of opportunistic networking and its associated challenges. The chapter starts with a discussion on wireless technologies that may suit the intermittent nature of opportunistic networks. A detailed account of how data is forwarded in opportunistic networks is given. Several protocols and their underlying principles to support intermittent data transfer are also highlighted in Chapter 5. Because there is an intermittent path between the source and destination, it is important for the network to be able to store some information. Memory elements called *repositories* are often employed to hold communication sessions in case a connection is suddenly lost [6]. Chapter 5 also covers a detailed discussion on repositories and their placement within an opportunistic network. Towards the end, Chapter 5 reports some interesting simulation results that may be useful for designing an opportunistic network.

1.3.1 Opportunistic Vehicular Communication

Vehicular communication is the information exchange between vehicles, as well as between vehicles and the roadside infrastructure. The information that these vehicles share may be related to on-road safety services or other infotainment services. Vehicular communication is an essential part of the intelligent transportation systems project, which focuses on improving passenger safety on roads and highways [8].

The notion of opportunity in vehicular communication stems from the fact that vehicles change their position very rapidly, and hence their ability to send data to their immediate neighbors also changes drastically. The length of time for which the vehicles are within each other's coverage range is an "opportunity" for them to exchange data. At two different instants in time, a vehicle may have completely different neighbors because of high mobility. As can be seen from Figure 1.2, vehicle V_1 has taken a turn that resulted in it losing the opportunity to communicate with its previous neighbors. Of course, it will encounter other vehicles on the way and become a part of another VANET.

Chapter 4 of this book starts by defining VANETs and comparing them with MANETs. It then explores various wireless technologies that may be used for

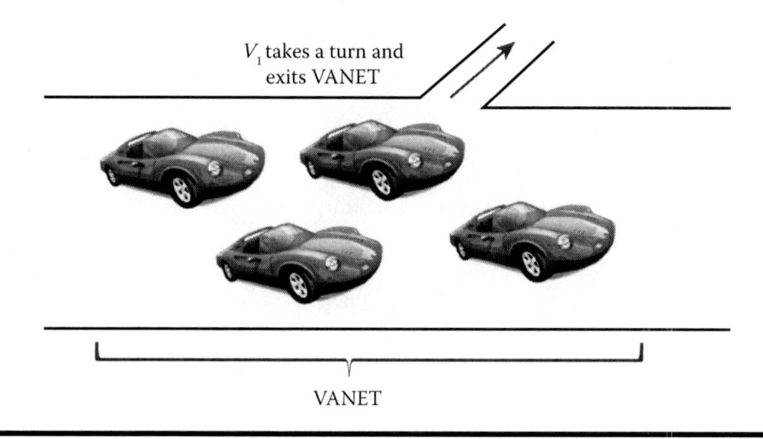

Figure 1.2 The concept of opportunity in vehicular communication.

enabling vehicular communication. Among several technologies that may be used, IEEE 802.11 has been used extensively in the literature. Therefore, a detailed account of this wireless technology and its application in vehicular environments is provided. The IEEE has standardized a modified version of legacy 802.11 for use in vehicular communication. It is known as 802.11p [9] and is discussed in Chapter 4. This chapter also outlines several ongoing projects around the world that explore the use of vehicular communication for passenger safety and infotainment services. Several other issues including handovers, routing, and predictive methods for determining the most probable path of a certain vehicle are also highlighted in this chapter. Among these issues, routing is one of the most important. This is because without customized routing protocols it is impossible to exchange information within the network.

1.3.2 Routing in Opportunistic Networks

In an ad hoc network, routing is concerned with delivering information from source to destination over multiple hops [10]. Routing becomes a big challenge in networks that rapidly change their topology. It has been pointed out that a rapidly changing topology is one of the main characteristics of opportunistic environments. A changing topology would mean that the number of hops between source and destination, as well as their spatial positions, will continue to change. Therefore, designing effective routing protocols for an opportunistic network requires careful consideration of a variety of factors [11]. Routing lies at the heart of a communication network, without which data exchange is not possible. Due to its significance, we provide a detailed discussion on routing in Chapter 5 with special regards to opportunistic networking environments.

The discussion in Chapter 5 begins with an introduction to routing in opportunistic environments. Routing in simple terms is the process of finding forwarding nodes that can relay information between a source and its destination. Chapter 5 highlights how context information is very helpful in making important routing decisions.

Context information may include users' behaviors, their mobility patterns, network demands, and so on. Context information can be useful for selecting an appropriate set of forwarders to achieve better routing services in a rapidly changing environment. Based on the use of context information, Chapter 5 classifies routing protocols into three categories and comparatively analyzes them. Several examples are drawn from each category to explain the main concept behind routing and its use of context information. Towards the end, this chapter introduces a few recent projects that aim to analyze the routing issues pertinent to opportunistic networking scenarios.

The end of Chapter 5 also concludes the first half of this book, which describes the basic theory and operation of new networking paradigms. The second half of this book is dedicated to exploring the application of these concepts in a number of scenarios.

1.4 Applications

The use of state-of-the-art communications technology can help raise the standard of living for people across the world. Following this theme, the second half of this book considers the use of technology in a number of daily life setups. The book discusses the notion of smart environments in the context of our homes, cities, hospitals, and so on. Smart agriculture is also discussed in this half of the book, as a large number of countries are driven by their agricultural products. The topics covered in the second half are outlined in the following subsections.

1.4.1 Smart Environments

Smart environments first emerged in the form of automatic and remote controlling of different appliances. An early example of a "smart" environment—probably the most common example—had automatic lights with remote controlled entry and exit. By definition, smart environments consist of an intelligent agent (or group of agents) that respond to different events [12]. A general diagram depicting some of the entities of a typical smart environment is shown in Figure 1.3. The figure will become more populated with advances in technology; at this point it serves to define the interconnection between different smart entities. With advances in technology, smart environments have become increasingly complex but also increasingly reliable. The underlying principles of the application of smart environments stem from the improvements in sensing technology and short-range communications [13].

Complete system designs that convert an ordinary environment into a smart environment have been proposed and tested [14]. The main idea is to enable communication between all user appliances and create a central portal from which everything can be controlled. For example, a user can switch on a personal heater using a web application when it is getting cold. The heater can even be automated to start itself if the concerned sensor feels a drop in temperature. Smart environments

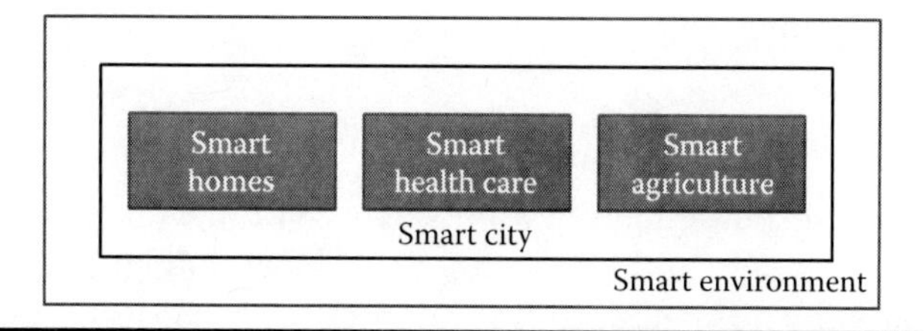

Figure 1.3 A conceptual block diagram describing the hierarchy of smart entities.

can provide a number of services, the most common of which is localization. A number of daily-life setups as well as emergency scenarios need to know a person's position. Therefore, the bulk of research focuses on the challenges of localization within smart environments [15].

Chapter 6 of this book examines different techniques for providing positioning services in an indoor environment. The authors of this particular chapter take into account trilateration and other proximity-based localization services. However, the main point of discussion centers on passive Radio Frequency Identification (RFID) technology [16] and its use in positioning. The discussion on RFID is backed with experimental results that endorse the use of this technology in smart environments. Although RFID seems to be a good technology for localizing a device (or a person), it still faces a number of issues. For example, the main parameter of interest in RFID techniques is received signal strength intensity (RSSI). RSSI itself is a highly fluctuating parameter that varies almost unpredictably [17]. Using such an unpredictable variable to provide reliable services is a huge research challenge that is well covered in Chapter 6.

1.4.2 Smart Homes

A smart environment is composed of a number of smart entities, one of which is a smart home. A smart home offers a residential setting with automated appliances that can react to changes in their surroundings [18]. Designing a smart home is a good starting point towards designing a smart environment. Chapter 7 of this book focuses entirely on smart homes and more specifically on their design.

The authors of Chapter 7 have considerable experience in designing smart homes, which they put to use to explore good design practices. The design practices highlighted in Chapter 7 can be extended to other smart environments as well. A good design is generally centered on optimizing a predefined set of requirements. For example, a smart home intended for offering more comfort to residents will have completely different specifications than one designed for energy-efficient living. Designing a smart home, regardless of its purpose, essentially has two design tasks: one related to hardware (e.g., which sensor to install, etc.) and the other related to software (e.g., when will a sensor actuate what to do). Chapter 7 covers reflections from both hardware and software standpoints. It also presents a case study of a purpose-built intelligent home infrastructure.

1.4.3 Smart Cities

A typical smart city is envisaged as a system that ensures better living standards by benefitting from a set of components working together. A smart city is often a software platform that establishes links between different social services, for example health care, agriculture, and so on, as shown in Figure 1.3. Hence, the "smartness" of a city is linked with the smartness of its constituent components. Some of the most common constituents of a smart city include its citizens, economy, environment, communication infrastructure, and so on. Because a smart city is an amalgamation of a number of subfunctions, the communication technologies that link these functions together are of key importance. Strides in the world of telecommunication help enable the implementation and planning of smart cities. For example, as highlighted in Ref. 19, key enablers like IPv6 are important for a plausible implementation plan. Of course, there are a number of issues that still require careful consideration, for example, standardization and specification.

An important constituent of the smart city is smart agriculture, which is the focus of the following section.

1.4.4 Smart Agriculture

Given that agriculture is the driving force for a number of countries across the world, its effective use is crucial for the success of various world economies. Interesting milestones achieved by electronic and wireless technology have also affected agricultural activities. Chapter 8 of this book covers an interesting discussion on smart agriculture, its underlying principles, and good practices.

The authors of Chapter 8 start by defining the term *smart agriculture* and move on to advocating the use of wireless sensor networks (WSNs) to make it more effective. The authors emphasize that WSNs are at the heart of data collection for agricultural fields and that the data thus collected can help farmers take necessary steps at appropriate times [20]. Figure 1.4 shows a sample smart farm that has sensors spread across the land and also mounted on farm animals [21]. The sensors collect data and send it to the central management system through the gateway. The management system runs analytical functions to make observations and possibly predict any future threats (or trends). The information generated is then passed on to the user.

The authors of Chapter 8 mention some detailed applications of smart agriculture and introduce a generic architecture for such systems. The chapter then provides some guidelines for deployment of smart agriculture solutions and how existing hardware from different manufacturers could be deployed for building such a system. The authors also mention some examples of existing solutions that adopt different approaches to the domain and make projections about the future of smart agriculture.

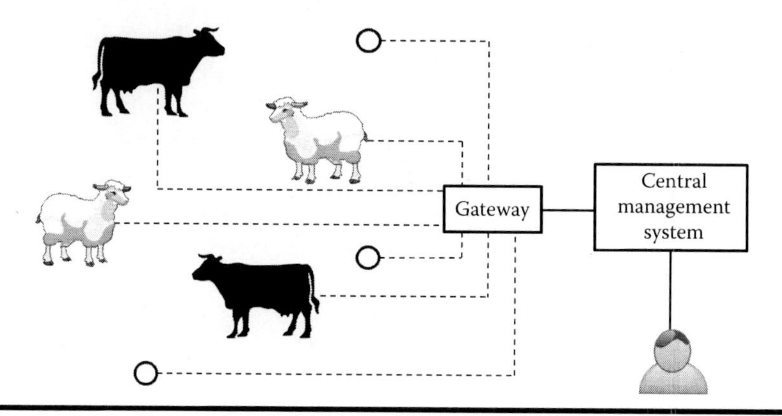

Figure 1.4 A smart farm that comprises sensors deployed on land and mounted on animals. All sensors send data to the central server via the gateway, where it is processed.

1.4.5 Smart Health Care

Continuing with the discussion about smart environments, Chapter 9 of this book examines the use of technology in the health care sector. In Chapter 9, the authors introduce the concept of cognitive radio networks (CRNs) [22] and demonstrate its application in hospitals using a case study. Other applications of CRNs, for example in military and homeland security, are also highlighted.

Whereas all wireless networks use a fixed frequency spectrum to exchange data, cognitive networks search for a spectrum hole and use that white space for wireless transmission (and reception). Users of CRNs are referred to as *secondary users*, in that they do not own the spectra that they use. The authors argue that CRNs can be employed in body area networks (BANs). A BAN is the wireless interconnection of a number of medical devices that monitor the state of a patient often in a critical condition [23]. The current state of BANs uses unlicensed technologies, for example Wi-Fi and Bluetooth, which are prone to interference from other users. The authors of Chapter 9 propose that short-range communication between medical devices can be managed using a CRN. The idea is that the devices would determine the best possible spectral space and transmit data over that space.

1.4.6 Emergency Communications: An Introduction to Never-Die Networks

Chapter 10 is an invited chapter on never-die networks (NDNs) contributed by the original proposers of the concept [24]. As experienced in the March 2011 earthquake in Japan, big natural disasters may cause the breakdown of network communication infrastructure. This leaves the people in the affected areas isolated from the rest of the world and also from their family, friends, and peers. As a result, important information—particularly regarding potential precautions against further deterioration of the situation, evacuation, and rescue—cannot be exchanged. NDNs propose a framework for offering minimal connectivity to the general population in such situations. Such a network could ensure minimal or short-term exchange of information for a large number of people simultaneously in a disaster scenario. As explained by the authors, the design of an NDN has the following features [25]:

- An NDN should provide real-time services.
- The number of users connected to an NDN is very large.
- Data exchange is meant to support services of short duration only.

One distinguishing aspect of Never Die Network is that in order to ensure connectivity during disaster situations when the whole or part of the communication infrastructure has been damaged, it proposes a layered communication architecture comprising wired infrastructure as well as wireless connectivity including WiFi, balloon mounted base stations, ship mounted base stations, satellite links etc,. However, NDN is not only about hardware. Rather, it also incorporates software, communication methods, and protocols. As such, the concept of Never Die Network includes the concepts of both Network Fault Tolerance System and Resilient Overlay Network.

In addition to the conceptual background, the authors of this chapter describe their experience of the situation after the Great East Japan Earthquake and Tsunami in 2011. They also provide an outline of ongoing research activities and experiments on NDNs.

1.5 The Road Ahead

Chapter 2 commences the discussion on basic working and principles of ad hoc networks. Issues such as routing in MANETs and the fundamentals of opportunistic communication are covered. Readers who already have a background in these topics can skip ahead to Chapter 6, which covers the application part of this book. The use of communication technology in several important sectors of daily life is highlighted in the second half of this book.

References

1. A. Zemlianov and G. de Veciana, Capacity of Ad Hoc Wireless Networks With Infrastructure Support, *IEEE Journal on Selected Areas in Communications*, 23(3), 2005.
2. P. Li, Y. Fang, J. Li and X. Huang, Smooth Trade-Offs between Throughput and Delay in Mobile Ad Hoc Networks, *IEEE Transactions on Mobile Computing*, 11(3), 2012.
3. Y. Lu, X. Li, Y.-T. Yu and M. Gerla, Information-Centric Delay-Tolerant Mobile Ad-Hoc Networks, IEEE INFOCOM Workshops, 2014.

4. Y. Ma and A. Jamalipour, Opportunistic Geocast in Large Scale Intermittently Connected Mobile Ad Hoc Networks, 17th Asia-Pacific on Communications, 2011.
5. A. Elmangoush, R. Steinke, M. Catalan, A. Corici, T. Magedanz and J. Oller, Interconnecting Standard M2M Platforms to Delay Tolerant Networks, International Conference on Future Internet of Things and Clouds, 2014.
6. M. Pitkanen, T. Karkkainen and J. Ott, Opportunistic Web Access via WLAN Hotspots, IEEE Pervasive Computing and Communications, 2009.
7. S. F. Hasan, N. H. Siddique and S. Chakraborty, *Intelligent Transport Systems*, Springer, New York, 2012.
8. S. F. Hasan, Vehicular Communication and Sensor Networks, IEEE Potentials, 2013.
9. B. Bellalta, E. Belyaev, M. Jonsson and A. Vinel, Performance Evaluation of IEEE 802.11p-Enabled Vehicular Video Surveillance System, *IEEE Communications Letters*, 18(4), 2014.
10. I. T. Haque, On the Overheads of Ad Hoc Routing Schemes, *IEEE Systems Journal*, 9(2), 2015.
11. Z. Zhang and R. Krishnan, An Overview of Opportunistic Routing in Mobile Ad Hoc Networks, IEEE Military Communications Conference, 2013.
12. M. Roscia, M. Longo and G. C. Lazaroiu, Smart City by Multi-Agent Systems, International Conference on Renewable Energy Research and Applications, 2013.
13. D. –M. Han and J. –H. Lim, Smart Home Energy Management System using IEEE 802.15.4 and Zigbee, *IEEE Transactions on Consumer Electronics*, 56(3), 2010.
14. O. Evangelatos, K. Samarasinghe and J. Rolim, Syndesi: A Framework for Creating Personalized Smart Environments using Wireless Sensor Networks, IEEE International Conference on Distributed Computing in Sensor Systems, 2013.
15. S. A. Kharidia, Q. Ye, S. Sampalli, J. Cheng, H. Du and L. Wang, HILL: A Hybrid Indoor Localization Scheme, 10th International Conference on Mobile Ad Hoc and Sensor Networks, 2014.
16. F. Manzoor and K. Menzel, Indoor Localization for Complex Building Designs using Passive RFID Technology, 30th URSI General Assembly and Scientific Symposium, 2011.
17. S. F. Hasan, N. H. Siddique and S. Chakraborty, Extended MULE Concept for Traffic Congestion Monitoring, 63(1), 2012.
18. C. Yang, B. Yuan, Y. Tian, Z. Feng and W. Mao, A Smart Home Based on Resource Name Service, IEEE International Conference on Computational Science and Engineering, 2014.
19. J. V. den Bergh and S. Viaene, Key Challenges for the Smart City: Turning Ambition into Reality, International Conference on System Sciences, 2015.
20. F. Mehdipour, Smart Field Monitoring: An Application of Cyber-Physical Systems in Agriculture (work in progress), International Conference on Advanced Applied Informatics, 2014.
21. K. Taylor, C. Griffith, L. Lefort, R. Gaire, M. Compton and T. Wark, Farming the Web of Things, IEEE Intelligent Systems, 2013.
22. S. Sengupta, J. Jay and K. P. Subbalakshmi, Open Research Issues in Multi-Hop Cognitive Radio Networks, IEEE Communications Magazine, 2013.
23. P. Honeine, F. Mourad, M. Kallas, H. Snoussi, H. Amoud and C. Francis, Wireless Sensor Networks in Biomedical: Body Area Networks, 7th International Workshop on Systems, Signal Processing and their Applications, 2011.

24. N. Uchida, K. Takahata, Y. Shibata and N. Shiratori, A Large Scale Robust Disaster Information System based on Never Die Network, IEEE International Conference on Advanced Networking and Applications, 2012.
25. N. Uchida, K. Takahata, Y. Shibata and N. Shiratori, Evaluation of Never Die Network for a Rural Area in an Ultra Large Scale Disaster, 6th International Conference on Complex, Intelligent and Software Intensive Systems, 2012.

Chapter 2

Opportunistic Networking: An Application

Abdolbast Greede and Stuart M. Allen

Contents

Recent years have seen an explosive growth in data-hungry mobile services, including social networking, video conferencing, photos, and streaming video, motivating the design of more efficient networking technologies. Opportunistic networking (OppNet) has emerged as a potential solution; however, there remain several different research issues that need to be addressed before mainstream adoption is practical. In an OppNet, mobile nodes are able to communicate without an available end-to-end path or explicit knowledge of the network topology. Instead, mobile devices carry data until an opportunity arises to forward it to a peer via a short-range wireless link. The start of this chapter discusses wireless mobile communication in general, before giving an overview of OppNets and their applications in the context of general wireless communications. The specific problem of efficient data delivery in OppNets is then considered, addressing the limitations of current approaches and what is needed to overcome these. Finally, repository-based data dissemination (RDD) is introduced as a proposed approach to improve data delivery in practical OppNets.

2.1 Wireless Mobile Communication

The use of wireless mobile devices continues to enjoy rapid growth as a direct result of recent technological advances in both the capability of devices and wireless communication technology. These developments have led to lower prices, higher data rates, and a greater proliferation of wireless communication between mobile users. There are two fundamental approaches for enabling wireless communication between mobile devices, depending on whether they involve a central network infrastructure.

2.1.1 Infrastructure-Based Communication

In infrastructure-based communication, devices communicate with each other via a base station, which represents a key component of the network backbone (e.g., cellular network). This kind of network is limited to the regions where such an infrastructure is deployed and by the bandwidth provided by the network. The second approach is infrastructureless wireless communication, wherein mobile devices communicate with each other either directly (when a pair of devices are within mutual transmission range) or indirectly (a multihop network). Most wireless mobile devices nowadays can support both types of wireless networks, as can be seen in smartphones that are equipped with a Global System for Mobile communication (GSM) [1] (infrastructure), Wi-Fi, and Bluetooth (infrastructureless) wireless networks.

The most widely used infrastructure wireless networks are cellular. When the first-generation (1G) cellular network was commercially introduced in the United States in 1983 [1], voice communication was the main service provided. However, the great advances in wireless technologies have taken mobile devices beyond this classic use as a voice communication tool to also encompass data communication. In fact, mobile users' demand for content is increasing, driven particularly by streaming multimedia and the growth of user-generated content, which has led to cellular networks shifting toward data-oriented services. This direction is clearly seen in the third-generation (3G) cellular network standard, which has a higher data rate and data-oriented services, compared to the second-generation (2G) standard, which has a low data rate and voice-oriented services [1,2].

Although the bandwidth provided by 3G to some extent solves the problem of mobile users' data demands, continuous advances in smartphones, the fast-growing number of users for new social network applications such as Facebook and Twitter, and an increase in user-generated content (i.e., more sharing between users) have led to increased demand for data. There is thus a need for a new cellular network standard that provides a better data rate. For this reason, in October 2009, six technologies were submitted to the International Telecommunications Union (ITU) seeking approval as the fourth-generation (4G) communications standard [3]. In January 2012, two of the six technologies—namely LTE Advanced and WirelessMAN Advanced (known as *WiMAX*)—were approved as official platforms for 4G. 4G represents a big shift in data rate, which can peak at 1 Gb/s compared to 10 Mb/s for 3G. Although the 4G standard could solve the data rate problem, it still requires the mobile user to be attached to the network infrastructure to enjoy the benefits of the new standard. In addition to this infrastructure requirement, cellular network providers usually apply expensive charges for the use of their bandwidth. Although the cost problem may be solved at some stage, the infrastructure requirement is still hard to guarantee.

2.1.2 Infrastructureless Communication

There are several scenarios in which a network infrastructure could be unavailable for mobile users. First, the infrastructure may simply not exist, as seen in many developing countries or in some rural areas of the developed world. In the second scenario, the infrastructure has been temporarily disabled, as can happen during any natural disaster, such as an earthquake or tsunami. Third, mobile users may find themselves incapable of reaching the network infrastructure, even if it exists, as seen in the so-called Arab Spring countries in 2011 (Libya, Egypt, and Tunisia), where their dictators shut down the network operators, leaving infrastructureless networks as the only options for users. Because of the extensive use of the Internet by smartphone users, other problems start to appear, such as mobile phone providers running out of capacity to handle all data requests. This can be particularly problematic in a scenario such as the Olympic Games, where hundreds of thousands of people come from abroad with their smartphones and tablets to temporarily join the local network providers.

In classic infrastructureless wireless networks, mobile devices communicate with each other either directly or indirectly. In a multihop network, every device can behave as a router, forwarding packets on behalf of other devices in the network. This type of network is known as a *mobile ad hoc network* [4]. Although there are a few implementations of ad hoc networks in the military sector, the end-to-end connectivity requirements make successful implementation in the civilian sector harder.

Delay-tolerant networking (DTN) is another infrastructureless network paradigm that is capable of providing communication, relaxing the need for end-to-end connectivity [5]. This is facilitated by information being stored at nodes and subsequently shared with peers when an opportunity to forward arises. When DTN was initially introduced, the popularity and advances of mobile devices were not at the stage they are today. Therefore, DTN was not specifically designed for mobile phones but in a generic form for heterogeneous wireless mobile devices that move from one place to another and employ that movement to deliver data.

In the last 5 years, a new infrastructureless network paradigm has emerged, known as *OppNets* [6]. In an OppNet, wireless mobile devices are clearly defined as homogeneous mobile devices carried by individual people. Therefore, by distinguishing OppNets from DTNs through the mobile devices and their carrier, OppNets could be considered as a subset of DTNs. However, there is another distinguishing characteristic between OppNets and DTNs. In a DTN, the disconnection point is usually known in advance, and sometimes it is even known how long this disconnection will last. By contrast, in OppNets these occur as a consequence of an individual's movement and they do not exhibit any clearly defined patterns, which is why OppNets are called *opportunistic*. Hence many researchers see OppNet as a general form of DTN.

A recent ITU report estimated the number of mobile phone users at 6.8 billion subscriptions in 2013 [7]. The fact that the population of this planet is only 7.1 billion means that there is nearly one mobile phone for each human on the planet. The number of smartphone users is expected to continue to grow very rapidly in the coming years; conversely, the number of non-smartphone users is expected to shrink. Smartphones are often equipped with short-range wireless connectivity, making OppNets a good solution for sharing data between smartphone users.

2.2 Opportunistic Networks

OppNets share the same fundamental concept with DTNs in that neither requires the sender and recipient to be connected at the same time. This is facilitated by information being stored at nodes and subsequently shared with peers when the opportunity to forward arises. The simplest form of opportunistic networking can be described as follows. If a node has information to be sent, this node stores the information until it encounters another node, when it forwards this information to it. If the recipient node is not the target node, the recipient node stores the information, to be forwarded when it encounters other nodes. This forwarding process between encountering nodes is continuous until the information is delivered to its destination.

The OppNet uses a store-carry-forward communication paradigm. First a node stores data to be sent; it then uses its mobility to carry the data and, when another node is encountered, it forwards the data. Figure 2.1 shows an example of OppNets between people, highlighting the use of wireless communication (in forwarding the data) in addition to user mobility (in carrying the data) for data delivery.

2.2.1 Characteristics of Opportunistic Networks

Although an OppNet shares some concepts of other wireless infrastructureless networks, it has some special characteristics:

1. The network is comprised of mobile devices carried by humans.
2. The mobility patterns for the devices are the same as those for their carriers—independent and uncontrollable movement patterns.

Figure 2.1 An example of an opportunistic network (OppNet).

3. The devices comprising the network have limited resources of battery, storage, and networking.
4. Communication between devices is through the network, using only infrastructureless wireless networks.
5. All wireless devices comprising the network have short-range wireless capability with the same radio technology—that is, it is a homogeneous network.
6. Wireless devices are capable of communicating with each other directly if they are in mutual range.
7. Communication between devices that are not in mutual range does not require end-to-end connectivity. This is facilitated by using the store-carry-forward communication paradigm explained above.

From these characteristics, there are some differences between DTNs and OppNets. In DTNs the wireless device is not clearly defined so it could be a bus, airplane, spaceship, or any other wireless device, whereas in OppNets it is clearly defined as a mobile device carried by humans. In DTNs, the place of possible disconnection is usually known in advance, but OppNets are more unpredictable. Furthermore, in DTNs the duration of disconnection is usually known whereas OppNets are designed assuming the duration is more unpredictable. Therefore, if OppNets were distinguished from DTNs by the nature of the mobile device carrier, OppNets could be classified as a subset of DTNs. However, if they were characterized by the mobility pattern of the mobile device, DTNs could be seen as a subset of OppNets, because the disconnection point and time in a DTN is often known in advance. By contrast, in an OppNet the mobile devices are carried; hence their mobility is usually hard to predict. Because of the use of mobile device in OppNets, some researchers employ the term *pocket switched networks* (PSN) as seen in Refs. [8–10].

An OppNet enables users to share data among themselves without the need to worry about the cost and it doesn't need the end-to-end connectivity required by classic mobile ad hoc networks. However, the design of efficient data delivery protocols for OppNets is a complicated task due to the absence of knowledge about the topological evolution of mobile users. Although in the last few years, much research has been conducted in designing the data delivery protocol, the implication of the human aspect on the networks has not gotten enough attention. This research gives more consideration to the implications of human aspect.

2.3 Data Delivery in Opportunistic Networks

Data delivery in OppNets is a problem because of two key challenges: the unpredictable mobility patterns of mobile users and the resource constraints of mobile devices. Therefore, most of the work to date in OppNets is on the design of data delivery protocols.

Data delivery protocols can be classified into data forwarding and data dissemination protocols. This classification is based on knowledge of the destination address. In a data forwarding protocol, the data that needs to be delivered has a specific destination address; in a data dissemination protocol, the data objects that need to be delivered have no specific destination address. Instead of using the physical address, a more meaningful identifier could be used, such as the user name (as used in the Haggle Project).

Data forwarding can be defined as the process of moving data from the source node to the specific destination node. The forwarding process could be single hop or multihop to reach the destination node. Data dissemination can be defined as the process of moving data objects from the source node to other nodes that are interested in the data object.

Most of the research on OppNets has focused on forwarding protocols (also known as *routing*). Because the destination address is known in advance, the main target of a forwarding protocol is to move messages closer and closer to their destinations. Forwarding protocols usually choose the next hop toward the destination by estimating the probability of delivery to the destination node. Information such as node contact history and spatial information can be used to calculate or estimate these values.

In data dissemination protocols, the data needs to be delivered to any who are interested—that is, there is no specific destination. In this environment, the user generates data and wants to share it with others. The users who generate data are unaware of the nodes interested in their data.

2.3.1 Forwarding Protocols

The literature classifies forwarding protocols based on the amount of information used in predicting the topological change of the network [6] into *flooding-based* (also known as *dissemination-based*) and *context-based* protocols.

2.3.1.1 Flooding-Based Protocols

In flooding-based protocols, each node carrying a message forwards that message to all nodes it encounters in order to deliver the message to its final destination by a pure flooding policy. The idea behind this policy is that, in the absence of detailed information about a possible path to the destination, a message should be sent everywhere. If each node forwards the message to every node it meets, it will eventually reach the destination, but with a high overhead of unnecessary message transmissions. Protocols in this class usually yield the highest throughput and the lowest delay but suffer from high energy cost and resource consumption simply because of the high traffic overhead. Epidemic routing (EPR) [11] is an example of such a protocol.

Epidemic routing: EPR relies on the theory of epidemic algorithms [12] by passing messages between nodes as they come into contact with each other to eventually deliver messages to their destination. In EPR, each node maintains an index of all messages kept by that node, which is called a *summary vector.* When two nodes meet, they exchange summary vectors. The exchange of summary vectors is used to determine whether the source node has a message that was previously unseen to the destination. In that case, the node requests the messages from the other node. By applying this method, the messages will spread like a disease epidemic. In EPR, each message must contain a globally unique message ID to determine whether it has been previously seen. In addition to source and destination addresses, messages also contain a hop limit field, used to limit the number of hops it is allowed to travel to reach its destination. To limit message flooding in EPR, network-coding-based routing [13] has been introduced; it aims to limit message flooding by combining (encoding) blocks of data together.

2.3.1.2 Context-Based Protocols

Context-based routing exploits some information that aids in making decisions to identify suitable next hops towards the eventual destinations. Notable examples are the PRoPHET protocol [14], which exploits the observed contact history to select the next node. In other protocols, such as history-based routing protocol for opportunistic networks (HiBOp) [15], nodes use more information to choose the next hops. In this protocol, messages are forwarded to nodes that share increasing amounts of contextual data with the message destination. Other routing protocols, such as MobySpace [16] and meetings and visits (MV) [17], exploit historical information about node mobility patterns and the places nodes frequently visit.

Although the above protocols use only mobile nodes to forward messages, others such as Pan et al. [18] have investigated how the possible performance of OppNets can be improved by making use of whatever infrastructure is available. Pan et al. also looked at which conditions are necessary or useful for successful OppNet operation, and how much improvement it can provide over a purely infrastructure-based system. The infrastructure here is the Internet access point (AP). Their study concludes that there is a phase transition, as an initial deployment of APs had a significant impact on network performance. However, after a certain level, the benefits of additional AP deployments are minor and the use of the opportunistic communication system remains stable. Le and Liu [19] presented a similar approach (using infrastructure networks), showing that opportunistic routing can efficiently utilize infrastructure networks and achieve significantly higher throughput performance.

A different use of fixed infrastructure was introduced in the shared wireless infostation model (SWIM) [20]. In the SWIM model, there is only one final destination for all nodes. Suppose a node wishes to send a message to the final destination, it will deliver the message to the base station directly if within communication range; otherwise it delivers the message opportunistically to a nearby node that will eventually forward it to the base station when encountered. The base stations, which are distributed across the network region, are gateways to less challenged networks and the goal of the SWIM protocol is to deliver messages to the gateways only. In the approaches given in Refs. [18–20], the use of infrastructure networks is essential, which is not the way an OppNet should work.

2.3.2 Data Dissemination Protocols

Research on OppNet protocols has mainly focused on routing and forwarding protocols. Nevertheless, a number of approaches have been developed in recent years focusing on data dissemination. Yoneki et al. [21] presented a socio-aware protocol for data dissemination, which takes a clustering-based approach. Each cluster or community selects the closeness centrality node as a broker node. Closeness centrality is a measure of how long it will take data to spread to others in the community. The closeness of a vertex is the inverse sum of the distances to the other nodes. Socio-aware protocol supports the publish/subscribe paradigm, which usually contains three elements: a publisher who publishes messages, a subscriber who subscribes his interests to the system, and an event broker network to match and deliver the events to the corresponding subscribers. Therefore when a node has a publication, it sends it to its broker node within the community. When the broker node receives a publication, it matches it against its subscription list. If it matches, it floods the community with the publication. This operation may be performed as either a unicast or a broadcast. The socio-aware protocol needs a lot of traffic to keep up-to-date with which node is the broker node, simply because the network topology changes so often. When a broker node changes upon calculation of closeness centrality, the subscription list is transferred from the old one to the new one. Furthermore, the socio-aware protocol fails to deal with situations where mobile nodes refuse to work as brokers, which is expected to be quite common in an OppNet environment.

Another publish/subscribe paradigm approach is SocialCast, presented by Costa et al. [22]. SocialCast is based on the assumption that users with common interests tend to meet each other more often than other users. SocialCast consists of three phases: interest dissemination, carrier selection, and message dissemination. During interest dissemination, each node broadcasts a control message containing the list of its interests to its one-hop neighbors, along with a corresponding list of utility values. The utility is calculated using social information. During carrier selection, this utility is compared, for each interest, against the highest among those communicated by neighbors. If the utility of any neighbor is higher than that of the local node, it is selected as the carrier. In message dissemination, messages are forwarded to the interested nodes and the best carrier. This approach uses interest utility to select the carrier while it is ignorant of nodes willing to be carriers. Nevertheless, this approach is more advanced than the one presented in Ref. [21] because it exploits social information for message dissemination.

Boldrini et al. [23,24] present ContentPlace, which is a more refined approach using social information in data dissemination. Specifically, dissemination is driven by the social structure of network users, so that nodes store data items that are likely to be of interest to users they have social relationships with (and who, therefore, they are expected to be in touch with in the near future). Once again, this approach builds on the assumption that a mobile node is usually willing to relay message, which is arguable. Other data dissemination schemes for OppNets include that defined in the PodNet project [25], which does not exploit social information. Instead the dissemination process distinguishes two phases: during the first phase, the node satisfies the user's requests for the content of subscribed channels. When all available entries of the subscribed channels are downloaded, the device may use the remaining connection time to update and download content that it caches for public use. A different policy is used for public use, such as caching the most popular object, a less popular object, or a policy of no cache at all.

Wang and Wu [26] presented a social-tie-based information dissemination protocol. In this protocol, every mobile node maintains a tie strength table, which records social tie relationships with encountered nodes. This protocol including two phases: weak-tie-driven forwarding and strong-tie-driven forwarding. The idea of using weak ties comes from the American sociologist Mark Granovetter, who first highlighted the strength of weak ties [27]. In the weak-tie-driven forwarding phase, the message is first spread to more weak tie nodes, which leads to an intercommunity information spread. Hence, forwarding more tokens to nodes with more weak ties can increase the speed of spreading. After a while, the information will have been propagated to multiple communities. Hence, strong ties will play a more important role, which means that influential individuals (with many strong ties) can disseminate the information to many individuals in a short period. Therefore, the next stage is strong-tie-driven forwarding, which is like intracommunity information dissemination. One of the interesting things about this approach is that it runs counter to the popular argument that individuals with many strong ties catalyze information dissemination in society; this paper suggests that individuals with many weak ties are important. However, this approach is built around the assumption that all individuals are willing to spread a message to other individuals.

2.4 Repository-Based Data Dissemination System

The success of most of the proposed approaches depends on how much the mobile nodes (MNs) are willing to cooperate in facilitating data dissemination to other mobile nodes. In real-life scenarios, MNs usually are not willing to work as intermediate hops for others for the following reasons:

- Security concerns of MN carriers. Humans usually decline to offer relay services because they often associate this with some security risk. Li and Du [28] describe the types of active malicious behavior (such as supernova and hypernova) that can be present in OppNets.

- OppNets use resource-constrained mobile devices. Users usually try to make efficient use of device resources. Li and Du [28] also described other behavior that exists with MNs, such as free riders (where the MN refuses to serve as a relay for other MNs) and black holes (where the MN drops all of the relayed data without forwarding it to other MNs). MNs with free rider or black hole behavior negatively affect data transmission performance in OppNets. For example, MNs with free riders require less memory and energy (which is in the user's interest) than MNs without free riders, but the system has to bear the cost in terms of the overall data transmission performance.

For the above reasons, MNs may not be the right candidate as the exclusive relays in a practical scenario. Therefore, this section introduces RDD as a networking system aiming to provide practical data dissemination for OppNets. In RDD, the relaying of data objects to other mobile nodes is not solely reliant on mobile node cooperation. Instead, RDD uses carefully placed repositories to relay data objects. To describe the simplest form of the protocol, suppose a mobile node s wishes to send a data object to a set of other mobile nodes that are interested in this message. If s comes directly within the communication range of one of the interested nodes, it delivers the data object directly to that node. Otherwise, if s comes within communication range of a repository r, it delivers a copy of data object to r, which eventually forwards this message to interested nodes if encountered.

In RDD, a *repository* is defined as a standalone fixed wireless device with no access to fixed infrastructure that uses a short-range wireless technology such as ZigBee or Wi-Fi to communicate. Although RDD has not implemented a prototype, the repository envisioned in the model is practical, as a small wireless device, with the following characteristics:

1. It is a completely standalone wireless device and it is distinct from traditional APs, which are connected to the central network infrastructure.
2. The repository can easily be installed and moved.
3. It could be similar in hardware, size, and cost to the Intel iMote, a wireless device prototype, which was used in Ref. [8]. The iMote has been used to collect connection opportunity data and mobility statistics by some researchers at the University of Cambridge (Cambridge, UK). The iMote has an ARM processor, Bluetooth radio, flash RAM, and CR2 battery. The estimated cost of such a device is less than £100. Because the iMote is a compact device, it should be easy to install and remove.
4. The repository could use an AC power supply to eliminate power constraints.

In OppNets, mobile nodes represent two integrated parts: the mobile device and its user (as seen in Figure 2.2). As such, the data delivery process in OppNets exploits the wireless communication of mobile devices as well as user mobility; hence, the characteristics of the carrier's mobility play an important role

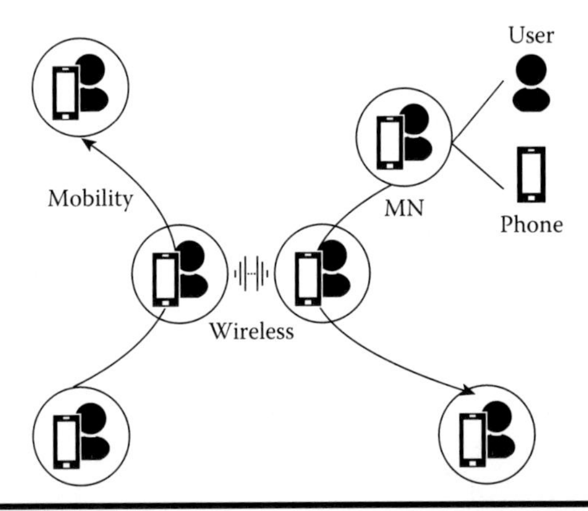

Figure 2.2 Two integrated components of MNs in OppNets.

in data delivery. Although the majority of existing protocols have focused on the development of the mobile devices themselves, the characteristics of user mobility patterns have not received the same attention. User mobility is characterized by a high degree of temporal and spatial regularity, where encounters between peers exhibit a regular structure. Based on this fact, RDD utilizes these characteristics to find the most commonly used routes or locations in the target environment to place the repository, to work most effectively as relay nodes.

Although the mobile nodes in OppNets are independent identity of nature, many protocols use a clustering-based approach where a special node in the cluster plays the major role in the dissemination process between other nodes (see Ref. [21]). In RDD, nodes operate completely independently in terms of which data objects to disseminate and which devices to disseminate them to. In the push-based dissemination system, nodes neither advertise their data object's availability nor send requests for their needs to other nodes. Instead, RDD employs only node preferences to deliver data objects to their interested users.

2.4.1 Repository-Based Data Dissemination Design Assumptions

The assumptions underlining the design of RDD are as follows:

1. Mobile nodes are carried by individuals walking in environments constrained by obstacles, such as a city. This implies that there are areas more commonly visited.
2. Mobile nodes are not aware of their absolute geographical location. This assumption is based on currently available location technologies, where keeping track of the position of a device (e.g., via GPS) requires a prohibitively high cost in battery resources.

3. Mobile nodes do not maintain lists of other nodes or their preferences—that is, apart from during encounters, each mobile node is completely unaware of who requires individual data objects and this information is not stored for later use.
4. The target environment of the network is the size of a university campus or region of a city.
5. Data objects are relevant over the period of a single day, and it is assumed that the buffer size is unlimited.

2.4.2 Repository-Based Data Dissemination Architecture

RDD is composed of four parts. The first part is data object management, which is responsible for managing data objects. The second part is user preference management, which is responsible for managing users' subscriptions to channels. The third part is data dissemination, which is responsible for pushing data objects to intended mobile devices. The fourth part is repository placement, which is responsible for placing the repository in the most suitable place. The following subsections discuss these parts in details.

2.4.2.1 Data Object Management

Mobile nodes form the backbone for an OppNet. Here, the characteristics and nature of mobile nodes used in RDD are presented before describing how mobile nodes manage data objects and channel subscription.

Mobile nodes in OppNets are composed of two parts: the mobile wireless device itself and the carrier of the device. The nature and characteristics of these parts are now discussed in detail.

Wireless devices: The type of mobile wireless device envisaged in this study is a typical smartphone or tablet with short-range wireless radio connectivity. In this research, the approach is not restricted by the type of wireless technology used; however, all devices are assumed to be equipped with the same technology (i.e., a homogeneous network). This technology could be Bluetooth, ZigBee, Wi-Fi, or any other short-range wireless technology operating in unlicensed bands. The mobile wireless devices are usually small. The device size, portable nature, and user lifestyle place a number of constraints on these devices. Managing these constraints is the major challenge in designing and implementing data dissemination protocols for OppNets. The following points summarize these constraints.

1. *Limited memory storage:* Most mobile devices have limited storage, which has a strong impact on their capability to store data on behalf of other mobile nodes.
2. *Limited battery life:* Batteries are also important for the mobility and portability of mobile devices. Batteries run down quickly in most handheld devices.

Therefore, adding traditional opportunistic networking where devices are required to work as relays for other nodes will add an extra load to the batteries of these devices.

3. *Limited wireless range:* The nature of the wireless technology used in OppNets is short range, usually stretching from 10 to 100 meters.

Device carriers: In the definition of OppNets used in this chapter, the mobile devices are carried by people; therefore, the device carrier has all the mobility characteristics of human movement. Because OppNets exclusively employ device mobility, in addition to wireless media, to achieve data communication between devices, the carrier's movement plays an important role in data communications. RDD exploits the characteristics of human mobility in choosing the repository location, summarized as follows:

1. Carrier mobility patterns show a high degree of temporal and spatial regularity.
2. Carriers choose destinations based on their personal preferences.
3. Carriers take the shortest route to reach their destinations.
4. Carriers travel along selected pathways, meaning they have the ability to avoid environmental obstacles and share the paths or locations with others.

In the design, the data objects are small in size, such as short text messages, small pictures, or brief video clips. One interesting aspect of smartphones currently is the ability to create such data objects easily. In RDD, it is assumed that most data objects injected into the networks are created by individual mobile nodes. Data objects could be generated by other parties, but mobile nodes are the only parties responsible for injecting data objects into the network.

Each node maintains a buffer consisting of all the data that it holds. When a mobile node m creates a new data object, this data object is added into the buffer. Assume that each data object fits into one packet. Each data object packet d is composed of two parts (Table 2.1): the data to be delivered (or *payload*) and a header composed of five fields, as follows:

1. $c(d)$: Channel of the data object d, which classifies the content and is discussed in detail in the next section.
2. $m(d)$: Source node.
3. $Id(d)$: Message sequence number, as created by the source node. Hence each message is defined by a unique pair <source node Id, message Id>.
4. $ct(d)$: Data object creation time.
5. $ttl(d)$: Time to life for data object d.

Let $D_t(m)$ be the set of all data stored at a mobile node m at time t. Define $D_t^{wt}(m)$ as a data object waiting to be sent by node m at time t and $D_t^{rc}(m)$ as data objects

Table 2.1 Packet Structure for a Data Object (*d*)

Header					*Payload*
Id(d)	*c(d)*	*m(d)*	*ct(d)*	*ttl(d)*	Data object content

that were received by node m before time t; hence $D_t(m) = D_t^{wt}(m) \cup D_t^{rc}(m)$. Let $D_t(r)$ be the set of all data stored at a repository node r at time t and $D_t(r) = D_t^{wt}(r)$, where $D_t^{wt}(r)$ are data objects waiting to be sent by node r at time t.

To keep buffer size low, RDD removes any data objects that reach their expired time. At time t, the expiration time of d (denoted as $x(d)$) is defined in Equation 2.1.

$$x(d) = ct(d) + ttl(d) \tag{2.1}$$

Note that the buffer may contain more than one data object for a given subscription channel. As there is a limit on how many messages can be passed in each interaction, finding the right prioritization for pushing these data objects is a crucial element. In this design, the data objects are ranked such that those due to expire sooner have a higher priority than others. At time t, the priority $p(d)$ of data object d is defined as follows:

$$p(d) = \begin{cases} \dfrac{1}{x(d) - t} & \text{if } x(d) > 1 \\ 0 & \text{otherwise} \end{cases} \tag{2.2}$$

Data objects are organized into different categories, a conventional way to access required data. A similar approach can be seen in Refs. [25,29], and a simpler form of such classification is used on local advertisement websites such as Gumtree, where the ads are classified into different channels. RDD organizes the data objects the protocol is intended to disseminate into specified channels according to content. The assumption is that there are agreed or standard classification methods that are used by all mobile nodes to classify their data objects.

Let C be a set of channels, where $C = \{c_1, c_2, c_3, \ldots, c_{n_c}\}$. Each created data object d is assigned to exactly one channel on creation time and this channel belongs to the set C—that is, $c(d) \in C$ and $|c(d)| = 1$.

2.4.2.2 User Preference Management

Each mobile node m subscribes to a set of channels $S(m)$, where $S(m) \subseteq C$. The subscription process is assumed to be carried out manually by the device users according to user needs, although this could be automated based on the history of previous content consumption. Therefore in a real system, this assumption could be easily relaxed in practical implementation.

2.4.2.3 Dissemination Protocol

The dissemination protocol of RDD is responsible for pushing data objects to the intended mobile nodes. RDD is capable of delivering data objects both directly between MNs (i.e., without storing them in the buffers of relay nodes) and indirectly (i.e., by means of a store-and-forward technique). If the source node and an interested node encounter each other, the message is delivered using an underlying direct data dissemination protocol. Otherwise, it is delivered using an underlying indirect data dissemination protocol, which is responsible for forwarding the messages to a repository, ideally with a high chance of successful delivery.

At the beginning of an encounter, a new connection is established if the two nodes are both free from any ongoing communication. During a connection, each node acts as both a data pusher and a data receiver, for itself (mobile node only case) or on behalf of other nodes (repository node case). A connection is completed when both nodes finish receiving the data objects in which they are interested. In many cases, a connection can terminate before that, simply because the two nodes have moved out of range or one of them has run out of battery power. When a new connection is established, the two nodes first exchange their subscribed channel set (S)—that is, they tell each other what channels they are currently subscribed to.

2.4.2.3.1 Direct Data Dissemination Algorithm

When two mobile nodes m_i and m_j encounter each other, they exchange their channel subscription sets $S(m_i)$ and $S(m_j)$. This exchange enables nodes to advertise the channels they are subscribed to, and this information is used to determine which data objects should be pushed for other nodes. The pseudo-code of direct data dissemination algorithms is shown in Algorithm 2.1. Figure 2.3 shows the communication process between two mobile nodes (m_i and m_j).

2.4.2.3.2 Indirect Data Dissemination Algorithm

As mentioned above, the repository is defined as a standalone fixed wireless device with no access to a fixed infrastructure. Under RDD, the repository completely controls multihop communication.

When a repository r encounters a mobile node m, the repository initiates the communication and m pushes its channel subscription $S(m)$. The repository uses $S(m)$ to determine which data objects from its buffer will be pushed to mobile node m. At the same time, the mobile node uses this interaction to push any data objects in its waiting buffer to the repository, to be pushed to other subscribed mobile nodes that it encounters. The pseudo-code of indirect data dissemination algorithms is shown in the Algorithm 2.2.

Initialization connection;

t=Current time;

Send $S(m_i)$ from m_i to m_j;

Send $S(m_j)$ from m_j to m_i;

Let the list A be a copy of the elements of $|D_t^{wt}(m_i)|$ ordered by decreasing expire time;

Let the list B be a copy of the elements of $|D_t^{wt}(m_j)|$ ordered by decreasing expire time;

```
while Connection true do
    A_rec=false; B_rec=false;
    while Not A_rec and|A| >0 do
        d = Pop(A);
        if x(d) < t then
            if c(d) ∈ S(m_j) then
                if d ∉ D_t^rc(m_j)) then
                    Push d from m_i to m_j; D_t^rc(m_j) = D_t^rc(m_j) U {d}; A_rec=True;
                end
            end
        else
            D_t^wt(m_i) = D_t^wt(m_i) – {d}; A = A–{d}
        end
    end
    while Not B_rec and |B| >0 do
        d = Pop(B);
        if x(d) < t then
            if c(d) ∈ S(m_i) then
                if d ∉ D_t^rc(m_i)) then
                    Push d from m_j to m_i; D_t^rc(m_i) = D_t^rc(m_i) U {d}; B_rec = True;
                end
            end
        else
            D_t^wt(m_j) = D_t^wt (m_j) – {d}; B = B–{d}
        end
    end
end
```

Algorithm 2.1 Direct data dissemination algorithm.

2.4.2.4 Repository Placement

Placing the repository in the most suitable place is a challenging research problem and will depend on knowledge of mobile node movement and communication patterns. Under RDD, the repository is the only node responsible for message relay, so it completely controls multihop message delivery. Therefore, the repository controls all indirect communications between MNs. As the repository location changes,

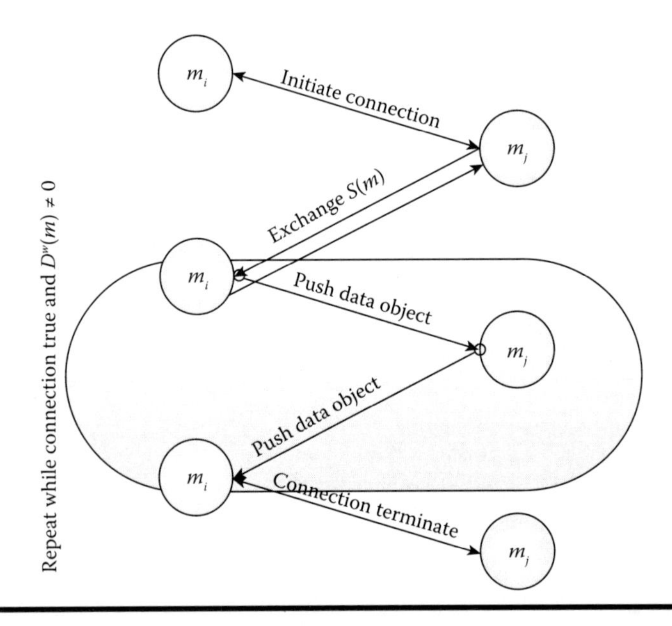

Figure 2.3 Diagram showing the communication process between mobile nodes.

the number of MNs with the opportunity to contact the repository may change. An increase in these contact opportunities should lead to an increase in message relay rate. Consequently, the repository performance strongly depends on its location, so optimizing the location for the repository is very important in the system design.

The last few years have seen the emergence of location-based social networking such as Foursquare [30], GyPSii [28], and Gowalla [31], which provide many services such as finding nearby friends and letting friends know you are "checking in" to a location. Although the Gowalla service recently shut down, Foursquare and GyPSii are still providing services. These applications employ the mobility information of users that mobile cellular networks can provide. The same mobility information can easily be used to generate traffic predictions according to historical patterns as well as to find the most crowded road or location, which is likely to be a suitable location for a repository; hence, the most crowded locations are usually the most suitable for placing a repository. However, in RDD two strategies are proposed—betweenness (*B*) and social betweenness (*SB*)—to address the repository location problem. These two strategies are discussed in the following subsections.

2.4.2.4.1 Betweenness

The fact that the map is a graph means that the centrality of any vertex in that graph may be calculated by using the betweenness measure of how important a vertex is by counting the number of shortest paths that it is a part of. Vertices that occur

```
Initialization connection;
t=Current time;
Send S(m) from m to r;
Let the list A be a copy of the elements of |D_t^wt(m)| ordered by decreasing expire time;
Let the list B be a copy of the elements of |D_t^wt(r)| ordered by decreasing expire time;
while Connection true do
    A_rec=false;
    B_rec=false;
    while Not A_rec and|A| >0 do
        d = Pop(A);
        if x(d) < t then
            if d ∉ D_t^wt(r)) then
                Push d from m to r; D_t^wt(r) = D_t^wt(r) U {d}; A_rec = True;
            end
        end
        D_t^wt(m) = D_t^wt(m) – {d}; A = A – {d}
    end
    while Not B_rec and|B| >0 do
        d = Pop(B);
        if x(d) < t then
            if c(d)∈ S(m) then
                if d ∉ D_t^rc(m)) then
                    Push d from r to m; D_t^rc(m) = D_t^rc(m) U {d}; B_rec = True;
                end
            end
        else
            D_t^wt(r) = D_t^wt (r) – {d}; B = B – {d}
        end
    end
end
```

Algorithm 2.2 Indirect data dissemination algorithm.

on many shortest paths between other vertices have higher betweenness than those that do not [32]. Because humans usually choose the shortest path between their source and destination, the vertex with the highest betweenness in a graph should be expected to be a good place for the repository in that graph, simply because a lot of people pass that vertex. Let G be the graph, and let V be the set of vertices in G. The betweenness $B(v)$ for a vertex v is defined in Equation 2.3.

$$B(v) = \sum_{\substack{v_s \neq v \neq v_t \in V \\ v_s \neq v_t}} \frac{\sigma_{v_s v_t}(v)}{\sigma_{v_s v_t}} \tag{2.3}$$

where $\sigma_{v_s v_t}$ is the number of shortest paths from v_s to v_t, and $\sigma_{v_s v_t}(v)$ is the number of shortest paths from v_s to v_t that pass through a vertex v.

2.4.2.4.2 Social Betweenness

While the betweenness strategy makes it possible to determine the highest centrality ranking vertex, this ranking is based on the assumption that MNs have equal preferences for all locations, which is not representative of real life. Therefore, despite a vertex being considered the highest betweenness vertex in the graph, it may not be passed often by mobile nodes because it is on a route to some locations not preferred by mobile nodes. An alternative strategy is needed to determine the repository location. In real life, users usually choose their destination according to their spatial relationships.

The preferences SpR for particular destinations can be taken into account by extending *betweenness* to give the new measure proposed, referred to as the *social betweenness* (SB) measure. Social betweenness measures how likely a vertex is to be used by mobile nodes, based on knowledge of user preferences as well as location of the vertex. The SB centrality of a vertex v in graph G is calculated from the probability of use of the shortest paths between each pair of vertices of graph G passing by vertex v. Equation 2.4 defines the SB for vertex v.

$$SB(v) = \sum_{v_t \in V} \sum_{m \in M} \frac{\sigma_{h(m)v_t}(v).SpR(m, v_t)}{\sigma_{h(m)v_t}} \tag{2.4}$$

where $\sigma_{h(m)v_t}$ is the number of shortest paths from $h(m)$ to v_t, and $\sigma_{h(m)v_t}(v)$ is the number of shortest paths from $h(m)$ to v_t that pass through a vertex v.

Social betweenness thus motivates a placement strategy that employs knowledge of user preferences as well as geographical location in determining the suitability of a location.

2.5 Repository-Based Data Dissemination for Opportunistic Networking

A new simulator was implemented to evaluate the performance of RDD. Mobility models such as the random walk, random waypoint [28], and community-based models [10] are often used to simulate the movements of real mobile nodes.

However, they may fail to capture the key features of human mobility; hence in this simulation, a new model—the human mobility model (HMM)—is used. The movement in this mobility model is based on a graph representation of discrete locations and the potential movement between them. This mimics how humans use memorized road maps. In the simulation such a graph is generated by distributing a chosen number of vertices randomly in the plane, connected using Delaunay triangulation. The home locations for the mobile nodes are chosen from a set of further vertices, which are randomly added, connected by a single edge. Such a graph mimics the main characteristics of real environments, where there are roads, paths, or corridors to follow and obstacles to avoid. To mimic a real-life scenario, MNs choose their destination using Zipf's power law, with the ranking of locations randomly rotated among MNs.

Given a graph, the HMM works as follows. At the beginning of the simulation, the nodes are placed in their home location (defined as one of the terminals). Once the simulation starts, the nodes choose their destination according to the Zipf distribution. MNs reach their destination by following the shortest path in the graph. In each step of the simulation, each node moves along this path at constant speed. If the node arrives at its destination, it pauses for some specified time steps and then returns to its home location by the reverse path. The node repeats this process until the end of the simulation time steps.

Data objects (messages) are created by MNs according to a Poisson distribution. The channels of a created data object are decided by the MN at generation time. In these simulation experiments, a list of 10 channels the MNs can subscribe to is defined. It is known that the intercontact time of human mobility exhibits a strong power law.

2.5.1 Performance Metrics

Many metrics are commonly used to measure the performance of network protocols. This section presents the metrics used in the evaluation process.

1. *Data delivered* (*DD*): This metric gives an indication of the ability of a protocol to deliver data to its destinations (i.e., how many messages are successfully delivered to interested nodes). *DD* is defined as the total number of messages successfully delivered to its destinations for all mobile nodes in the entire duration of the simulation time.
2. *Data sent* (*DS*): This metric gives an indication of the traffic generated by a protocol and is the total number of messages sent by all nodes in the entire duration of the simulation time.
3. *Data redundancy* (*DR*): This metric gives an indication of the number of messages sent without immediate benefit for final delivery (i.e., the traffic overhead created by the protocol). *DR* is the total number of data objects sent by all mobile nodes minus the number of data objects delivered by all nodes.

2.5.2 Evaluation of Repository Location Strategies

Two repository location strategies were introduced, betweenness (*B*) and social betweenness (*SB*), which will be evaluated in this section. In the investigation, data delivered (DD) is used as the main metric to determine the performance of the proposed strategies.

2.5.2.1 Experimental Scenarios

To study the performance of the proposed strategies, a series of simulation experiments were carried out. The following experiments examine the two repository location strategies (SB and B) in four simulation scenarios (Table 2.2). The simulation setting used the default value presented in Table 2.3.

The first scenario is considered as small and dense, the second scenario is small and sparse, the third scenario is large and dense, and the fourth scenario is large and sparse. The evaluation process investigated the performance of each scenario using three different maps, and for each map the experiment was run in seven different mobility patterns. Therefore for each scenario, 21 experiments were carried out.

2.5.2.2 Test Results

This experiment examines the effect of the repository location on the DD. The repository is placed at each $v \in J$, and the DD evaluated in each time. Figures 2.4 through 2.7 show the performance comparison between using *B* and *SB* strategies in four simulated — scenarios. In Figures 2.4 through 2.7, the x-axis represents the repository location ranking according to B for the betweenness graph and according to SB for the social betweenness graph. The ranking starts with the highest (ranking 1 in x-axis) to the lowest, which equals $|J|$ used in the corresponding test.

Despite the different simulated scenarios (1–4), Figures 2.4 through 2.7 show that the performance of the DD often improves as the repository location B and SB value increases; nevertheless, using B as a location strategy, DD sometimes shows unpredictable change, representing the inconsistency of this strategy. In the four tested scenarios (1–4), the results show that using *SB* as a location strategy for the repository provides more consistent and smooth results compared to *B*, leading to the conclusion that SB should be used as a repository location strategy. From this point on, all results presented in this thesis are based on a single repository located in the best SB location.

2.5.3 Repository-Based Data Dissemination Performance

To evaluate RDD, simulation experiments were performed comparing RDD with EPR. EPR is a popular protocol for data delivery in OppNets. EPR relies on the flooding concept. The main idea of this protocol is as follows. If a carrier of a message encounters another node, it will deliver all the messages it carries that the other node does not already have. This approach can achieve the highest possible delivery

Table 2.2 Simulation Scenarios for the Evaluation of R Location Strategies

	Scenario 1	*Scenario 2*	*Scenario 3*	*Scenario 4*
$\|M\|$	10 nodes	10 nodes	25 nodes	25 nodes
$\|J\|$	20 vertices	50 vertices	50 vertices	100 vertices
$\|R\|$	1 node	1 node	1 node	1 node

Table 2.3 Simulation Parameters

Parameter	*Value*
Simulation steps	86,400 steps
Communication range	2 meters
Speed	1 meter/step
Average messages created	Three messages
Environment size	1000 * 700 meters
Minimum off time	7200 steps
Maximum off time	21,600 steps
Number of channels	10 channels
Messages allowed to pass	Two messages
Data handling priority	$x(d)$
$ttl(d)$ minimum time	21,600 steps
$ttl(d)$ maximum time	43,200 steps
Buffer size	Unlimited
$\|M\|$ Number of mobile nodes	25
$\|R\|$ Number of repositories	1
$\|J\|$ Number of junctions	25
$H(v)$	1
Density of MNs with $W(m) = 1$	0%
Density of MNs with $S(m) = C$	100%

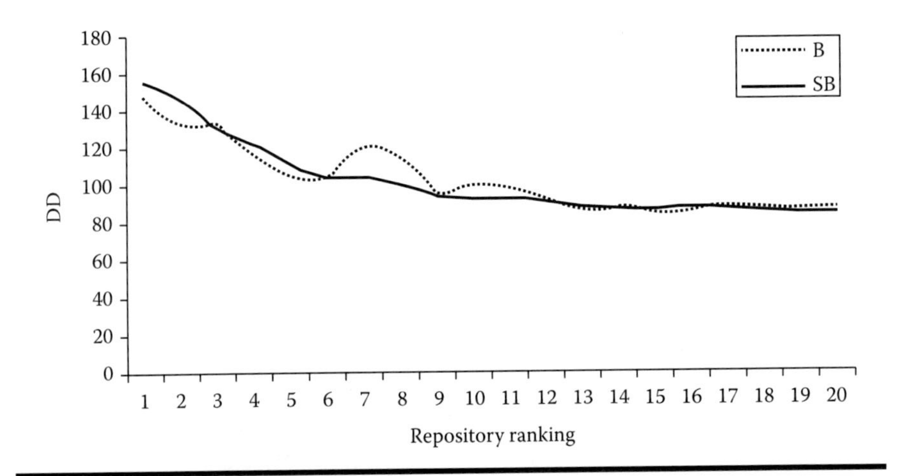

Figure 2.4 Comparison of B and SB performance in Scenario 1. B, betweenness; SB, social betweenness.

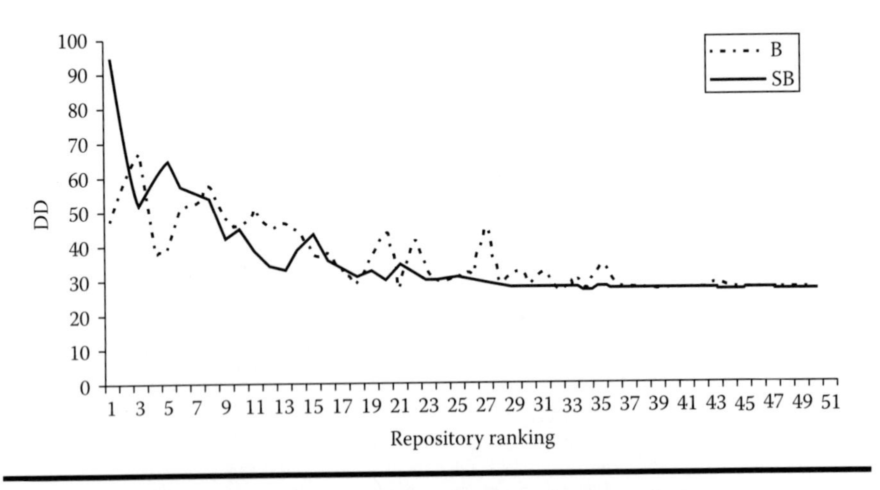

Figure 2.5 Comparison of B and SB performance in Scenario 2.

ratio if it is provided with infinite bandwidth and buffer resources. Therefore, it is considered as a good benchmark for comparison with alternative protocols.

Because EPR was designed as a forwarding protocol (data delivery to a specific destination), some modification to this protocol was necessary so that it could be used as a data dissemination protocol (where data is delivered to all nodes encountered). Here, EPR has been implemented with three settings for the hop limit. The hop limit for a message is the maximum number of hops that the message is allowed to travel to reach its destination. These settings are a two-hop limit (2h), a four-hop limit (4h), and unlimited hops (Uh).

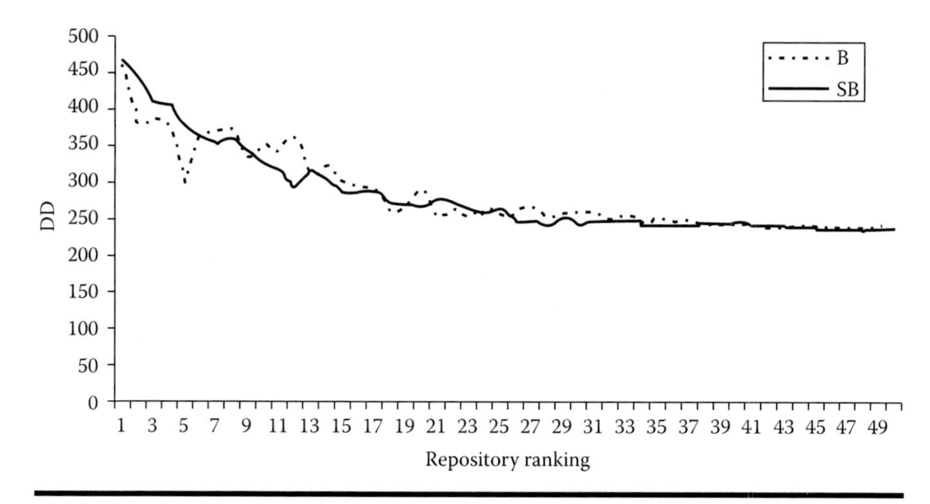

Figure 2.6 Comparison of B and SB performance in Scenario 3.

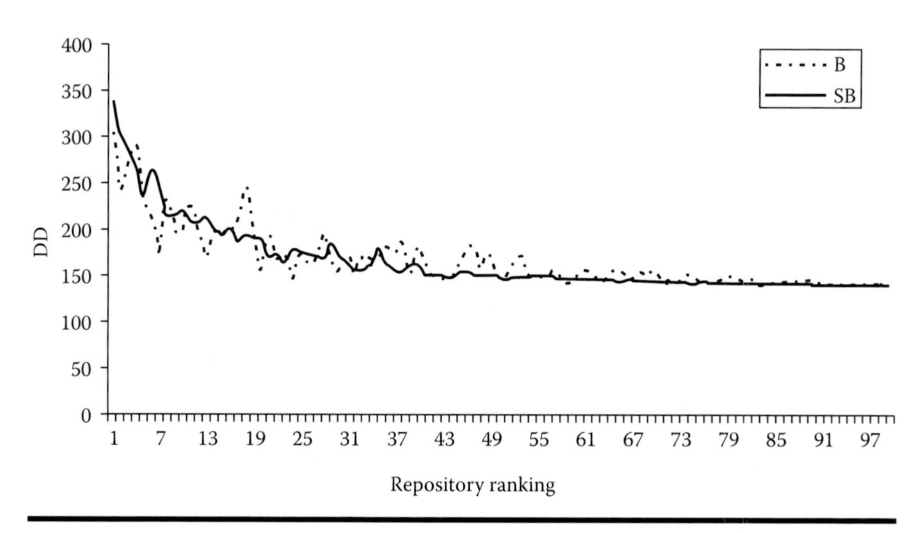

Figure 2.7 Comparison of B and SB performance in Scenario 4.

2.5.3.1 Experiment Setting

The simulation setting used the values presented in Table 2.2, whereas the protocol (RDD and EPR) settings are presented in Table 2.4.

Figure 2.8 shows that the EPR approach provides better DD despite the hop limit. Because the density of MNs with $S(m) = C$ is 100%, there is no redundancy in EPR; however, there is some redundancy associated with RDD. This occurs because messages routed through the repository require more than one hop, and some messages forwarded to a repository may never reach their final destination.

Table 2.4 Experiment Setting

Parameter	*RDD*[a]	*EPR*[b]
Hop limit	n/a	Two, four, and unlimited hops
Repository location strategy	SB	n/a
$\|M\|$	25	25
$\|R\|$	1	0
$\|J\|$	25	25
$H(v)$	1	1
Density of MNs with $W(m) = 1$	0%	100%
Density of MNs with $S(m) = C$	100%	100%

[a] Repository-based data dissemination.
[b] Epidemic routing.

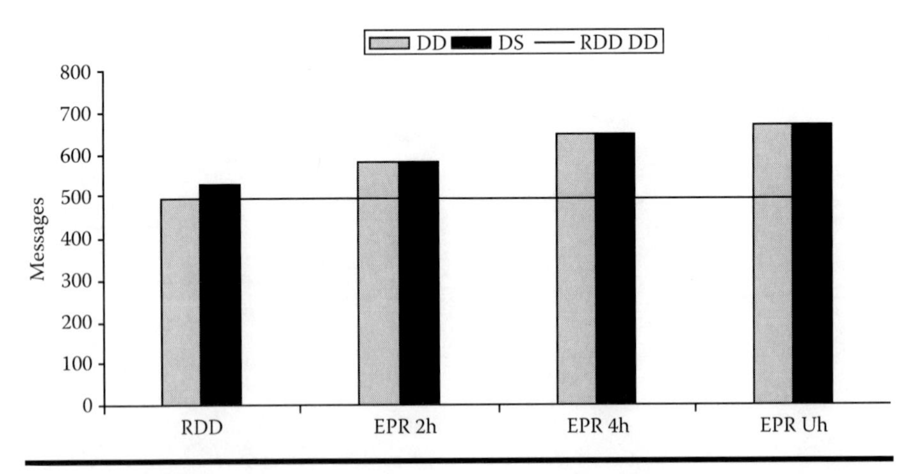

Figure 2.8 RDD vs. EPR performance. RDD, repository-based data dissemination; EPR, epidemic routing.

Further experiments were carried out with varying densities of mobile nodes. Define the mobile node density as the number of mobile nodes $|M|$ divided by the number of junction vertices $|J|$ in the graph used the corresponding scenario. Table 2.5 presents the densities used in these experiments.

For a graph with $|J| = 25$, Figure 2.9 shows that when the density of mobile nodes equals 48% or lower, RDD performance is better than EPR, whereas when the mobile node density is 100% or 200% the performance of EPR is better. For graph with $|J| = 50$, Figure 2.10 shows that when the density of mobile nodes is equal

Table 2.5 Mobile Density Scenarios

Scenario	*1*	*2*	*3*	*4*	*5*	*6*	*7*	*8*
$\|J\|$	25	25	25	50	50	50	50	50
$\|M\|$	12	25	50	15	25	35	40	50
$H(v)$	1	1	2	1	1	1	1	1
EPR[a] hop limit	Unlimited							
Mobile node density	48%	100%	200%	30%	50%	70%	80%	100%

[a] Epidemic routing.

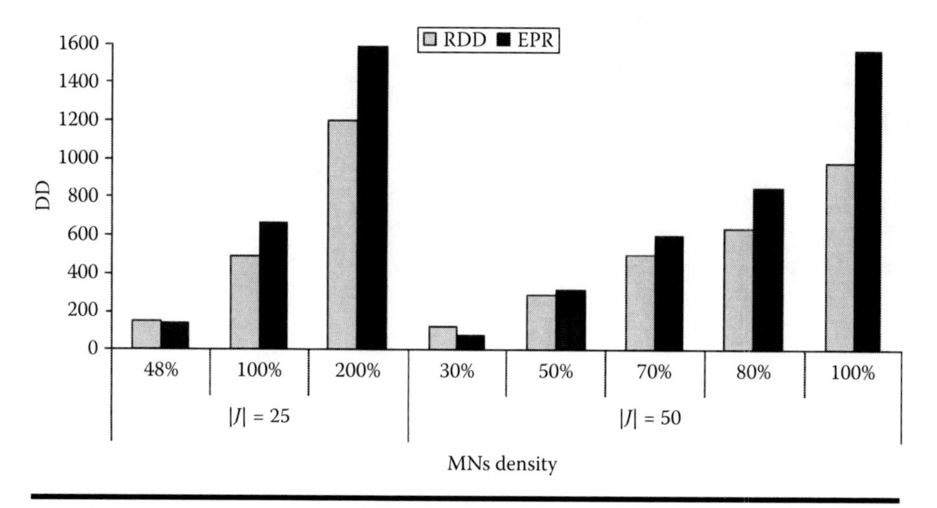

Figure 2.9 Impact of mobile nodes' density on protocols performances.

to 30%, RDD performance is better than EPR, whereas when mobile node density is 50% EPR starts to perform better than RDD. This leads to the conclusion that RDD will perform better than EPR in low-density scenarios despite the hop limits and that EPR will perform better than RDD in medium- or high-density scenarios. This finding goes with the fact that more mobile nodes mean more relay for EPR, so the EPR performed better with a high density of mobile nodes.

2.5.4 Repository-Based Data Dissemination in a Partially Cooperative Environment

One of the arguments for using a repository in data dissemination is that MNs are not usually willing to work as a relay, or if they are willing it may be inefficient for them to work as a relay. This test compares the performance of RDD against

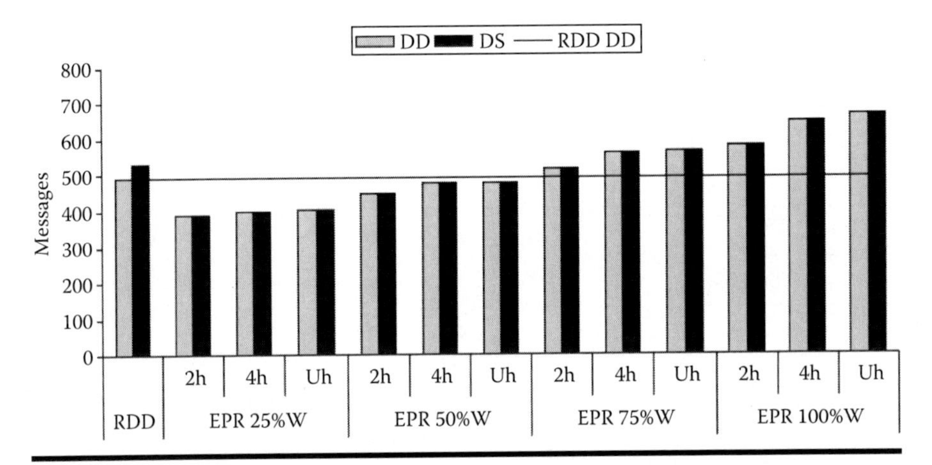

Figure 2.10 Impact of density of MNs with *W*(*m*) = 1 on the performances.

EPR in an environment where the density of MNs with $W(m) = 1$ is varied. Note that RDD requires no willingness—that is, the repository is guaranteed willing to relay. The main point of this test is to investigate the performance of both protocols in an environment closed to real-life environment where mobile nodes are not always willing to carrying messages to the other nodes. Note that throughout the evaluation process, 25% *W* refers to the density of mobile nodes willing to cooperate being 25%.

2.5.4.1 Experiment Setting

The simulation setting used the same default values presented in Table 2.2; the protocol (RDD and EPR) settings are presented in Table 2.6. For this test, four scenarios were defined where the density of mobile nodes willing to relay varied as shown in Table 2.6. When this density was less than 100%, the selection of nodes that were willing to relay was made uniformly random.

Figure 2.10 shows that if the percentage of MNs that are willing to relay is less than 50%, RDD performs better despite the hop limits used by EPR. When the density reaches 75%, EPR performs better. The improvement EPR gains is up to 15% when 75% of mobile nodes are willing to cooperate and 17%–36% when 100% of MNs are willing to cooperate.

2.5.5 Repository-Based Data Dissemination in Partial Subscription Scenarios

In real-life scenarios, a mobile node channel subscription could have the following setting: $S(m) = \varnothing$ or $S(m) = C$ or $S(m) \subset C$. As stated above, at this stage the assumption is that the mobile node m can be either subscribed to all channels,

Table 2.6 Experiment Setting

Parameter	*RDD*	*EPR*
Hop limit	n/a	Two, four, and unlimited hops
Repository location strategy	SB	n/a
Number of repository	1	n/a
$\|M\|$	25	25
$\|R\|$	1	0
$\|V\|$	25	25
$H(v)$	1	1
Density of MNs with $W(m) = 1$	0%	25% (Scenario 1)
—	—	50% (Scenario 2)
—	—	75% (Scenario 3)
—	—	100% (Scenario 4)
Density of MNs with $S(m) = C$	100%	100%

that is, $S(m) = C$, or not subscribed, that is, $S(m) = \varnothing$. This investigation studies the performance of RDD compared with EPR in different scenarios, where the density of subscribed mobile nodes varies. Note that throughout the evaluation process, 25% S refers to the density of subscribed mobile nodes being 25%.

2.5.5.1 Experiment Setting

The simulation used the same default values presented in Table 2.2, and the protocol (RDD and EPR) settings are presented in Table 2.7. This test defines three scenarios where the density of mobile nodes with $S(m) = C$ is set to 25%, 50%, and 75%. However, $W(m) = 1$ for all mobile nodes in these scenarios. The subset of MNs with $S(m) = C$ are selected uniformly randomly.

Figure 2.11 shows that if the density of MNs with $S(m) = C$ is 25%, EPR provides the better DD rate. The improvement provided by EPR in DD is approximately 40% compared to RDD. However, this improvement comes with a high cost in DR, with an average of 900% (over 2h, 4h, and Uh scenarios) compared to RDD. When the density is 50%, EPR still maintains a higher DD with roughly the same percentage improvement but a lower cost in DR, with an average of 600% compared to RDD. When subscription density reaches 75%, the DR cost decreases to an average of 250% compared to RDD. Therefore, despite the improvement in DD provided by EPR, the high cost of DR makes RDD more practical and

Table 2.7 Tested Subscription Scenarios

Parameter	*RDD*	*EPR*
Hop limit	n/a	Two, four, and unlimited hops
Repository location strategy	SB	n/a
Number of repository	1	n/a
$\|M\|$	25	25
$\|R\|$	1	0
$\|I\|$	25	25
$H(v)$	1	1
Density of MNs with $W(m) = 1$	0%	100%
Density of MNs with $S(m) = C$	25% (Scenario 1)	25% (Scenario 1)
—	50% (Scenario 2)	50% (Scenario 2)
—	75% (Scenario 3)	75% (Scenario 3)
—	100% (Scenario 4)	100% (Scenario 4)

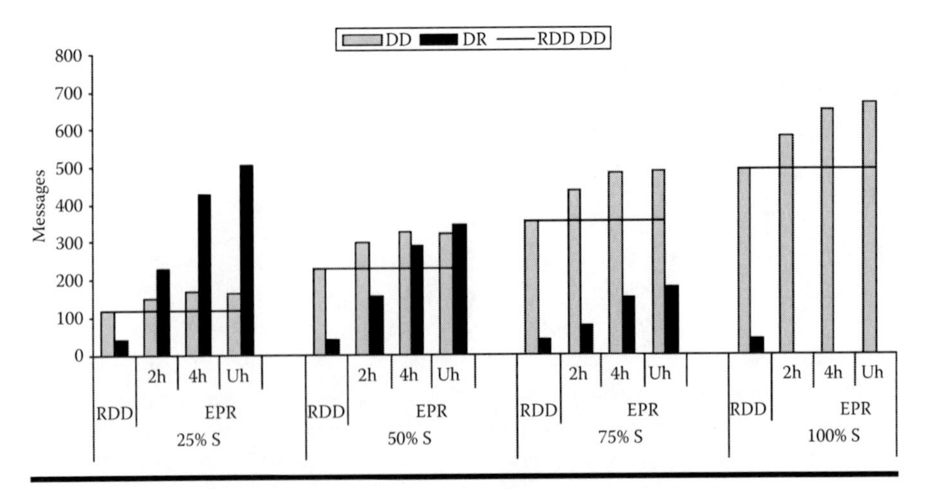

Figure 2.11 Impact of density of MNs with *S*(*m*) = *C*.

efficient to use. It is notable that the DR decreases in EPR when the percentage of MNs subscribed increases.

The results in Figure 2.11 show that in practical scenarios, where not all users require information, RDD is able to significantly reduce overhead, with only a small impact on delivery.

2.6 Conclusion

This chapter presented a short overview of wireless communications before focusing on performance in *OppNets*. The main problem associated with opportunistic networking is how to design an efficient and realistic data delivery mechanism. The literature review in this chapter concludes that the success of most proposed approaches depends on how willing MNs are to work as a relay for others. As discussed in detail in this chapter, in real-life scenarios, MNs are not usually willing to work as intermediate hops for such reasons as security and resource constraints. The proposed RDD system addresses this problem through a different approach, employing standalone fixed wireless devices to relay data objects instead of using mobile nodes. The first part of the presented evaluation process compared repository placement strategies in different scenarios, concluding that considering social information improves the consistency of performance. RDD performance was then compared across a range of metrics to the EPR protocol. Even with a single, well-placed repository, RDD was able to provide better performance when the density of mobile nodes willing to cooperate was 50% or lower. Another important finding is that RDD was able to significantly reduce resource consumption. The investigation also showed that when the density of mobile nodes was low, RDD provided better performance in terms of data delivery. As such, the results clearly indicate that RDD is a promising way to facilitate data dissemination in OppNets, especially in real-life scenarios where other approaches fail due to the noncooperative nature of MNs in carrying data objects to other MNs.

References

1. Timo Halonen, Javier Romero, and Juan Melero. *GSM, GPRS and EDGE performance: Evolution towards 3G/UMTS*. Wiley, 2004.
2. L.S. Ashiho. Mobile technology: Evolution from 1G to 4G. *Electronics for You*, 2003.
3. Ian F. Akyildiz, David M. Gutierrez-Estevez, and Elias Chavarria Reyes. The evolution to 4G cellular systems: LTE-advanced. *Physical Communication*, 3(4):217–244, 2010.
4. Subir Kumar Sarkar, T.G. Basavaraju, and C. Puttamadappa. *Ad Hoc mobile wireless networks: Principles, protocols and applications*. Auerbach Publications, USA, 2007.
5. Kevin Fall. A delay-tolerant network architecture for challenged internets. In *The 2003 Conference on Applications, Technologies, Architectures, and Protocols for Computer Communications*, pages 27–34, 2003.
6. L. Pelusi, A. Passarella, and M. Conti. Opportunistic networking: Data forwarding in disconnected mobile ad hoc networks. *IEEE Communications Magazine*, 44(11):134–141, 2006.
7. International Telecommunication Union. *The world in 2013: ICT facts and figures*. 2013.
8. A. Chaintreau, A. Hui, J. Crowcroft, C. Diot, R. Gass, and J. Scott. *Pocket switched networks: Real-world mobility and its consequences for opportunistic forwarding*. Technical Report UCAM-CL-TR-617. Computer Laboratory, University of Cambridge, Cambridge, 2005.

9. Pan Hui, Augustin Chaintreau, Richard Gass, James Scott, Jon Crowcroft, and Christophe Diot. Pocket switched networking: Challenges, feasibility and implementation issues. *Autonomic Communication*, 3854:1–12, 2006.
10. Hui Pan, Chaintreau Augustin, Scott James, Gass Richard, Crowcroft Jon, and Diot Christophe. Pocket switched networks and human mobility in conference environments. In *ACM SIGCOMM Workshop on Delay-Tolerant Networking*, pages 244–251, 2005.
11. Amin Vahdat and David Becker. *Epidemic routing for partially-connected ad hoc networks*. Technical Report CS-200006. Duke University, Durham, NC, 2000.
12. Alan Demers, Dan Greene, Carl Hauser, Wes Irish, John Larson, Scott Shenker, Howard Sturgis, Dan Swinehart, and Doug Terry. Epidemic algorithms for replicated database maintenance. In *Sixth Annual ACM Symposium on Principles of Distributed Computing*, pages 1–12, ACM, 1987.
13. Jorg Widmer and Jean-Yves Le Boudec. Network coding for efficient communication in extreme networks. In *The 2005 ACM SIGCOMM Workshop on Delay-Tolerant Networking*, pages 284–291, 2005.
14. Chen Ling Jyh. An evaluation of routing reliability in non-collaborative opportunistic networks. In *International Conference on Advanced Information Networking and Applications*, volume 0, pages 50–57, 2009.
15. Chiara Boldrini, Marco Conti, and Andrea Passarella. Design and performance evaluation of ContentPlace, a social-aware data dissemination system for opportunistic networks. *Computer Networks*, 54(4):589–604, 2010.
16. Timur Friedman, Vania Conan, and Jeremie Leguay. Evaluating MobySpace-based routing strategies in delay-tolerant networks. *Wireless Communications and Mobile Computing*, 7(10):1171–1182, 2007.
17. B. Burns, O. Brock, and B. Levine. MV routing and capacity building in disruption tolerant networks. In *INFOCOM 2005, 24th Annual Joint Conference of the IEEE Computer and Communications Societies*, volume 1, pages 398–408, 2005.
18. Hui Pan, A. Lindgren, and J. Crowcroft. Empirical evaluation of hybrid opportunistic networks. In *Communication Systems and Networks and Workshops*, pages 1–10, 2009.
19. Tan Le and Liu Yong. On the capacity of hybrid wireless networks with opportunistic routing. *EURASIP Journal on Wireless Communications and Networking*, volume 1, 2010. DOI: 10.1155/2010/202197.
20. Small Tara and J. Haas Zygmunt. The shared wireless infostation model: A new ad hoc networking paradigm (or where there is a whale, there is a way). In *The 4th ACM International Symposium on Mobile Ad Hoc Networking & Computing*, pages 233–244, 2003.
21. Eiko Yoneki, Pan Hui, ShuYan Chan, and Jon Crowcroft. A socio-aware overlay for publish/subscribe communication in delay tolerant networks. In *10th ACM Symposium on Modeling, Analysis, and Simulation of Wireless and Mobile Systems*, pages 225–234, 2007.
22. P. Costa, C. Mascolo, M. Musolesi, and G.P. Picco. Socially-aware routing for publish-subscribe in delay-tolerant mobile ad hoc networks. *Selected Areas in Communications, IEEE Journal on Selected Areas in Communications*, 26(5):748–760, 2008.
23. Chiara Boldrini, Marco Conti, and Andrea Passarella. ContentPlace: Social-aware data dissemination in opportunistic networks. In *11th International Symposium on Modeling, Analysis and Simulation of Wireless and Mobile Systems*, pages 203–210, 2008.

24. Chiara Boldrini, Marco Conti, Jacopo Jacopini, and Andrea A. Passarella. HiBOp: A history based routing protocol for opportunistic networks. In *The IEEE International Symposium on a World of Wireless, Mobile and Multimedia Networks, WoWMoM 2007*, pages 1–12, 2007.
25. Gunnar Karlsson, Vincent Lenders, and Martin May. Wireless ad hoc podcasting. In *Sensor, Mesh and Ad Hoc Communications and Networks, SECON'07*, pages 273–283, 2007.
26. Yunsheng Wang and Jie Wu. Social-tie-based information dissemination in mobile opportunistic social networks. In *14th International Symposium on a World of Wireless, Mobile and Multimedia Networks*, 2013.
27. Mark S. Granovetter. The strength of weak ties. *American Journal of Sociology*, 1360–1380, 1973.
28. Min Li and Zhenjun Du. Dynamic social networking supporting location-based services. In *International Conference on Intelligent Human-Machine Systems and Cybernetics, 2009. IHMSC'09*, volume 1, pages 149–152, 2009.
29. Gunnar Karlsson, Vincent Lenders, and Martin May. Delay-tolerant broadcasting. In *The 2006 SIGCOMM Workshop on Challenged Networks, CHANTS '06*, pages 197–204, New York, 2006.
30. M.J. Chorley, G.B. Colombo, M.J. Williams, S.M. Allen, and R.M. Whitaker. Checking out checking In: Observations on foursquare usage patterns. In *International Workshop on Finding Patterns of User Behaviors in NEtwork and MObility (NEMO)*, 2011.
31. Colleen Cuddy and Nancy R. Glassman. Location-based services: Foursquare and Gowalla, should libraries play? *Journal of Electronic Resources in Medical Libraries*, 7(4):336–343, 2010.
32. A. Keränen. *Opportunistic network environment simulator.* Technical Report. Department of Communications and Networking, Helsinki University of Technology, Helsinki, 2008.

Chapter 3

Mobile Ad Hoc Networks: Rapidly Deployable Emergency Communications

Niaz Chowdhury and Stefan Weber

Contents

This chapter describes Beatha, a transport protocol for rapidly deployable emergency communications. The Irish Gaelic word *beatha* means "life." In keeping with the meaning of its name, the proposed protocol introduces a lifelike quality to end-to-end data delivery in emergency response, where nodes are not the first point of communication but rather their services are. Beatha makes use of characteristic-based network architecture to route packets from source to destination and along its network layer counterpart forms a new network paradigm that replaces the Internet Protocol–based network in wireless environment.

3.1 Introduction

Over the past two decades, researchers across the globe have extensively investigated wireless communications. The mobile ad hoc network (MANET) is one of the outcomes of that research. The community has defined the MANET in many ways:

- A MANET is a collection of wireless mobile nodes that cooperatively form a network without infrastructure [1].
- A MANET is a dynamically reconfigurable wireless network that does not have a fixed infrastructure [2].
- MANETs consist of a group of mobile and autonomous nodes that are directly interconnected to each other. These networks are independent of any communication infrastructure such as access points or base station and rely entirely on node cooperation for relaying information from data source to intended destination [3].
- A MANET is a group of mobile wireless nodes that cooperatively form a network independent of any fixed infrastructure or centralized administration [4].
- A MANET is a network that works under dynamic routing with a multihopping mechanism and has no centralized body to govern the network under which the network has to communicate [5].

From the above definitions, it is clear that for the most part nodes in MANETs are wireless and mobile. The network is dynamically reconfigurable, does not have a fixed infrastructure, and relies on node cooperation for delivering data from one

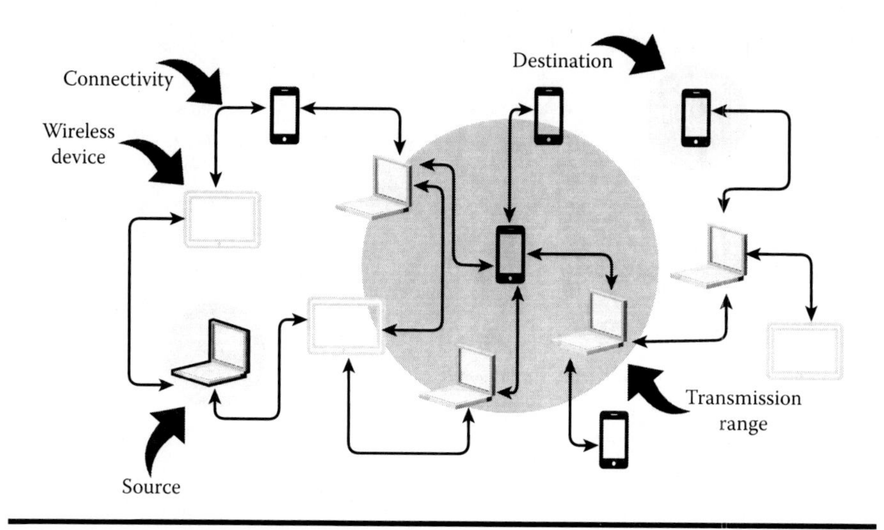

Figure 3.1 Architecture of mobile ad hoc network.

node to another (Figure 3.1). Although IEEE 802.11 is the most frequently used link layer protocol for MANETs, other protocols such as IEEE 802.15.4 (ZigBee) or IEEE 802.16 (WiMAX) standards can also be used as underlying medium access control (MAC) and physical layer protocol [6].

The infrastructureless nature and being able to operate without central administration control make MANETs very effective in playing an increasingly important role in extending the coverage of traditional wireless networks. This technology is widely used to deploy decentralized communication in remote places [6]. There are many applications that can be built around MANETs. Military operations and emergency police response are already in use. Vehicle-to-vehicle communication is another dimension derived from this network that helps to prevent accidents and make communication available in next-generation vehicles. Nevertheless, in recent years emergency communication for disaster recovery has become one of the most demanding applications where MANETs can be best fitted [7].

Communication over wireless networks, however, is currently managed at higher layers based on the use of Internet Protocol (IP) and its transport layer counterpart Transmission Control Protocol (TCP). Wired networks assume that nodes are interconnected by immobile elements and individual streams of communication do not interfere with each other. However, a rising number of ubiquitous systems with wireless communication functionality removes the basis for these assumptions, and the nature of communication between nodes in a given area becomes inherently sequential in order to avoid the interference of concurrent transmissions. This has led to poor performance of transport protocols used in wired networks in their wireless MANET counterparts. Another prominent difference between wired and wireless networks is the change in topology. Topology seldom changes on wired networks. Arguably, the only reason for altering the topology of a wired network is to have a router down. However, in wireless networks, especially in MANETs, a number of causes bring changes to the topology. Because MANETs are comprised of mobile nodes with differing velocities, the chance that a node will move away is very high. Moreover, in MANETs each node acts as a router and because of its decentralized nature a node may join or leave at any time. This feature, however, is also responsible for creating changes in the topology. These frequent changes eventually cause two problems that largely hamper end-to-end communication in MANETs, particularly in emergency situations where nodes move frequently. First, they cause frequent link failures while routing data from the source to the destination and, second, a node serving a particular task may move away, resulting in a permanent service failure.

It is therefore eminent that protocols designed for wired networks are not adequate to serve the purposes of wireless networks, and the necessity for an alternative approach to an IP network model is at its peak now. In recent years, several approaches have been proposed from different parts of the world. Some proposals have come in the form of cross-layer implementation, whereas some proposals have dealt with multipath routing to enhance transport layer performance. However, initiatives that replace both IP and TCP are still rare, and reliable communication over such an alternative paradigm has not yet been established.

This chapter stresses two main points: first, the importance of switching to an alternative paradigm in order to emphasize service over provider and, second, the need for a more suitable transport layer capable of adjusting the delay in emergency communications. Both motivations have their roots in the drawbacks that occur when protocols designed for wired networks operate in a dynamic wireless environment.

The justification for switching to an alternative paradigm lies in the nature of MANETs. Because MANETs are highly mobile and nodes change positions frequently, a fixed identifier like an IP address is not useful in a wireless environment. For example, a disaster recovery mission, where rescuers search for victims in a relatively broad area, might have a situation where someone requires a digger. In such a case, that person would be happy to have the service available for use; it would not be necessary to know the person providing the service. As a result, tracking nodes with a fixed identifier in these applications often do not play an important role. This example makes it clear that in emergency communication MANET services should take priority over service providers. A provider may remain visible or invisible but should not be taken into account for communication purposes, and the identity of nodes needs to be revealed based on the services they offer. In recent years, a number of network platforms have been proposed that keep this principle in mind. Content centric networks [8], attribute-based routing [9], and Publish–Subscribe Architecture [10] are a few examples that provide alternative communication based on data or services.

The need for a more suitable transport layer is particularly important because in emergency situations, networks face long delays and sometimes might get disconnected for awhile. Because the basic assumptions of existing transport protocols designed for wired networks do not include properties found in wireless networks, they exhibit poor performance while operating over an IP network. The design of the new transport protocol takes three main issues into account:

1. Congestion can be the principal reason for data loss in wired network but in wireless ad hoc networks mobility or link failure could be other reasons for this problem.
2. Low-bandwidth wireless networks, in contrast to wired networks that have dedicated bandwidth, are not suitable for window-based flow control.
3. Because of the collision avoidance feature of medium access control protocols, segment-by-segment acknowledgment is an obstacle for achieving desired throughput.

3.2 Rapidly Deployable Emergency Communications

The nature of emergency communications is different. This kind of communication is rapidly deployed in the event of a natural disaster, accident, war, or similar situation, where using an infrastructure-based network is difficult or sometimes impossible. It is evident from the type of scenarios that the users and the data they exchange are also very different from regular data being exchanged on the Internet.

Let us consider a real-life scenario as illustrated in Figure 3.2. A disaster recovery team has been trying to rescue victims at a site. This team would possibly be comprised of rescuers, diggers, emergency medical technicians (EMTs), an ambulance, firefighters, a base camp, gateway, and so on. These entities need to communicate with each other to support the rescue mission as a team. For example, a rescuer might seek assistance from an EMT to give immediate medical support or a digger might report on the state of a victim by sending photographs. Again, diggers or ambulances might need to communicate to decide where to provide service at any given moment. One noticeable attribute common to all communication in emergency response is service dependency. A rescuer is not concerned with the person serving as a digger, EMT, or firefighter as long as he or she is receiving the service requested. Thus communication occurs between services instead of nodes.

The existing architecture, however, does not support this concept. At the inception of the Internet, the IP-based communication model became the *de facto* standard for establishing communication at higher layers. Wireless networks, which are the backbone of establishing emergency communications, require their own protocols at the transport and network layers to provide adequate performance in the described scenarios. Although medium access control in wireless network is

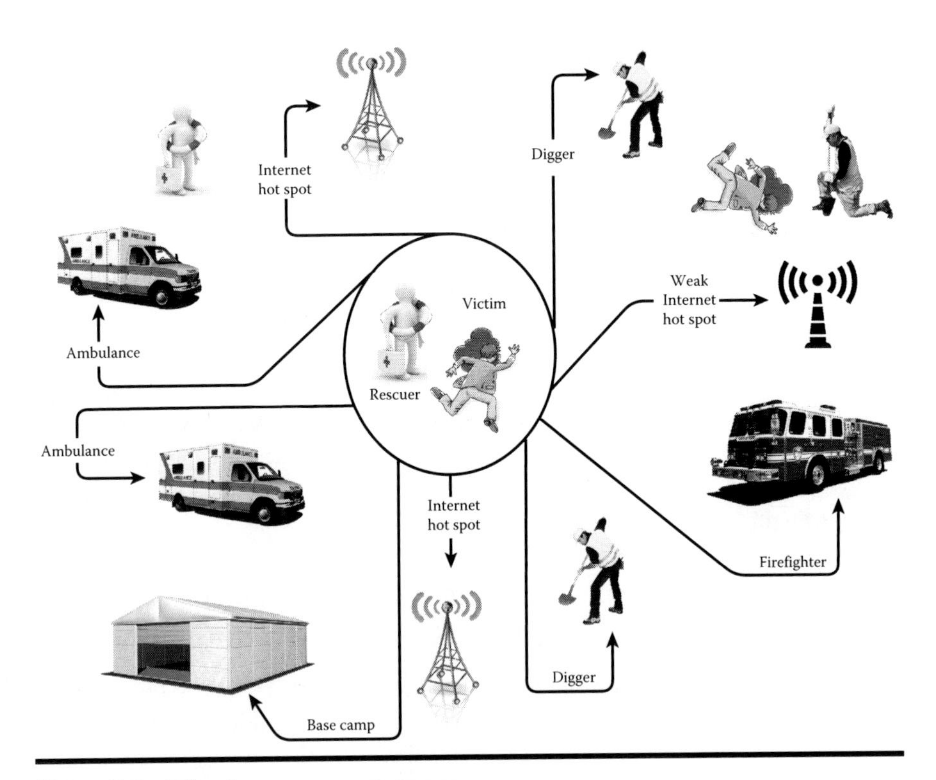

Figure 3.2 Disaster recovery program.

managed by specialist protocols, for example IEEE 802.11, higher layers continue to be managed by IP protocols. Despite the fact that this old practice has been the cause of poor performance at both the network and transport layers, until recently sufficient initiatives have not been undertaken to explore alternative approaches.

Over the past two decades, a number of redesigns have been proposed for the end-to-end approach in distributed systems [11] and conventional Internet structure [12]. Nevertheless, it was only when Wi-Fi became popular that researchers from around the world started undertaking initiatives to design alternative protocols to IP. Content-Based Networking [13] and Content-Based Routing [8] were two of those early initiatives. In recent years two other protocols, Adaptive Attribute-Based Routing [9] and Publish–Subscribe Architecture [10], have been proposed. These protocols, however, are designed for application-specific network layers, except content-based routing, which has some degree of functionality to provide unreliable transport layer support, but they are not a successful alternative to the TCP/IP paradigm. One of our main challenges in this work was to find a suitable routing protocol that works with Beatha set up on top of it. After exploring the existing protocols in the literature, we ended up choosing Uisce, which is an anonymous routing protocol and replaces IP in the communication protocol stack.

3.3 Alternative Paradigm

Because of the highly mobile nature of MANETs, nodes change their positions frequently. This behavior suggests that a fixed identifier such as an IP address does not have a great impact on controlling the network and nodes need to keep changing their service directory. Let us take a more technical look at the example described in the previous section. Figure 3.3 shows that Node 3 is a rescuer with multiple nodes around it. At the network layer, only their IP addresses are visible. Suppose this rescuer finds a victim and wants to communicate with an ambulance, ask for help from a digger, or seek Internet access. In order to do that, he must know which node is providing what services—that is, he needs to go through a service discovery procedure (Figure 3.4).

This can be achieved in two possible ways. The interested node (in this example, Node 3, the rescuer) may keep track of the IP addresses of every neighboring node and maintain a table of services associated with each IP address. While looking for a specific service, it can just look up the table and get the IP address of the desired node. Unfortunately, this solution is unrealistic due to the uncertainty of node movement in MANETs. An alternative approach can also be used. The interested node may go to every neighboring node and ask whether it is an ambulance or a digger or such. This second approach will significantly increase overhead in the network.

This example points out one important aspect: IP addresses may not be very helpful in a highly mobile environment, and therefore a network protocol is

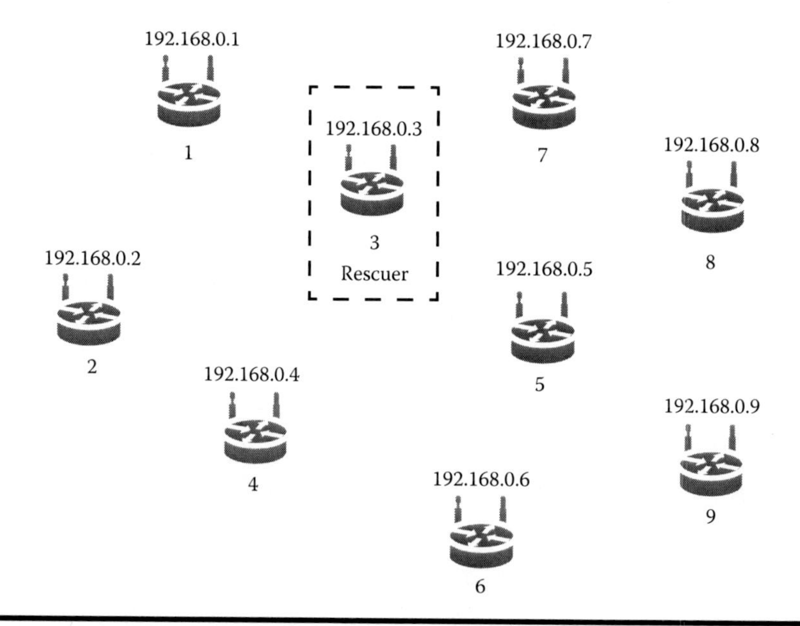

Figure 3.3 Network prior to service discovery.

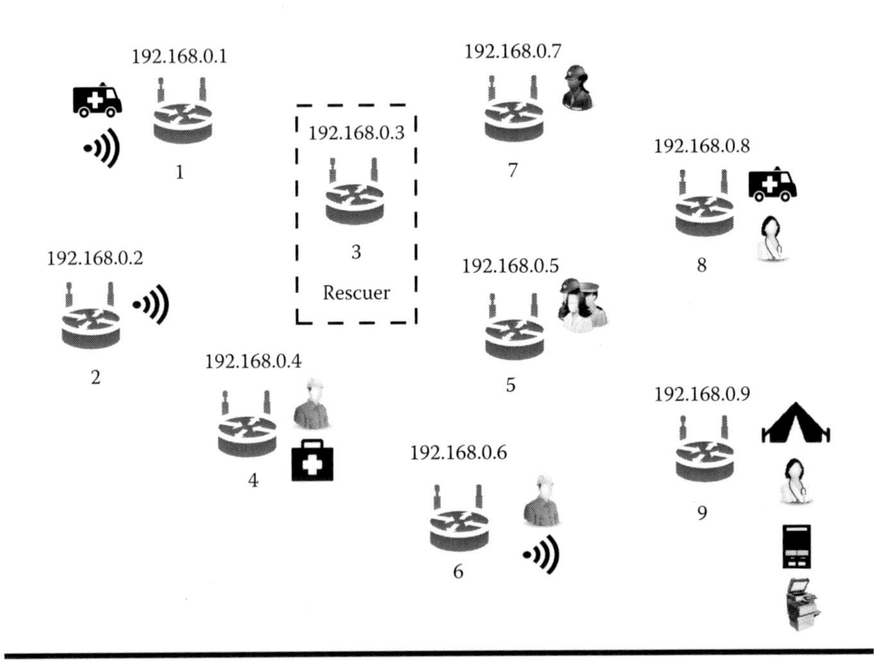

Figure 3.4 Network after service discovery.

required that omits the concept of addressing nodes with fixed identifiers and finds routes based on the services available at each node. As a result, the demand for an alternative approach was intense and Uisce grabbed attention by meeting that need.

Uisce is a characteristics-based routing protocol for MANETs designed and developed by Guoxian and Weber [14]. In this protocol, the term *characteristic* refers to a service. From that point of view, Uisce is very similar to other service-oriented routing protocols. However, unlike IP routing protocols it does not identify nodes with a fixed identifier. It even cannot differentiate nodes from one another. It only communicates based on characteristics, overlooking nodes that retain that characteristic. Because of this special property Uisce is called an *anonymous routing protocol*. Note that this chapter does not try to improve the performance of Uisce; rather the proposed transport protocol works on top of it with a view to forming an alternative paradigm.

The real-world analogy used for characteristic dissemination in Uisce is a natural river. The flow of water in a river is driven by gradients in the landscape, from high to low altitudes, and multiple rivers merge at a confluence to form a new river, which flows on with a combined water level. Uisce associates a characteristic with a potential value representing the water level of the river. At the source node of a characteristic, the potential describes the capacity of the characteristic. The dissemination of characteristic information is performed through broadcasting. A host node broadcasts its characteristic information periodically. It helps neighboring nodes to keep their knowledge up to date. One of the most significant issues in

Uisce is that a node only has information about the characteristics of its one-hop neighbors; the rest of the network is anonymous to it.

Uisce operates based on three functions and a parameter to control the potential changes in values during the propagation. A cost function takes the current potential as input and calculates a reduction of the potential. For each characteristic, all nodes in a network use a uniform specification of this function. A cost compensation function is applied to reduce the potential value cost along the hop. It provides individual nodes with the flexibility to affect the characteristic flow that runs through them. Finally, the third function, known as the *fusion function*, calculates the potential of multiple flows merging into one. In addition to these functions, Uisce has a control parameter called w_{dif}. A node that contains a characteristic with potential w accepts an incoming flow of the same characteristic only if the incoming potential is not lower than $(w - w_{dif})$.

The specifications of these functions may vary from characteristic to characteristic. In order to have them operate correctly, a condition is given that confirms the stability of the characteristic and precludes any reverse flow. Uisce sets the maximum value of the cost compensation function to c_{max} and the minimum value for a characteristic to run further as *threshold*. Hence, the condition stands as follows:

$$w_{dif} + c_{max} < threshold \tag{3.1}$$

In short, the higher the capacity value of a characteristic, the further it may propagate and the closer to a characteristic source, the higher capacity could be sensed. To achieve an optimal path, data messages in Uisce are forwarded towards the direction of potential gradient that balances route selection between the distance and the source node.

Uisce has an on-demand and temporary characteristic known as the *trace characteristic* that is created by the source node. This temporary characteristic enables a return path between destination and source for each data packet. This special characteristic is, however, allowed to propagate only within a limited region and its lifetime is also very short. Despite the fact that Uisce is an autonomous routing protocol, the proposed transport protocol will use this characteristic to send acknowledgment by making the communication bidirectional.

3.4 Challenges

The design aspect of a transport protocol for a rapidly deployable mobile environment involves many challenges. This involves (if not limited to) sequential access to the wireless medium, controlling the flow of data, and the consequences occurring due to link failure. These challenges often prevent the transport protocol from obtaining the best possible performance. Nevertheless, a trade-off between challenges and performance can be made if potential handles are observed, studied, and addressed during the design phase of the protocol.

Wireless networks are different compared to wired networks when they communicate with multiple stations simultaneously. For example, when multiple stations in the Ethernet try to access the medium at the same time, the MAC of the transmitting station identifies the collision using a collision detection mechanism and switches to a retransmission phase based on an exponential random back-off algorithm. Although this technique is very effective in a wired LAN, it is not a suitable solution for wireless. This is mainly for two reasons. First, implementing a collision detection mechanism would require implementation of a full duplex radio, capable of transmitting and receiving at the same time. This approach would increase the price significantly. Second, in a wireless environment, we cannot assume that all stations can hear each other, which is the basic assumption of the collision detection scheme. In wireless, when a station is willing to transmit and senses that the medium is free, it does not necessarily mean that the medium is free around the receiver area. In order to overcome these two problems, the 802.11 standard, instead of using a collision detection mechanism, utilizes a different approach known as *collision avoidance* together with a positive acknowledgment scheme. This collision avoidance inherently makes all communications sequential [15].

Figure 3.5 explains this scenario with more context. It shows a four-hop path with five wireless nodes and stations placed in such a way that there is a fixed distance between each of them. In this scenario, when Node A tries to access Node B, Node B cannot access its neighbor. Even Node C needs to remain idle to avoid interference. At this given time, only Node D can access its neighbor Node E or vice versa. This sequential nature of the wireless medium eventually creates dependencies between nodes and decreases accessibility of the whole path at a given time. Most transport protocols currently used in MANETs are built for wired networks and do not account for this sequential nature of 802.11 MAC. In a multihop network, such transport protocols inappropriately interpret delivery delay caused by this sequential medium access and trigger remedies that are meant for something else.

This medium access problem is also responsible for obstructing reverse flow on the path. Because acknowledgment is an integrated part of any reliable transport protocol, naturally a flow gets created opposite to the flow of the data packet. Most transport protocols including TCP have segment-by-segment acknowledgment, and for each individual or group of segments the receiver node sends a message acknowledging successful delivery of the respective segments. This acknowledgment message creates a reverse flow. As described earlier, the wireless medium is

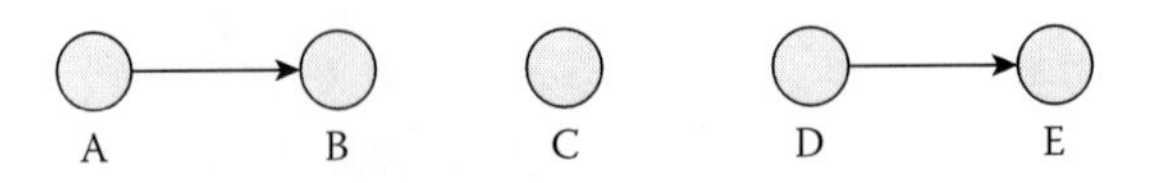

Figure 3.5 Sequential access to the wireless medium.

sequential in nature; while accessing nodes, this reverse flow becomes an obstacle for the forward flow and vice versa.

The process of identifying congestion in the network is also an important task for any reliable transport protocol. Often segments get lost in the network due to congestion, and transport protocols need to retransmit the missing segments. In the course of this process, some segments that are assumed lost may arrive at a later time. TCP usually suffers from this kind of problem when routing at the network layer is performed over a long multihop path in a wireless network. This delayed arrival of segments is interpreted by TCP as out-of-order data and it resends acknowledgment for the last ordered segment. The TCP receiver node continues to do this every time it receives such a delayed segment and, in response, the TCP sender node moves into the fast retransmission phase after receiving at least three acknowledgments. This special phase of TCP flow and congestion control assumes that the segment that immediately followed the delayed segment was also lost and therefore retransmits it without waiting for the timers to expire. This mechanism is useful in wired networks, as it helps maintain the flow even if the network is congested. However, in a wireless network that is not the case, and this misinterpretation unnecessarily triggers congestion control.

Moreover, round trip time (RTT) and retransmission time-out (RTO) in TCP are designed based on assumptions obtained from wired networks. However, in a wireless environment an unexpected situation often occurs and deceives TCP into calculating RTT and RTO inappropriately. We will revisit some of these challenges in the next section and see how they affect TCP flow control and end-to-end throughput if they are not properly taken care of.

3.5 TCP Flow Control

TCP, the most common transport protocol, is not suitable for a rapidly deployable wireless environment in emergency rescue missions mainly for two reasons. First, we have already seen that TCP is designed based on assumptions obtained from wired networks, and it therefore fails to achieve the performance it usually demonstrates in wired networks in a wireless environment. Second, applications that are used in rescue missions are different and largely differ from the kind used in everyday life. As a result of these combined factors, TCP flow control often triggers inappropriate actions that result in poor performance. In this section, we are going to review those inappropriate behaviors exhibited by TCP and pave the path for a potential solution to be presented in the next section.

TCP follows window-based flow and congestion control mechanisms [16,17]. Its flow control is closely tied to its congestion control and operates in a synchronized fashion. One of the blunt decisions these flow and congestion controls make is to identify any delay as congestion. Although this assumption is correct for wired networks, in a wireless environment delay could occur for a number of reasons, including link failure, temporary isolation, rapid movement, poor signal, and so on. TCP is completely unaware of these situations and triggers congestion control as soon as it identifies any delay [18].

TCP congestion control is divided into four phases: *slow start*, *congestion avoidance*, *fast retransmit*, and *fast recovery*. Among these four phases, slow start and congestion avoidance are responsible for slowing down data flow. Slow start is the initial phase of any TCP connection—it slowly probes the network and starts increasing the pace of the data flow. It holds two variables, namely congestion window (*cwnd*) and slow start threshold (*ssthresh*). When a new connection is established, TCP sets *cwnd* to one segment size (a segment size is either the size announced by the other end or 512 bytes by default) and *ssthresh* to 65,535 bytes. Every time an acknowledgment (ACK) is received, the *cwnd* is increased by one segment. This mechanism limits the sender to transmitting up to the minimum of the *cwnd* and advertised window (*rwnd*). On receiving each ACK, the TCP sender increases *cwnd* once, creating a provision for sending more data next time. This process, in fact, gives slow start an exponential breakthrough over RTT.

Slow start continues until *cwnd* is less than or equal to *ssthresh*; otherwise, it gets phased out and TCP flow control enters into congestion avoidance. Every time an ACK is received, this phase increases *cwnd* by $segsize^2/cwnd$, where *segsize* is the segment size. Thus *cwnd* grows linearly compared to slow start, which grows exponentially. During this phase, the increment of *cwnd* gets fixed by a maximum of one segment regardless of the number of ACKs received.

In event of a time-out, TCP assumes that congestion has occurred and invokes congestion control by resetting the *cwnd* size to one segment and triggering slow start again. Nevertheless, this time slow start continues until TCP reaches half of the size of the window when congestion occurred and then it swiftly moves to congestion avoidance. During this phase, TCP window growth will be much more conservative and, if no losses are detected, *cwnd* will grow no more than one segment per round trip. As TCP performs these phases assuming that network is congested, it unnecessarily slows down the data flow in event of link failure, temporary isolation, and so on and requires a longer period to get back to normal mode, even if new paths are established or nodes get out of temporary isolation quickly. This observation suggests that window-based flow control is not suitable for wireless networks. Link failure or situations involving temporary isolation can separate one or more nodes from other parts of the network for a specific period. The transport protocol needs to know when such events occur and when nodes are getting back to normal business. Rate-based flow control can be more effective in performing that job.

3.6 Overview of Beatha

Beatha is designed to overcome the challenges that transport protocols encounter in a rapidly deployable wireless environment. The primary objective of this protocol is to provide reliable end-to-end transport layer support for applications used in disaster recovery and rescue operations over characteristics-based anonymous communication. The following subsections describe components of Beatha in brief.

3.6.1 Connection Management

Transport protocols with reliable support do not communicate only on an end-to-end basis. They operate in a process-to-process manner through connection-oriented communication for specific streams. Although the use of Uisce benefits Beatha in many ways, in turn it creates a new problem for Beatha. Facilitating an anonymous underlying routing protocol to work with a connection-oriented continuous data stream seems difficult. Because Uisce is only able to see nodes within a one-hop distance, it is blind to the network and can recognize only its immediate neighbors.

Figure 3.6 gives more insight into this matter by illustrating an abstract model of how packets are seen from Uisce's point of view. It shows Node A transferring two sets of data, represented by the white and the grey packets, while Node C is transferring just one set, represented by the black packets. In this example, a set of data indicates a stream that may simply contain a file of multiple segments. Although segments from two different files are handed over to Uisce at Node A, it cannot identify those segments separately. To make things worse, it also cannot report which segment belongs to who while delivering those segments. For example, Node F receives packets from both A and C that Beatha receives in random order from Uisce.

This problem can easily be solved in TCP/IP with the help of creating a socket that involves an IP address and a port number. This combination is unique for any stream at a particular time and helps keep track of segments of the same stream. In the proposed alternative paradigm, however, the characteristic is the only tracing

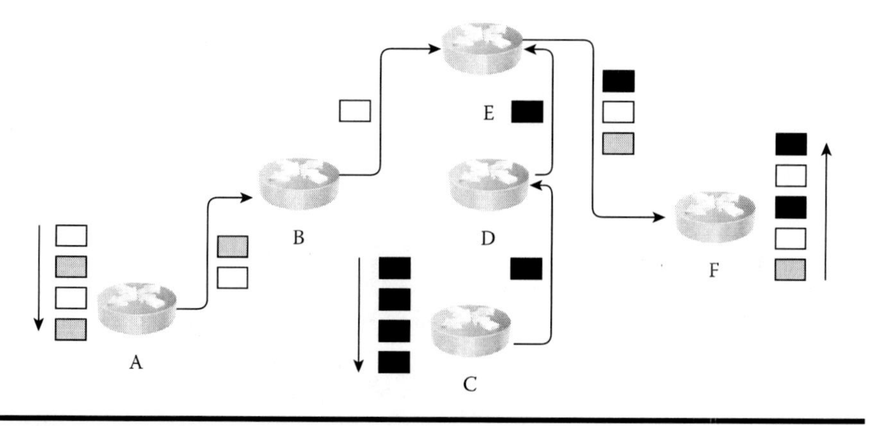

Figure 3.6 Packets from Uisce's point of view.

element to discover a node from the network. In order to take advantage of this in a wireless environment, earlier we deliberately gave up the uniqueness of any tracing element that makes it possible to have the same characteristics in multiple nodes. Therefore, a transport protocol operating over anonymous networks cannot correlate the concept of IP address–based connection management and requires a different mechanism to distinguish nodes and their streams from each other.

Beatha connection management consists of three elements: connection frame (CF), active connection table (ACT), and active reception table (ART). Each connection is associated with a set of specific variables that controls the procedure. Amongst those variables, members of CF act as the identification property of each stream. CF is comprised of a unique combination of three different variables: requested characteristics (RC), source tracker (ST), and destination tracker (DT), where RC is the characteristics requested by the application, ST is a randomly generated number for the source, and DT is another randomly generated number for the destination. Note that ST and DT are generated by the source and destination nodes, respectively.

In Beatha, ACT and ART are two tables maintained by the sender and receiver nodes, respectively. The ACT contains the CF of all initiated requests, whereas the ART contains the CF of all accepted requests. The entries in these two tables are finite and the maximum number of connections that can be initiated and accepted is controlled by a connection threshold called C_{thresh}. When Beatha receives a communication request from its application layer, it generates a CF and creates an entry in the ACT but leaves the DT field empty. It then sends a connection request with RC. Uisce, the underlying network protocol, is responsible for delivering this request to a node that holds the requested characteristic. When the node on the other end receives this request, it accepts it by generating a connection acceptance message with a DT. It also creates an ART entry for this connection. A connection is said to be established when the source Beatha receives a completed CF from the receiver and acknowledges it with a connection establishment message.

Once a connection gets established between two nodes, they enclose a CF within each data or control message to differentiate communication streams from each other. Despite the fact that the underlying network layer is anonymous, because of the presence of this connection frame, a receiver node can explicitly identify segments of appropriate streams.

3.6.2 Segment Management

Conventionally, transport protocol receives a data stream from the application layer and chunks them into small pieces called *segments*. These segments are the smallest unit of data at the transport layer. Each segment contains the necessary transport protocol header in front of its payload. When the transport layer passes its segments to the network layer, those segments are then encapsulated into a packet and routed to other nodes. This work maintains this convention of data flow at different layers.

For each connection, Beatha has a data buffer to hold its segments and keep track of associated variables. There are two types of buffers: the source node, which is responsible for initiating connections and sending data, maintains the outgoing segment buffer, and the receiver node, which acts as the destination, maintains the incoming segment buffer (ISB).

3.6.3 Acknowledgment

Beatha is designed especially to avoid reverse flow in data communication. Unlike TCP, Beatha does not have a segment-by-segment acknowledgment scheme. In fact, it never sends an acknowledgment to the sender in order to ensure successful delivery. It only sends an acknowledgment when it does not receive a segment. Specific to Beatha, this is called *negative acknowledgment* (NAK).

In Beatha, segments are organized in chronological order and the sender node sends them sequentially. It helps the receiver node to understand the arrival process more effectively. When a segment arrives, if Beatha finds that one or more segments have a smaller sequence number than the most recently received segment but have not arrived yet, it starts individual timers for all of them and waits for a specific period. If the timer expires but those segments are still due to arrive, the receiver node sends a NAK requesting the sender to retransmit missing segments.

This acknowledgment procedure works based on three variables that are located in the ISB control information. These variables are *isReceived*, *isSequenced*, and *isTimerSet*. Beatha initializes these variables with *false* at the time of establishing the connection. Subsequently, the values of these variables get updated and help the sender maintain the retransmission process smoothly. Briefly, when a receiver node receives a segment Beatha checks three things. First, it checks whether the segment has been received or not. If not, then it marks the associated *isReceived* variable *true* and checks whether the immediately previous segment is sequenced—that is, whether all the segments above the immediately previous segment have been received. If that segment is sequenced, Beatha marks *isSequenced* of the received segment *true*. However, if the immediately previous segment is not sequenced, there are two possible situations: either that segment has not been received yet or at least one segment above is yet to arrive. So Beatha starts checking *isTimerSet* for all the segments upwards that have not yet been received. If any such segment is found for which *isTimerSet* is false, Beatha starts a timer for it. If the respective segment does not arrive before the timeout, the receiver node sends a NAK to the sender to retransmit the segment and sets another timer for it. The details of these two timers are discussed later.

3.6.4 Retransmission Timer Management

Unlike the general convention for retransmission timer management, which is located at the sender's end, the retransmission timer in Beatha is controlled from the receiver's end. As described in Section 3.6.3, when the receiver experiences a time-out of an expected segment, it sends a NAK to the sender, requesting it to send the segment again. The receiver's end maintains a timer for this NAK and sends another NAK if it does not receive the data segment before the time-out. Note that Beatha also uses the same timer when it waits for an expected segment.

Most transport protocols including TCP use Jacobson's algorithm to calculate RTT and retransmission time-out (RTO). This algorithm was first introduced for TCP in the Request for Comments (RFC) 2581 [16]. However, later calculation of RTT and subsequently RTO was modified through an updated version of this algorithm described in RFC 2988 [17]. This new version introduced RTT variance instead of a fixed RTT calculation.

As per this algorithm, once the first RTT measurement (say, R) is made, the host calculates two variables as follows:

$$\text{RTT} = R \tag{3.2}$$

$$\text{RTTVAR} = \frac{R}{2} \tag{3.3}$$

The subsequent RTT measurement will then maintain the following equations:

$$\text{RTTVAR}_{\text{new}} = (1-\beta) \times \text{RTTVAR}_{\text{old}} + \beta \times |\text{RTT}_{\text{old}} - R| \tag{3.4}$$

$$\text{RTT}_{\text{new}} = (1-\alpha) \times \text{RTT}_{\text{old}} + \alpha \times R \tag{3.5}$$

In the above equations, R is the original RTT of the last segment, RTT is the measured round trip time, and RTTVAR is the RTT variance. α and β are two constants, defined as 1/8 and 1/4, respectively.

However, Jacobson's algorithm proposed for TCP involves operations based on feedback from previously travelled segments. As Beatha does not send regular acknowledgments, it is not possible to use this algorithm directly. Nevertheless, in order to adapt this algorithm to Beatha some necessary modifications are introduced in this chapter.

Beatha requires a timer based on the RTT when it sends NAKs for missing segments and waits for retransmissions from the sender's end. One of the difficulties in the RTT calculation in Beatha is its unconventional acknowledgment approach. Because the receiver never acknowledges any segments, it is not possible to get an RTT from data segments. Hence, RTT in Beatha is not calculated based on a real round trip; instead, it maintains a virtual return time to incorporate Jacobson's algorithm within it. This calculation is performed as follows: the Beatha sender includes a time stamp in the header of each segment it sends to the receiver's end. On receiving a segment, the receiver extracts this time stamp and deducts it from the current time to get the travel time (say tt) from the sender's to the receiver's end. Soon after, it calculates R_v using the following formula to replace R in Jacobson's algorithm.

$$R_v = 2.5 \times tt \tag{3.6}$$

There is a justification behind calculating R_v in this way. We saw earlier that Beatha segments receive a privilege when they access the medium. They do not face a reverse flow created by acknowledgments. However, when a NAK passes through a long multihop path, it faces a delay due to the constant flow of data segments. In order to adjust this delay, the segment travel time is multiplied 2.5 times instead of just doubling it.

Hence, by replacing R by R_v in Equations 3.4 and 3.5, we get

$$\text{RTTVAR}_{\text{new}} = (1-\beta)\times \text{RTTVAR}_{\text{old}} + \beta \times |\text{RTT}_{\text{old}} - R_v| \tag{3.7}$$

$$\text{RTT}_{\text{new}} = (1-\alpha)\times \text{RTT}_{\text{old}} + \alpha \times R_v \tag{3.8}$$

The RTT obtained from the above equations is used in Beatha to calculate the RTO. In this process, the first RTT is straightforward R_v and will be calculated just after receiving the connection request. Once the connection is established, Beatha starts calculating the RTT based on Equations 3.7 and 3.8.

Earlier, we saw that Beatha sets a timer on two different occasions: it sets a timer for expected segments and also sets a timer after sending NAKs. Although these timers are set for two different reasons, one single retransmission timer is sufficient to serve the purpose. The following expression as a function of RTT describes RTO calculation in Beatha.

$$\text{RTO} = fx(\text{RTT}) \tag{3.9}$$

where

$$fx(\text{RTT}) = 1, \text{ if RTT} < 1.0$$

$$fx(\text{RTT}) = \text{RTT} + 4 \times \text{RTTVAR}, \text{ if } RTT \geq 1.0$$

The calculation is carried out in this manner in order to adjust the unpredicted delay that the segment may face due to numerous reasons produced by the wireless environment. Sometimes it happens that a segment or group of segments face delay due to link failure or temporary isolation. However, this does not indicate that the segments are lost. During the design phase of Beatha, it was observed that an RTO having more than a 1-second period allows the network enough time to find a new route. Nevertheless, when the RTO is very small, it results in a time-out. This time-out triggers a retransmission request for a segment or group of segments that have not yet been lost. In order to reduce this kind of duplicate transmission, Beatha sets a 1-second minimum threshold for its retransmission time-out.

3.6.5 Flow Control

The flow control is a process of managing the data transfer rate between the sender and the receiver. The amount of data a sender can send at a given time towards a particular receiver is decided by the flow control scheme of the sender. The flow control plays an important role in a rapidly deployable wireless network, as it needs to change the data rate on the fly. Unlike window-based TCP flow control, Beatha opted out of rate-based flow control because it helps establish more command over the data transfer.

A transport protocol views the network as a black box and sends segments to lower layers to deliver them to the other end. The general assumption of sending a group of segments to lower layers significantly differs in wireless networks compared to a wired network. In a wired network, when a transport protocol tries to send multiple segments to the receiver's end, it assumes that they will take almost equal time to reach the destination from the source node. However, we have already seen that in a wireless network, the medium can only provide adequate support for the first segment and the rest face an increasing delay to get access to the medium. It eventually delays their round trip and gives a false interpretation to the transport layer. When the transport layer assumes that the previously passed segment should have been delivered by this time and provides another bunch of segments, it creates buffer overflow at the MAC layer and results in a time-out for multiple segments. In order to cope with this problem, Beatha uses flow control that is completely unrelated to acknowledgment and follows its adaptive nature to adjust its rate with the change of network topologies.

This flow control process (Figure 3.7) sends two different types of data blocks. The first block is called *complete block* (c_b). The length of this block is constant. Each complete block is again divided into small partial blocks (p_b). Beatha sends a whole p_b at a time and waits for a period of delay known as *internal delay* (d_{int}) and sends another p_b with a view to eventually sending a whole c_b. In between two c_b Beatha

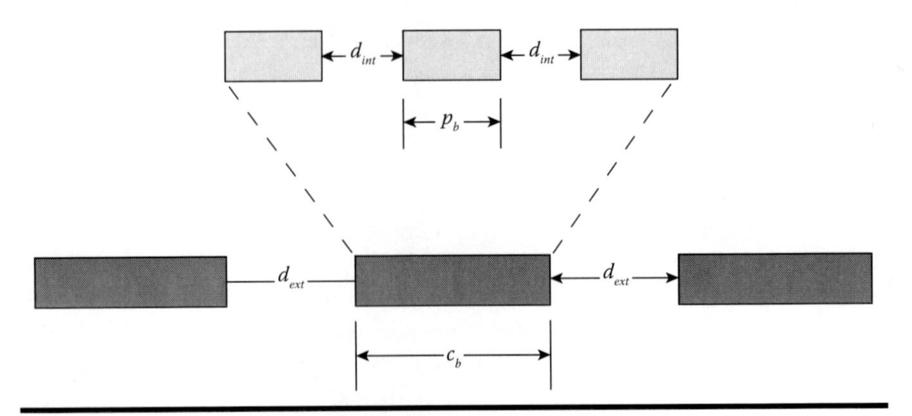

Figure 3.7 Beatha flow control basics.

uses another delay known as *external delay* (d_{ext}). Both d_{int} and d_{ext} are calculated based on the formulae explained later in this section.

Derivation of the Internal Delay (d_{int}): The reason for having internal delay is that it allows the p_b that has already been submitted to the lower layer enough time to get transferred to the other end. If T is the required time to transfer p_b from one end to another, it can be defined using the following formula:

$$T = TotalTrip \times TotalHop \times SingleHopTravelTime \tag{3.10}$$

In this equation, *TotalTrip* indicates the number of time Beatha needs to use the path from one end to another to transfer a p_b. *TotalHop* means the number of hops present on the path. Later it is denoted by *hop*. *SingleHopTravelTime* indicates the time a segment takes to travel over a single hop. It is later called t.

Before we move on, let us have a closer look at the capacity of a path. We know that a path having n hops cannot transfer n segments at a time [15]. Hence, the capacity of a path to simultaneously transfer segments is lower than the total number of free hops existing on that path. Figure 3.8 shows 10 different paths having different numbers of hops. It also shows the accessible hop on each path while the first hop is being used by a segment. A careful observation of this figure gives the idea that the capacity of a path is actually a series of 3. With an increment of every three hops, one extra available segment is created. Paths having one to three hops have the capacity to transfer one segment at a time, whereas paths with four to six hops have the capacity for two, those with seven to nine hops have three, ten to twelve hops have four, and so on. Therefore, the capacity of a path can be defined as follows:

$$\varsigma = \frac{hop}{3} \tag{3.11}$$

In this equation, the capacity is denoted as ζ and *hop* is the number of hops present on the path.

It was mentioned earlier that *TotalTrip* implies the amount of time Beatha needs to use a path from one end to another to transfer a p_b. The capacity of a path, however, varies (as described in Equation 3.11) and that causes *TotalTrip* to vary depending on the number of hops between two end points. Therefore, a path with ζ capacity requires the following number of trips to send a p_b:

$$TotalTrip = \frac{p_b}{\zeta} \tag{3.12}$$

The single-hop travel time (t) is a variable that Beatha updates on a regular basis. Upon receiving a NAK request, sender Beatha extracts the time stamp from the header to calculate the travel time (tt) for that particular NAK. Then it further

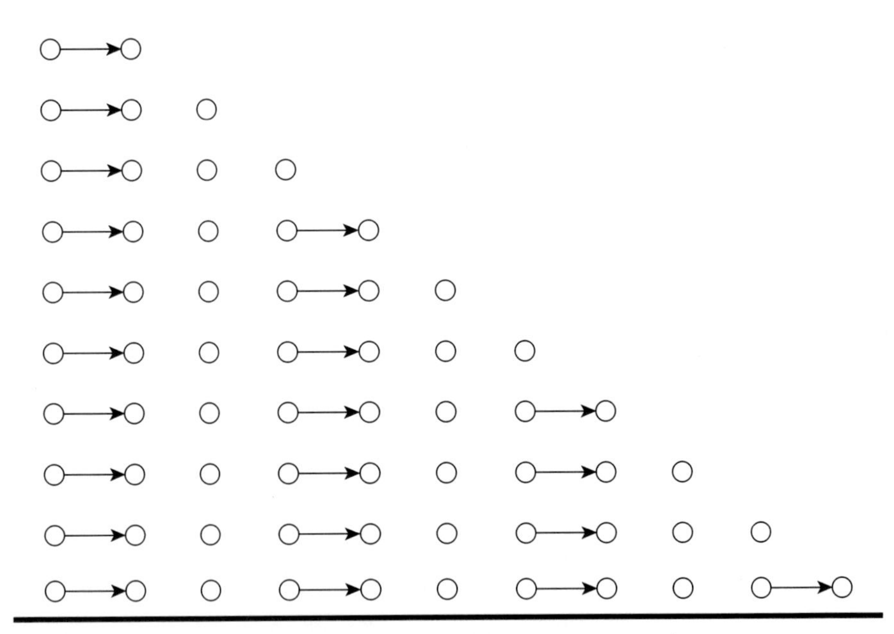

Figure 3.8 Number of segments travelling together on a path.

retrieves the number of hops that the NAK travelled on its way from the receiver's end to the sender's end. However, Beatha acquires this information from Uisce through a cross-layer information exchange. If receiver Beatha does not need to send a NAK for more than 5 seconds, it deliberately sends a message to update t at the sender's end.

$$t = 2 \times \frac{tt}{hop} \tag{3.13}$$

Replacing Equation 3.6 with Equations 3.7, 3.8, and 3.9, we get

$$T = \frac{p_b}{\varsigma} \times hop \times t \tag{3.14}$$

However, from the definition of d_{int} we know it is the time required to send a p_b. Therefore, T is actually d_{int}. So, finally we get

$$d_{int} = \frac{p_b}{\zeta} \times hop \times t \tag{3.15}$$

Derivation of the External Delay (d_{ext}): The main motivation behind the design approach of d_{int} was to estimate the delay over an uninterrupted path. However, in reality, nodes also act as intermediate routers. It may create further delay depending on the traffic on the path. To make the situation worse, in the event of link failure or temporary isolation, Beatha may need to slow down. Hence, its flow control process requires a feedback-based delay to control the rate adaptively and external delay (d_{ext}) serves that purpose.

The variable d_{ext} is designed in such a way that if a problem occurs and continues to give data transmission trouble, Beatha will be able to slow down its rate and gradually move back to the original flow once circumstances change. This delay incorporates three variables—*hop*, *t*, and *feedback*—and is defined as follows:

$$d_{ext} = \left(\left(hop - 1\right) \times t\right) + feedback \tag{3.16}$$

In Equation 3.16, *feedback* plays the role of adjusting d_{ext} with a change in the network topology. If Beatha suffers from segment loss, *feedback* increases itself to make d_{ext} larger, but as soon as the situation improves *feedback* starts decreasing itself to reduce d_{ext}. Because of its self-organizing nature, calculation of this variable is somewhat complicated. It works with the help of NAK and is controlled by two conditions: (i) it increases by 10% upon receipt of each NAK and (ii) it gets decreased by 50% every second. Let us consider that on an arbitrary second Beatha receives the first NAK. Thus, the change to *feedback* is as follows:

$$feedback_{new} = feedback_{old} + \left(feedback_{old} \times 0.10\right) \tag{3.17}$$

$$feedback_{new} = feedback_{old} \times (1 + 0.10) \tag{3.18}$$

If Beatha receives another NAK within that second, *feedback* looks like the following:

$$feedback_{new} = feedback_{old} \times (1 + 0.10) + \left(feedback_{old} \times (1 + 0.10) \times 0.10\right) \tag{3.19}$$

$$feedback_{new} = feedback_{old} \times (1 + 0.10) \times (1 + 0.10) \tag{3.20}$$

$$feedback_{new} = feedback_{old} \times (1 + 0.10)^2 \tag{3.21}$$

Therefore, the increment of *feedback* over a period of one second can be expressed as follows (where *n* is the number of NAKs received during that period):

$$feedback_{new} = feedback_{old} \times (1 + 0.10)^n \tag{3.22}$$

Nonetheless, according to the second condition, it gets decreased by 50% per second. Thus, this decrement can be expressed as follows:

$$feedback_{new} = feedback_{old} \times 0.5 \quad (3.23)$$

So, over a period of one second, the change in *feedback* can be expressed as follows:

$$feedback_{new} = feedback_{old} \times (1+0.10)^n - feedback_{old} \times 0.5 \quad (3.24)$$

$$feedback_{new} = feedback_{old} \times \left((1+0.10)^n - 0.5\right) \quad (3.25)$$

There might be a question as to why the increment per NAK received is 10% whereas the decrement per second is 50%. The answer lies in Equation 3.25. From this equation, it is clear that whether the value of *feedback* increases or decreases depends on n, the number of NAKs Beatha receives over a period of one second. During the development of Beatha, experiments showed that if segment loss occurs at a rate of up to four segments per second, Beatha can still operate without reducing the rate to achieve maximum throughput. However, if Beatha continues to do so for more than a four-segment loss, it bursts the network with excessive data. Based on this observation, the increment and decrement rates of feedback are set. In Equation 3.25, the multiplying factor with $feedback_{old}$—that is, $(1 + 0.10)^n - 0.5$—will remain smaller than 1 every cycle of a second for the value of n up to 4. It is always a new smaller value for $feedback_{new}$ that prevents d_{ext} from increasing. However, once n reaches 5 or more, the multiplying factor of $feedback_{old}$ grows to more than 1 and results in an increment in $feedback_{new}$. It eventually increases the value of d_{ext}. However, as soon as Beatha gets over with loss, *feedback* is exponentially decreased to trim d_{ext}. Because *feedback* is a delay, it is measured in seconds. Like all adaptive variables, *feedback* needs to be initialized before starting its operation. Its initial value is set to 0.1 second. It also maintains a lower threshold of 0.05 second—Beatha will not reduce *feedback* further once it reaches this threshold. The reason for having this lower bound is that if *feedback* ever reaches a value of 0, it will continue to be zero forever.

3.7 Experimental Setup

The proposed protocol was implemented and evaluated using the network simulator OPNET 14.0 [20]. This section describes the protocols, their configurations, scenarios, and the mobility patterns of Beatha.

3.7.1 Configurations

This simulator also provided the protocols used in the experiments except Beatha and Uisce. The implementation of Uisce was taken from an ongoing research project in the Distributed Systems Group of the Department of Computer Science at Trinity College Dublin, whereas Beatha was designed and developed by the authors. Microsoft Visual Studio.NET 2005 was used to compile the C++ implementation of all protocols used in this evaluation. The experimental hardware involved an Intel dual core 2.4 GHz machine with 3 GB RAM.

The OPNET simulator was configured with a 10×10 km^2 area and IEEE 802.11b was used as the MAC protocol. The following table shows the configuration of key parameters that were used in the experiments. The rest of the parameters were left on their default settings.

The goal of this evaluation is to compare the performance of Beatha to that of a conventional transport protocol. TCP Reno was carefully chosen as a representative of the conventional counterpart. A total of four pairs of protocols were used to conduct the experiments. Each pair was a combination of a network and a transport protocol. Beatha was paired with Uisce to form the Beatha–Uisce group, whereas TCP was paired with Dynamic Source Routing (DSR), Ad hoc On-Demand Distance Vector (AODV), and Optimized Link State Routing (OSLR) to form the TCP–DSR, TCP–AODV, and TCP–OLSR groups, respectively.

3.7.2 Simulation Scenarios

Three different scenarios were used in the experiments. The first scenario was called the *static line scenario*. This scenario was comprised of a number of paths with static nodes on them. Two end points on each path acted as the source and the destination. The number of hops on a path was varied in the experiments. Figure 3.9 shows the scenario where the nodes were placed with a gap of 70 m between each other. The transmission range of each node was set to 100 m, which is the most common range used in MANET simulations [21].

The second scenario was called the *moving line scenario*. It was comprised of eleven nodes placed as in the previous scenario; Node 1 acted as the source and Node 11 as the destination. Once the simulation began, each node started to move with a given speed ranging from 1 to 30 ms^{-1} in the same direction. When a node reached the edge of the simulation area, it hit the fence and changed its path depending on the deflecting angle.

The third and the last scenario was called the *mobility scenario*. It is illustrated in Figure 3.10, which shows how the initial placement was comprised of a 10-hop path surrounded by random nodes. On the path, two end nodes acted as the source and destination. Once the simulation began, the number of hops between the source and the destination started to change depending on the movement of the nodes.

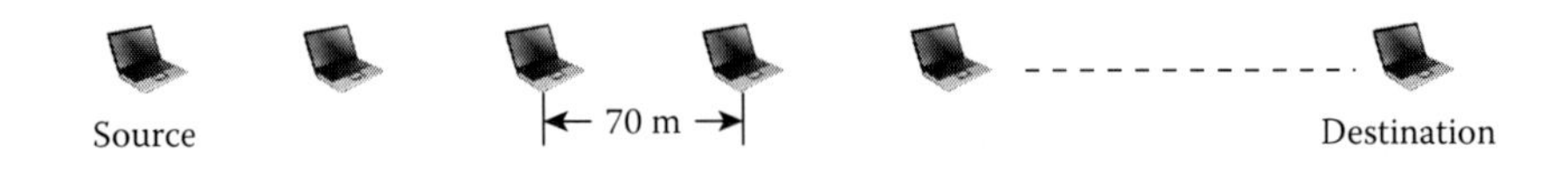

Figure 3.9 Static line scenario.

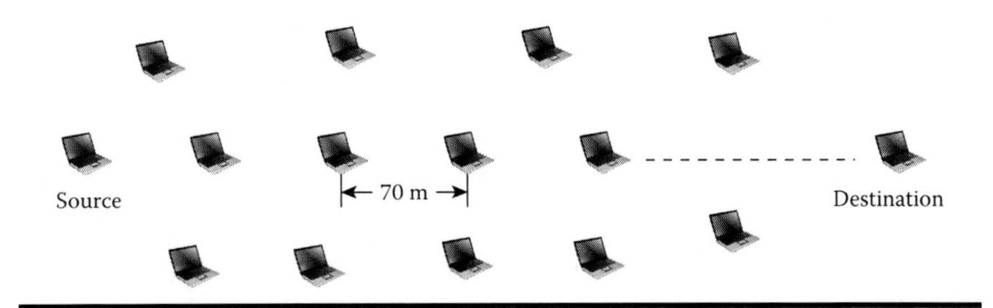

Figure 3.10 Mobility scenario.

Data rate (bps)	11 Mbps
Packet reception-power threshold (dBm)	–95
Short retry limit	7
Long retry limit	4
Max receive lifetime (secs)	0.5
Buffer size	256000
Transmit power (mW)	60

Figure 3.11 Parameter setting for IEEE 802.11b.

This scenario had two different sets of mobility patterns: the simple case, where nodes took a 90-degree turn after travelling every 100 m at a low speed between 5 and 15 ms^{-1}, and the complex case, where nodes took a 45-degree turn after travelling every 30 m at a speed of 20–30 ms^{-1}. In addition to these nodes in both cases, a total of 50 other nodes were placed in the simulation area that varied their speed between 5 and 15 ms^{-1} and changed direction randomly. Figure 3.11 shows an abstract view of this scenario.

3.8 Performance Evaluation

The objective of this performance evaluation was to analyze the Beatha protocol. Because this protocol was designed to address the problems transport protocols encounter when operating over a wireless network and quickly deploying emergency communication, this evaluation tried to demonstrate improvement by Beatha on particular matrices that were previously hindered by those problems.

3.8.1 Evaluation of Throughput

This part of the evaluation involved two experiments. The first experiment demonstrated measurement of throughput over long multihop paths when nodes are static. It showed the effect of hop increment on throughput. The second experiment took mobility into account and demonstrated how end-to-end throughput over long multihop paths changes in the presence of mobility.

The first experiment was conducted with the static line scenario. It involved a total of 10 paths with static nodes on them. The first path had only two nodes, separating the source and the destination by a single hop. The second path had three nodes, the third had four nodes, and the tenth had eleven nodes, separating the source and the destination by 10 hops. A file of 5 MB was transferred from the source to the destination to record throughput per second. It was obtained dividing the total size of the file by the time required to transfer it. This experiment was repeated 20 times and the average was taken to plot the graphs with positive and the negative errors.

It was previously mentioned that TCP exhibits poor performance over long multihop paths for a number of reasons, including sequential medium access of the wireless MAC protocol and window-based flow control. Figure 3.12 is evidence of this claim, showing a comparison of the performance of Beatha–Uisce to that of TCP–AODV, TCP–DSR, and TCP–OLSR over multihop paths. It is clear from the graph that all three combinations of TCP achieved decent throughput that started near 400 KB/s over the single-hop path. However, their performance worsened as soon as they started operating over long multihop paths and touched a throughput level near 20 KB/s on the 10-hop path. It is also notable that among the TCP groups, the TCP–AODV pair achieved the highest throughput (23.55 KB/s).

In order to improve transport layer performance over long multihop paths, a number of changes have been introduced in Beatha, as described earlier in this chapter. Its rate-based adaptive flow control is capable of adjusting rate with the change of topology, whereas its NAK-based acknowledgment reduces reverse flow significantly. As a result, Beatha performs better than TCP while operating on the same path. This experiment showed that Beatha achieved a throughput of 819.44 KB/s over the single-hop path. Later its performance degraded, but it still maintained a data rate of 151.51 KB/s on the 10-hop path, seven times higher than the average performance of the TCP groups.

The second experiment involved measurement of throughput in the presence of mobility. The goal of this experiment was to analyze the performance of Beatha over long paths with a verity of speed in the moving line scenario. Six speed patterns were used in this experiment: 5 ms^{-1}, 10 ms^{-1}, 15 ms^{-1}, 20 ms^{-1}, 25 ms^{-1}, and 30 ms^{-1}. For each instance, 20 readings were taken and the average was used to plot the graphs.

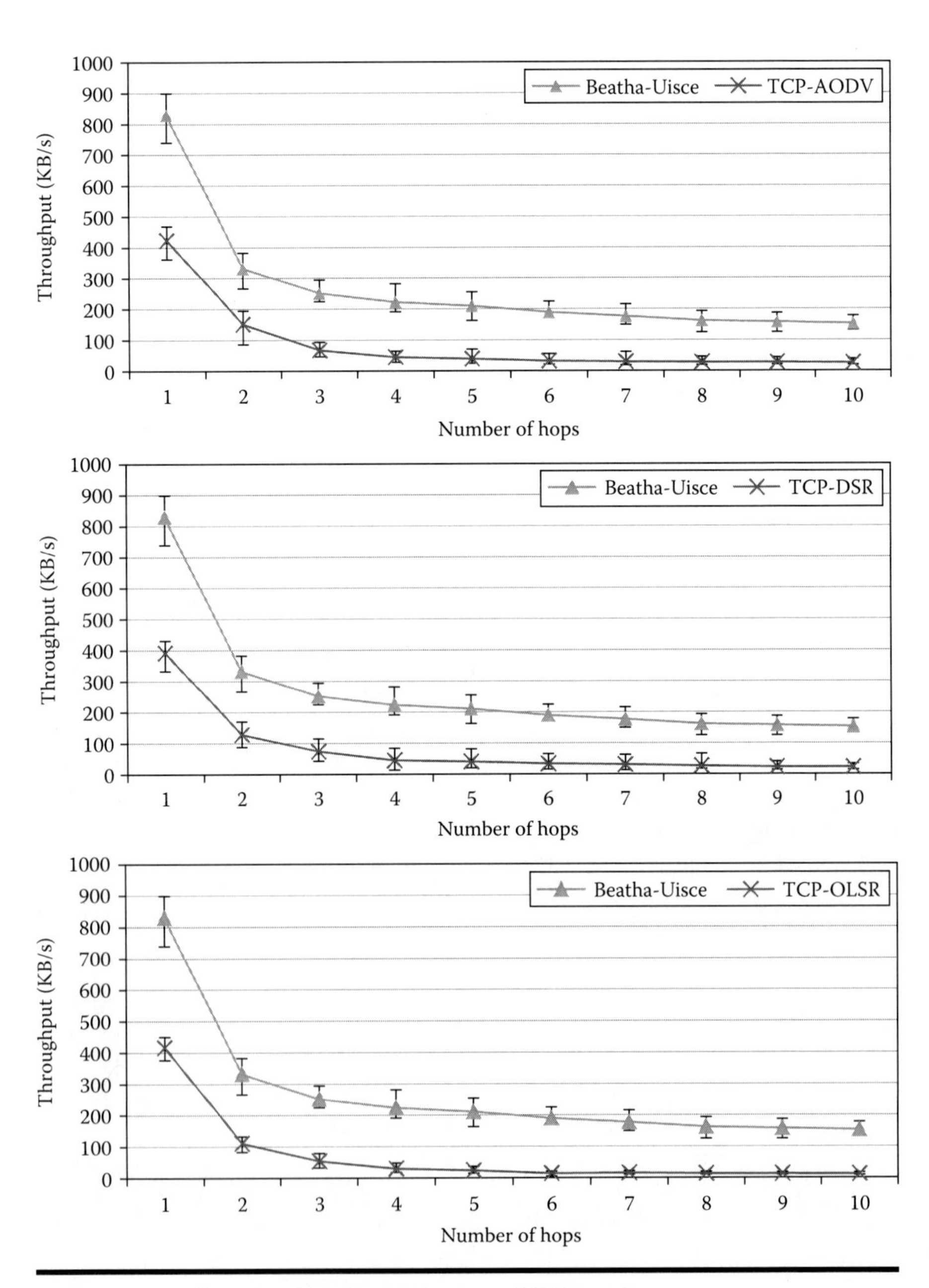

Figure 3.12 Throughput evaluation over multihop paths.

Transport protocols usually experience a drop in throughput when the nodes move with high mobility. In this kind of situation, TCP behaves differently depending on the routing protocols it uses. Figure 3.13 shows that at a lower speed the TCP–AODV pair performed better than the TCP–DSR and TCP–OLSR pairs. This is because OLSR is a proactive routing protocol that finds a path based on previously obtained information. As a result, in the event of path failure it cannot find a new path quickly. DSR is a proactive protocol, but it also uses some previously saved route caches to find paths between two end nodes. By contrast, AODV is a reactive and on-demand routing protocol that quickly recovers path failure. Because TCP performance over wireless networks is closely related to path failure [19], AODV served better than the other two pairs. Nevertheless, the performance of the TCP–AODV pair degraded at a higher speed (after 10 m/s) compared to that of the TCP–DSR pair. This is because at a higher speed, in the event of path failure, the DSR route cache frequently serves better than finding a new path.

However, this experiment clearly demonstrated that the performance of Beatha is not affected by these routing issues. Its performance is seven times better than the best TCP combination while operating with different degrees of mobility. It achieved a throughput level of 151.51 KB/s with no mobility and gradually became stable at 116.27 KB/s while operating at 30 ms^{-1} speed.

3.8.2 Evaluation of Retransmission

Two experiments were used to evaluate the retransmission performance of Beatha. These were conducted in conjunction with the previous experiments. While measuring throughput, measurements of retransmission were also made.

The first experiment was conducted in the static line scenario and the settings were identical to its experimental counterpart described earlier. It showed that TCP paired with OLSR provides the worst performance and that the TCP–DSR and TCP–AODV pairs act in a relatively similar way. However, the Beatha–Uisce pair retransmitted the least number of segments, boosting its performance to achieve the height throughput demonstrated in Figure 3.14.

The second experiment as conducted in conjunction with the second experiment described in the section "Evaluation of Throughput." The moving line scenario was used to evaluate the Beatha and TCP protocols. The settings of this experiment were the same as those described for the earlier experiment. This experiment demonstrated that the TCP–OLSR pair retransmitted more segments than any other pair but its performance was stable and did not fluctuate over the change of speed. The TCP–AODV pair suffered from more retransmissions when it operated at a speed of 10 ms^{-1} or more. However, Figure 3.15 demonstrates that Beatha performs better than any TCP pair with the least number of retransmissions while operating at different degrees of speed. Most importantly, this performance was stable regardless of the change of speed.

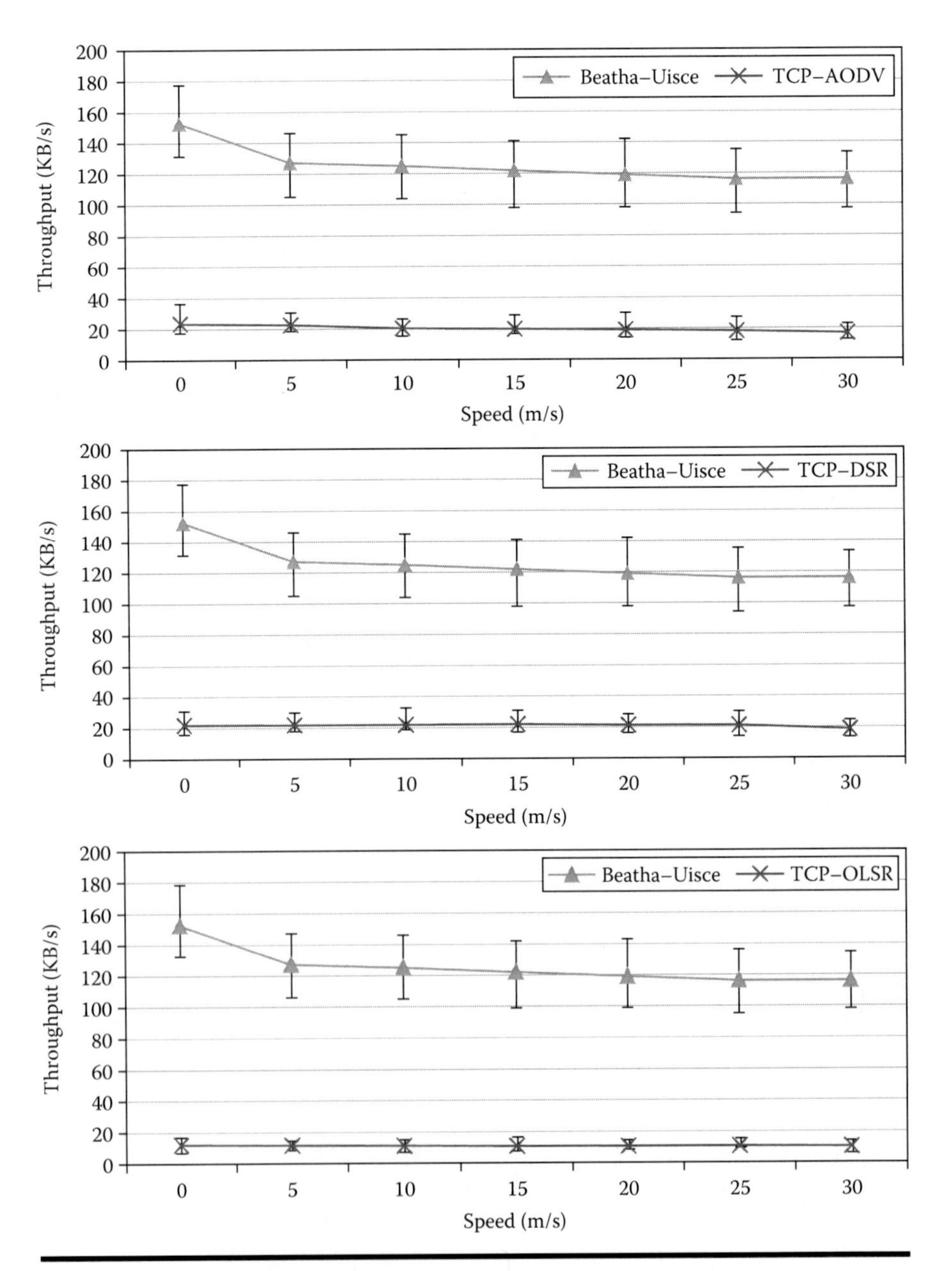

Figure 3.13 Throughput evaluation in the presence of mobility.

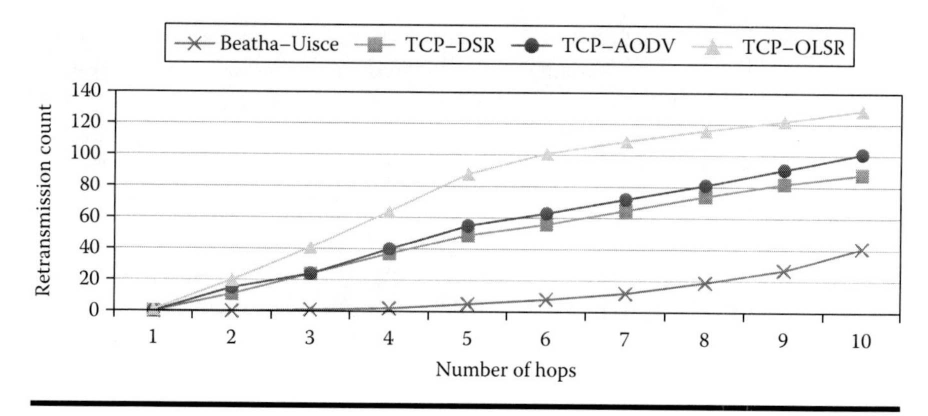

Figure 3.14 Retransmission of Beatha–Uisce and Transmission Control Protocol (TCP) with AODV, DSR, and OLSR over multihop paths.

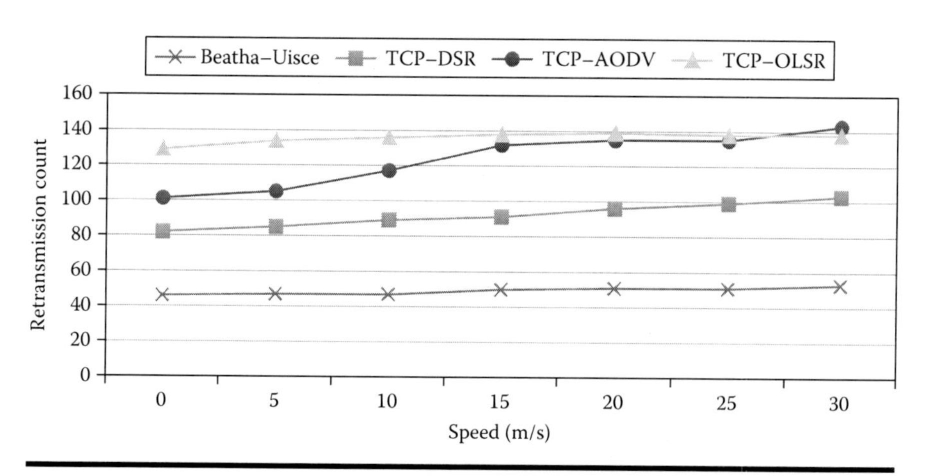

Figure 3.15 Retransmissions of Beatha and TCP in the presence of mobility.

3.8.3 Evaluation of Data Rate

This evaluation demonstrates the amount of data Beatha and TCP senders can deliver to the receiver per second. This evaluation is particularly important for demonstrating how TCP fluctuates over time while operating on long multihop paths. It also demonstrates the behavior of Beatha controlled by the flow control mechanism described earlier.

This evaluation took place in the mobility scenario. As described earlier, this scenario had two cases, one simple and one complex. There were separate experiments for each case. In order to demonstrate the exact picture, this time no average was taken; rather a first 30-second snapshot from both Beatha and TCP communication was taken to prepare the graph. The reason being, if the average were taken then the actual fluctuation in the data rate could not be shown. From the experiments demonstrated in the sections "Evaluation of Throughput" and "Evaluation of Retransmission," it is clear that the TCP–AODV pair performs better than the other two. Therefore, only this pair was used in this evaluation.

In the first experiment, the Beatha–Uisce pair was evaluated with the TCP–AODV pair. Figure 3.16 shows the amount of data both Beatha and TCP sent per second over a period of 30 seconds. It shows that TCP touched the ground a number of times, which means that on those occasions it completely failed to send data to the receiver. Moreover, TCP could not even reach half of the minimum level Beatha maintained while operating in the same environment. In contrast, Beatha showed stable behavior, always keeping the data rate between 120 and 160 KB/s.

The second experiment was conducted in order to demonstrate the deviation of performance in both Beatha and TCP when nodes change directions frequently. Figure 3.17 shows the amount of data Beatha and TCP sent in the complex case. It shows that TCP could hardly send data and touched the ground on several occasions. In this case the nodes moved randomly, causing frequent path failure. The results show that in the same situation Beatha maintained a relatively higher margin than TCP, though it also suffered from fluctuations. Beatha's performance can be interpreted with the help of Figures 3.18 and 3.19. Because Beatha flow control is engineered by its two delays, when those delays were increased,

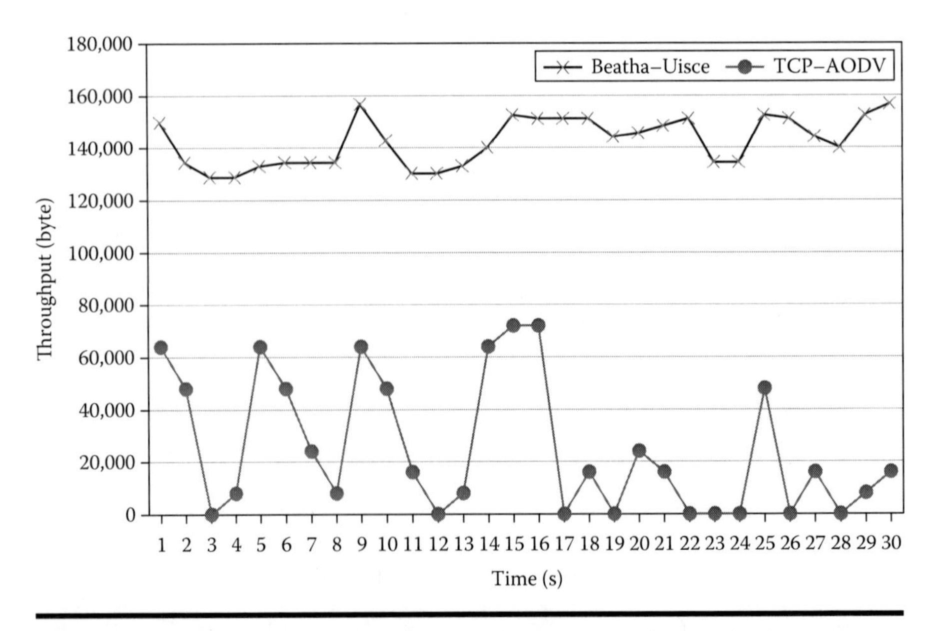

Figure 3.16 Data rate of Beatha–Uisce and TCP–AODV in the simple case.

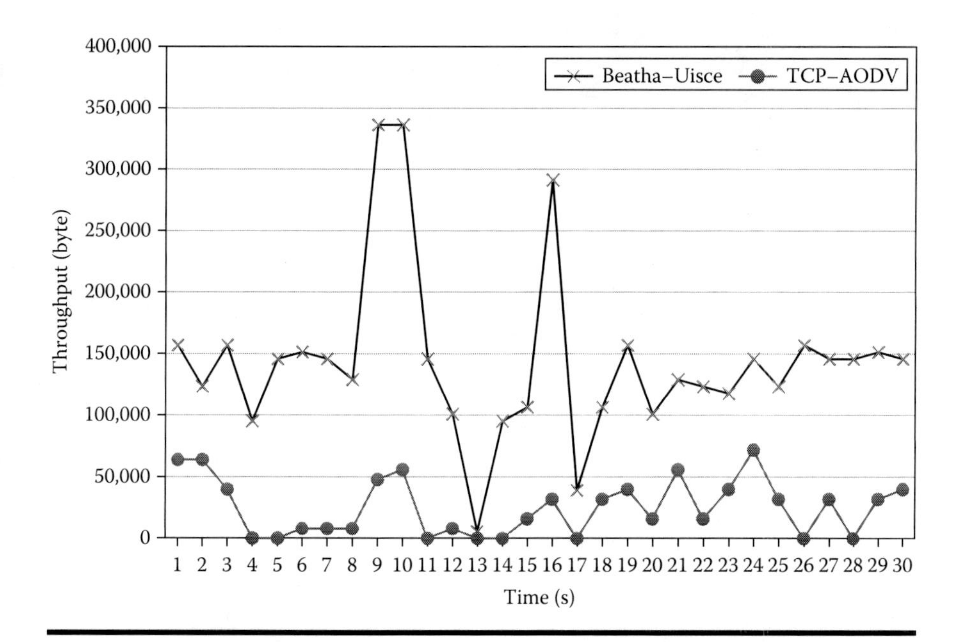

Figure 3.17 Data rate of Beatha–Uisce and TCP–AODV in the complex case.

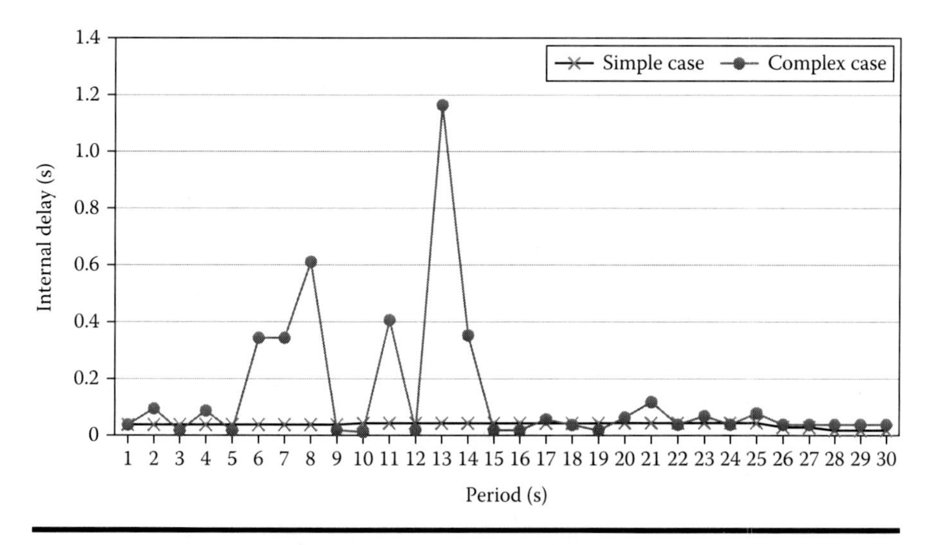

Figure 3.18 Internal delay of Beatha in the simple and complex cases.

the data rate decreased. In Figure 3.18, the internal delay reached 1.2 seconds at the 13th second of the simulation. During that period Beatha could not send any data and the data rate touched the ground (demonstrated in Figure 3.17). However, Beatha recovered from this situation very quickly and got back to a higher rate very soon because of its adaptively probing nature.

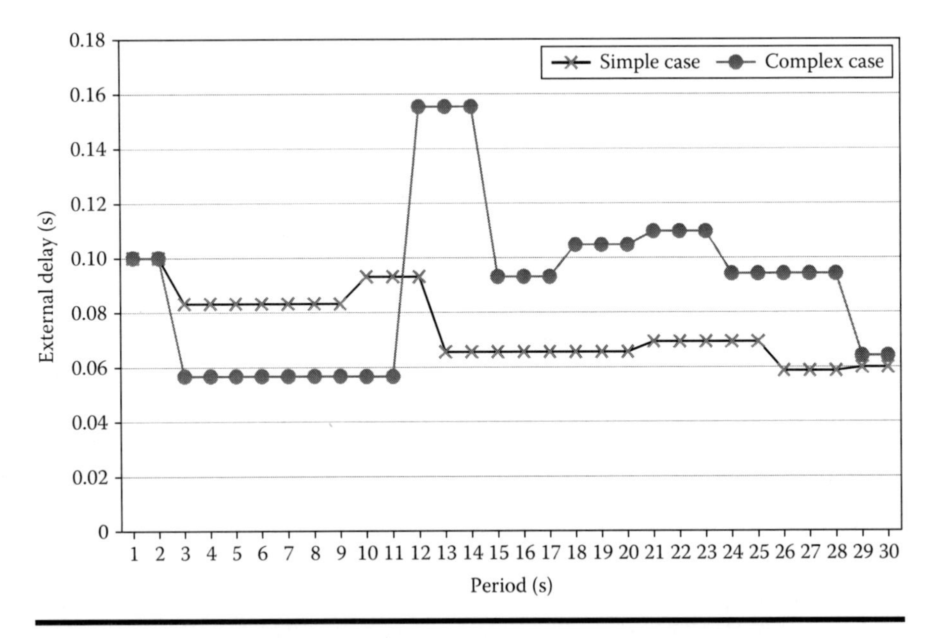

Figure 3.19 External delay of Beatha in the simple and complex cases.

3.8.4 Evaluation of Delay

This evaluation demonstrated the delay involved in delivering segments from the source to the destination node. Experiments for this evaluation took place in the mobility scenario and considered both the simple and complex cases to demonstrate the results. These experiments were conducted with the same configuration described in the section "Evaluation of Data Rate." The delay for all segments over a period of one second was accumulated first and then the average was taken by dividing it by the number of segments received during that period. As in the previous evaluation, the first 30 seconds from both the Beatha and TCP communications were taken to prepare the graphs. Figure 3.20 demonstrates that TCP exhibited mixed behavior with some fluctuations in the simple case. The graph representing TCP's performance is a broken graph, as at some points TCP failed to send any data whatsoever. However, Beatha demonstrated stable and continuous behavior without any breaks in the data transfer. Figure 3.21 demonstrates that in the complex case where nodes change their direction frequently, both Beatha and TCP exhibited poor performance. At some stage, the segment travel time in Beatha reached 1.8 seconds. Despite the fact that Beatha was having this fluctuation, its performance compared to TCP was better, as TCP failed to send data on a number of occasions. One key point in this graph is that the delay in TCP never reached especially high, but in Beatha it touched a couple of higher points. The reason for this behavior is that Beatha tries not to stop sending data even if the environment is hostile.

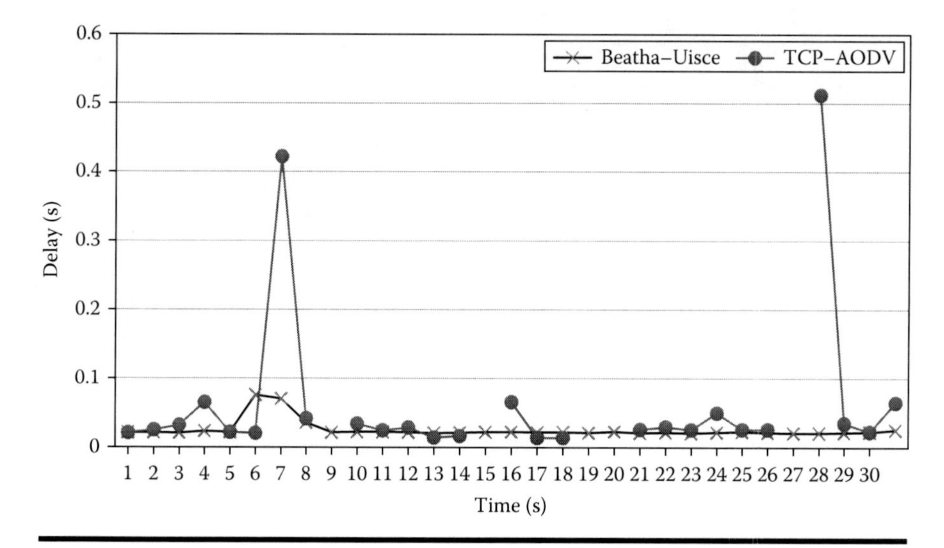

Figure 3.20 Beatha and TCP segment delay in the simple case.

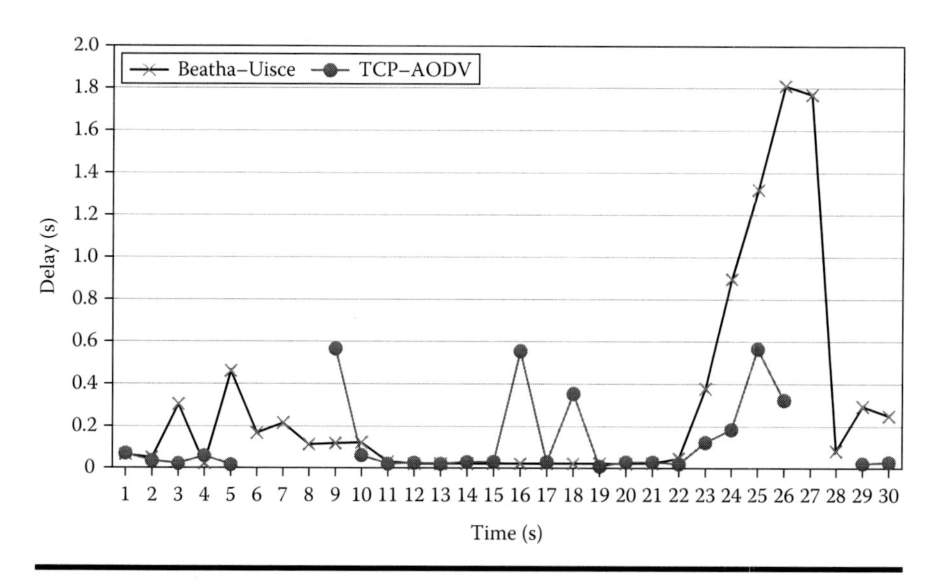

Figure 3.21 Beatha and TCP segment delay in the complex case.

3.8.5 Evaluation of Acknowledgment

This final evaluation demonstrates a comparison between Beatha and TCP based on the number of acknowledgments these two protocols sent over a particular communication. It was conducted in the static line scenario and used the same configuration as the first experiment described in the section "Evaluation of Throughput."

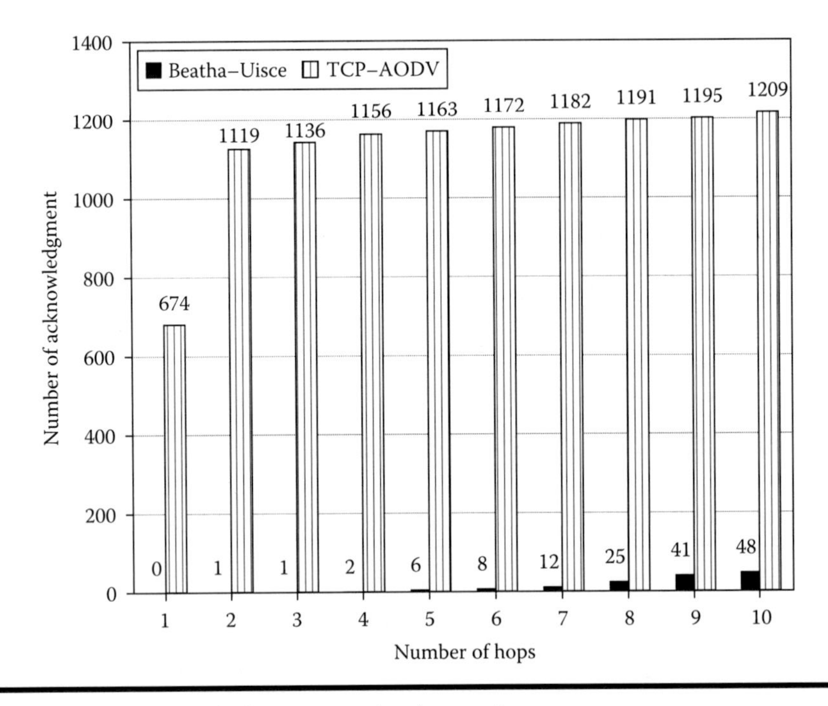

Figure 3.22 Acknowledgment sent by the receivers.

First a 5 MB file was transferred from the source to the destination, and later the number of acknowledgments required for maintaining reliability was recorded for both Beatha and TCP.

Figure 3.22 demonstrates that Beatha sent a very small number of acknowledgments for transferring a 5 MB file compared to TCP. On a single-hop path when TCP sent an average of 674 acknowledgments, Beatha sent nothing. In the worst case, on a 10-hop path Beatha sent an average of 48 acknowledgments, whereas TCP sent an average of 1209 acknowledgments.

3.9 Conclusion and Future Studies

The motivation for the work presented in this chapter arose from the observation that state-of-the-art research in supporting end-to-end reliable communication in quickly deployable wireless networks has failed to address the challenges of dynamic wireless environments where flow control needs to be adaptive and acknowledgment selective. As described in the section "Challenges," TCP exhibits poor performance in this kind of network because of its flow-control mechanism and behavior of triggering inappropriate congestion control in events where delay occurs for noncongressional reasons, mainly associated with the wireless environments. The proposed protocol addresses these problems and provides solutions to outperform those difficulties. The contribution of this chapter is twofold: it proposes a rate-based adaptive flow-control mechanism along with a NAK-based retransmission scheme that enhances end-to-end throughput in MANETs over long multihop paths and also co-founds a non-identifier-oriented architecture, an alternative to the TCP/IP architecture, for rapidly deployable emergency communication.

There are, however, some limitations in Beatha. Its flow control and acknowledgment mechanisms are designed to assume that clocks are synchronized. However, in reality that may not be the case and therefore a clock synchronization phase is required during connection establishment. Moreover, fairness and congestion control are two areas that remain completely unaddressed. Future study of Beatha also needs to pay attention to these gaps.

References

1. S. Kurkowski, T. Camp and W. Navidi, Two Standards for Rigorous MANET Routing Protocol Evaluation, in *Proceedings of IEEE International Conference on Mobile Ad Hoc and Sensor Systems*, Vancouver, BC, Canada, 2006.
2. N. Wang, Y. Huang and W. Liu, A Fuzzy-Based Transport Protocol for Mobile Ad Hoc Networks, in *Proceedings of IEEE International Conference on Sensor Networks, Ubiquitous and Trustworthy Computing*, Taichung, Taiwan, 2008.
3. T. Ramrekha and C. Politis, A Hybrid Adaptive Routing Protocol for Extreme Emergency Ad Hoc Communication, in *Proceedings of 19th International Conference on Computer Communications and Networks*, Zurich, Switzerland, 2010.
4. C. Fathy, T. El-Hadidi and M. El-Nasr, Fuzzy-Based Adaptive Cross Layer Routing Protocol for Mobile Ad Hoc Networks, in *Proceedings of IEEE 30th International Performance Computing and Communications Conference*, Orlando, FL, 2011.
5. A. Jangra, N. Goel and Priyanka, Efficient Power Saving Adaptive Routing Protocol (EPSAR) for Manets Using AODV and DSDV: Simulation And Feasibility Analysis, in *Proceedings of 2nd International Symposium on Intelligence Information Processing and Trusted Computing*, Hubei, China, 2011.
6. C. Murthy and B. Manoj, *Ad Hoc Wireless Networks: Architectures and Protocols*, Prentice Hall, Upper Saddle River, NJ, 2004.
7. B. Dilmaghani and R. Rao, Designing Communication Networks for Emergency Situations, in *Proceedings of IEEE International Symposium on Technology and Society*, New York, 2006.
8. M. Petrovic, V. Muthusamy and H. Jacobsen, Content-Based Routing in Mobile Ad Hoc Networks, in *Proceedings of 2nd Annual International Conference on Mobile and Ubiquitous Systems: Networking and Services*, San Diego, CA, 2005.
9. K. Wang, Adaptive Attribute-Based Routing in Clustered Wireless Sensor Networks, PhD thesis, Boston University, Boston, MA, 2006.
10. C. Rezende, B. Rocha and A. Loureiro, Publish-Subscribe Architecture for Mobile Ad Hoc Networks, in *Proceedings of ACM Symposium on Applied Computing*, Ceara, Brazil, 2008.
11. J. Saltzer, D. Reed and D. Clark, End-to-End Arguments in System Design, *ACM Transactions in Computer Systems*, 2(4), pp. 277–288, 1984.

12. M. Blumenthal and D. Clark, Rethinking the Design of the Internet: The End-to-End Arguments vs. Brave New World, *ACM Transactions on Internet Technology*, 1(1), pp. 70–109, 2001.
13. A. Carzaniga, M. Rutherford and A. Wolf, A Routing Scheme for Content-Based Networking, in *Proceedings of IEEE Conference on Computer Communications (INFOCOM)*, Hong Kong, 2004.
14. Y. Guoxian and S. Weber, Uisce: Characteristic-based Routing in Mobile Ad Hoc Networks, in *Proceedings of 10th IFIP Annual Mediterranean Ad Hoc Networking Workshop (Med-Hoc-Net)*, Sicily, Italy, 2011.
15. G. Holland and N. Vaidya, Analysis of TCP Performance over Mobile Ad Hoc Networks. Wireless Networks, in *Proceedings of 5th Annual ACM/IEEE International Conference on Mobile Computing and Networking*, 2002.
16. M. Allman, V. Paxson and W. Stevens, *RFC 2581: TCP Congestion Control*, NASA Glenn and Sterling Software, CA, 1999.
17. V. Paxson and M. Allman, *RFC 2988: Computing TCP's Retransmission Timer*, ACIRI and NASA GRC/BBN, CA, 2000.
18. J. Monks and P. Sinha, Limitation of TCP-ELFN for Ad Hoc Networks, in *Proceedings of the International Workshop on Mobile Multimedia Communication (MoMuC)*, Tokyo, Japan, 2000.
19. G. Holland and N. Vaidya, Impact of Routing and Link Layers on TCP Performance in Mobile Ad Hoc Networks, in *Proceedings of IEEE Wireless Communications and Networking Conference (WCNC)*, New Orleans, LA, 1999.
20. OPNET, *Modeling Concepts Reference Manual*, Technical report, OPNET Technologies, Inc., Bethesda, MD, 2008.
21. S. Kurkowski, T. Camp and M. Colagrosso, MANET Simulation Studies: The Incredibles. *Mobile Computing and Communications*, pp. 50–61, New York, 2005.

Chapter 4

Opportunistic Vehicular Communication: Challenges and Solutions

Ali Bohlooli

Contents

4.1 Introduction

Wireless applications have become significant in numerous fields [1] such as the auto industry. Indeed, the convergence of telecommunication, computation, wireless technology, and transportation technologies has contributed to the facilitation of our roads and highways as far as communications are concerned. This convergence in a sense is considered as a platform in intelligent transportation systems (ITS) where each vehicle is assumed to be equipped with devices as nodes in order to create contact with other nodes. Mobile ad hoc networks (MANETs) were introduced in Chapter 3. Because the features of a vehicle network are different from those of other types of MANETs, this network is called a *vehicular ad hoc network* (VANET) [2].

VANETs are an important part of the ITS, which has vast application potential and commercial value. It is considered one of the requirements for developing smart cities. A VANET is a special kind of MANET that supports cooperative driving among communicating cars on the roads. Vehicles act as communication nodes and relays, forming dynamic vehicular networks together with other nearby vehicles on the roads. As shown in Figure 4.1, in a VANET, packets are exchanged between mobile nodes traveling on constrained paths in the following order [3]:

Vehicle-to-vehicle (V2V) communications
Between vehicles and roadside access points (APs) (aka roadside units, RSUs) called vehicle-to-infrastructure (V2I) communications

VANETs share some common features with the general MANET. Both VANETs and MANETs are characterized by the movement and self-organization of the nodes. However, there are several features that distinguish VANETs from MANETs [3,4]. The first difference is in the networks applications. VANETs can support informative services, audio/video streaming, and generalized entertainment, which are mainly designed to improve the quality of transportation through time-critical safety and traffic management applications [5]. Moreover, vehicular applications demand strict

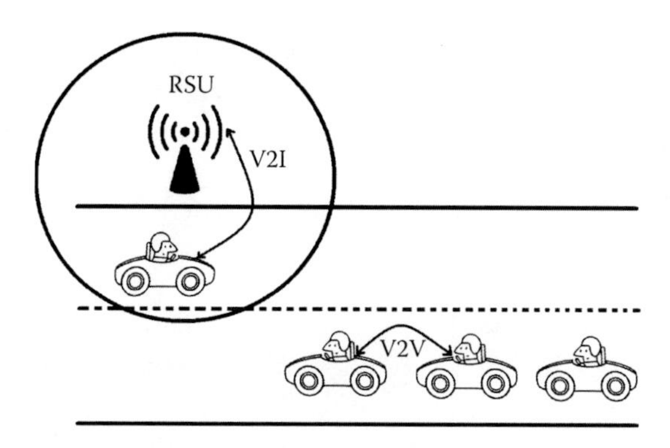

Figure 4.1 Vehicle-to-vehicle (V2V) and vehicle-to-infrastructure (V2I) communications in a vehicular ad hoc network. RSU, roadside unit.

communications performance (e.g., timely and reliable message delivery), which are not always needed in conventional wireless networks. MANET applications are identical (or similar) to those provided by the Internet network [6].

The second difference is that MANETs can contain many nodes that cannot recharge their power and have uncontrolled movement patterns, whereas the nodes in a VANET can be self-recharged frequently and their movement patterns are constrained by the road and traffic patterns. In this context, VANETs are characterized by highly mobile nodes and a potentially large network [1,3].

The third difference is related to the speed of nodes. Nodes in VANETs move at higher speeds (from 0 to 40 m/s) compared to nodes in MANETs (0–5 m/s). Data communication in vehicular networks is influenced by the high velocity of vehicles and their mobility patterns, which can cause rapid changes in the network topology. Environmental factors, such as physical obstacles or interferences, can limit communication ranges. These factors, when combined with node density and the distances/areas usually lead to frequent fragmentation and intermittent communications in the network. Naturally, these fragmentations disturb the end-to-end path between every pair of nodes [3].

The fourth difference is that VANETs have to cope with variable network densities induced by traffic conditions and by the mobility patterns of the vehicles: in this format vehicles do not move independently of one another, and this is related to constraints due to road topology, speed limits, and traffic lights [1].

The four mentioned differences between MANETs and VANETs expose some unique features regarding VANETs: the highly dynamic topologies caused by different vehicle speeds and mobility patterns (for example, vehicles traveling on the roadway in opposite directions, which leads to frequent link breakages that strongly hinder stable and durable V2V communications). Moreover, the limited infrastructure coverage, because of sparse RSU settling, may cause short-lived and intermittent V2I connectivity. All these unique features allow VANETs to fit well into the opportunistic network class [4,5].

Opportunistic networks are one of the most intriguing evolutions in MANETs. In these networks, mobile nodes are able to communicate with one another even when a connecting route is absent. Furthermore, due to the nature of the opportunistic network, nodes do not possess or acquire any knowledge about the network topology, something that is necessary in traditional MANET routing protocols. Routes are built dynamically, while messages are *en route* between their source and destination. Any possible node can be used as the next hop opportunistically, provided it is likely to convey the message closer to the final destination. In opportunistic networks, the assumption of a complete path between the source and destination is relaxed; mobile nodes are able to communicate with one another even if a connecting route is absent or frequently broken [6].

Traditional routing protocols for the Internet and MANETs [7] assume that an *end-to-end* path exists, something not true in opportunistic networks (e.g., VANETs). Opportunistic networking techniques allow mobile nodes to exchange messages by taking advantage of mobility and utilizing the store-carry-forward approach. According to this technique, a message can be stored in a node and forwarded through a wireless link as soon as a connection *opportunity* arises with a neighboring node, which is opportunistically used as the next hop toward the destination [8]. Messages that are cached in the network and waiting for an end-to-end path to be available can undergo a delay due to additional delivery time. This is why opportunistic networks are considered a special type of delay-tolerant networks (DTNs), allowing connectivity despite long link delays or frequent link breaks caused by nodes moving out of range, climatic changes, interference from other objects, and so on. In these networks, intermittent connectivity is common; however, there is a wide range of applications that are able to tolerate this fact [9,10]. The concept of opportunistic networking is widely applied on sparse and partitioned scenarios enabling non–real time services and applications.

The specific features of vehicular networks favor the development of efficient and challenging services and applications. Many researchers in this field consider the VANET to be one of the most prominent technologies for improving the efficiency and safety of modern transportation systems so far [2].

4.2 Wireless Technologies for Vehicular Communication Networks

There are several technologies involved in implementation of wireless communication in VANETs. These technologies include 3G cellular systems, long-term evolution (LTE), LTE Advanced, IEEE 802.16, and IEEE 802.11 [11].

IEEE 802.11 (Wi-Fi) has some special features that increase the motivation for using it in VANETs. These features are as follows [3]:

- IEEE 802.11 uses unlicensed spectrum.
- The operational costs of IEEE 802.11 are lower than other technologies.
- IEEE 802.11 access would be free or inexpensive for users.
- IEEE 802.11 would provide a bit rate of higher than 10 Mbps.

Because of these features, IEEE 802.11 is considered an appropriate wireless technology for application in VANETs. In this section the focus is on 802.11-based wireless communication in VANETs. First, we provide an explanation of IEEE 802.11.

4.2.1 IEEE 802.11

The first version of the IEEE 802.11 standard was published in 1997 and it specifies the medium access control (MAC) and physical layer (PHY) for wireless local area networks (WLANs). Over the years the standard has been continuously developed and has grown by annexing numerous amendments. The nature of these amendments differs in the presented technologies and achievements regarding speed. Some of these amendments are listed in Figure 4.2. WLANs based on 802.11 are starting to be deployed in office buildings, airports, hotels, restaurants, and campuses around the world [12].

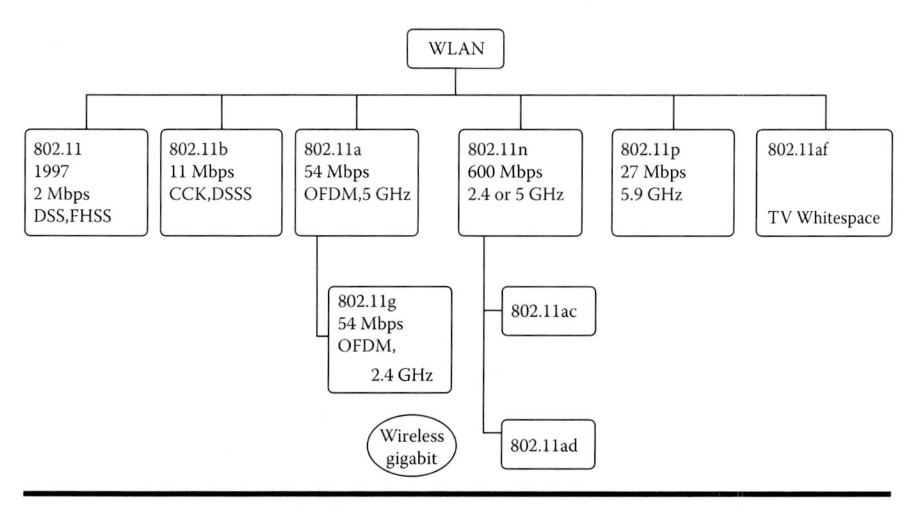

Figure 4.2 Amendments to the IEEE 802.11 standard.

4.2.1.1 IEEE 802.11 Components

IEEE 802.11 networks consist of four major components as follows [13]. These components are depicted in Figure 4.3.

4.2.1.1.1 Distribution System

When several APs are connected to form a vast coverage area, they must communicate with one another to track the movements of mobile stations. The system is a logical component of IEEE 802.11 applied to forward frames to their destination. 802.11 does not specify any particular technology for the distribution system. In most commercial components, the distribution system is implemented as a combination of a bridging engine and a distribution system medium, which is often called the backbone of the network used in relaying frames among APs. In nearly all components used commercially, Ethernet is used as the backbone network technology.

4.2.1.1.2 Access Points

An AP operates within a specific frequency spectrum and uses an 802.11 standard modulation technique. It informs the wireless clients of its availability and authenticity and its association with wireless clients to the wireless network. An AP coordinates the wireless clients' use of wired resources. There are several types of APs, including single radio and multiple radios, based on different 802.11 technologies.

4.2.1.1.3 Wireless Medium

To move frames from station to station, this standard uses a wireless medium. Here, several different physical layers are defined; the architecture allows multiple physical layers to be developed to support the 802.11 MAC. Initially, two radio frequency (RF) physical layers and a infrared physical layer are standardized, though the RF layers have proven more common [14].

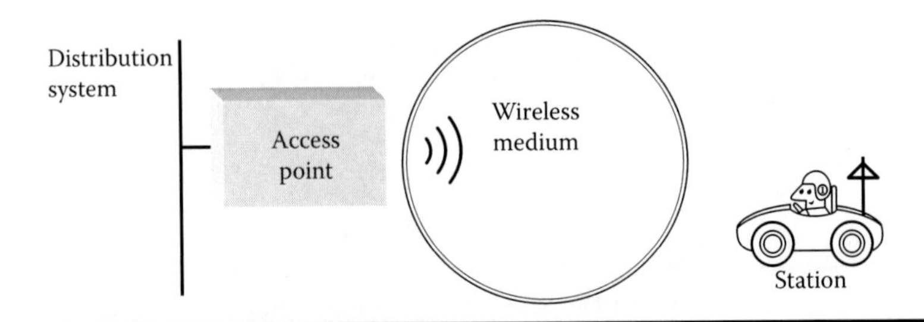

Figure 4.3 IEEE 802.11 components.

4.2.1.1.4 Stations

Networks are built to transfer data among stations where the computing devices are equipped with wireless network interfaces. Typically, these stations are mobile, though there is no reason why stations should be portable. In some environs, wireless networking is used to avoid a new wiring system, where desktops are connected by WLANs.

4.2.1.2 802.11 Working Modes

IEEE 802.11 has two modes of operation: infrastructure and ad hoc. In infrastructure mode, devices need to connect to an AP and all communications are transmitted through the AP. The AP and the devices associated with it form a basic service set (BSS) identified by a service set identifier (SSID). In ad hoc mode, devices within communication range of one another communicate directly without going through APs. These devices form an independent basic service set (IBSS), identified by an SSID as well [15].

4.2.1.2.1 802.11 Ad Hoc Mode

In Ad Hoc mode, the wireless network is relatively simple and consists of 802.11 network interface cards. The networked nodes communicate directly with one another without using an AP (Figure 4.4a). In this mode, there is no central control where stations compete for air time, just as they do with Ethernet. Stations in an IBSS communicate directly with one another; therefore they must be within direct communication range. The smallest possible 802.11 network is an IBSS with two stations. Typically, IBSSs are composed of a small number of stations set up for a specific temporary purpose. One common use of ad hoc mode is to create a short-term connection between two vehicles to exchange safety messages.

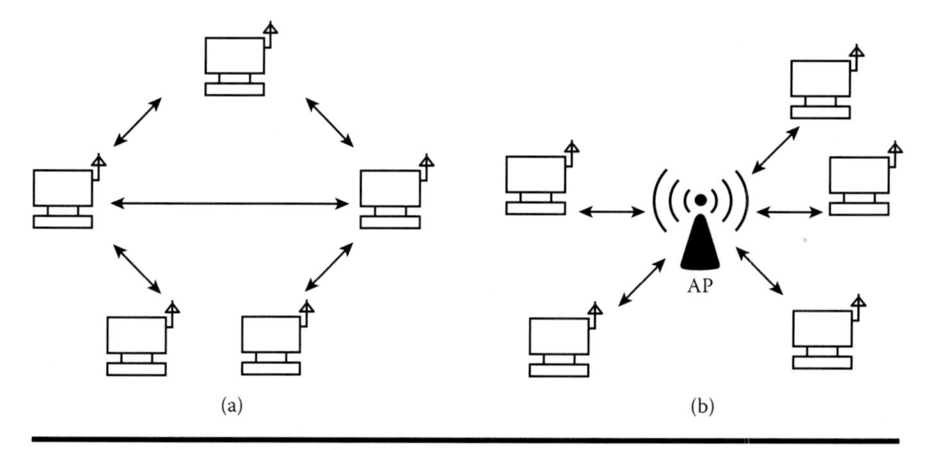

Figure 4.4 IEEE 802.11 working modes: (a) ad hoc mode; (b) infrastructure mode. AP, access point.

4.2.1.2.2 802.11 Infrastructure Mode

The use of an AP distinguishes the infrastructure networks (Figure 4.4b). If one mobile station in an infrastructure BSS needs to communicate with a second mobile station, the communication must take two hops: the first hop is the origination of the mobile station transferring the frame to the AP; the second hop is transferring the frame from AP to the destination station. The basic service area corresponding to an infrastructure BSS is defined by the points where transmissions from the AP can be received. This phenomenon embodies all communications relayed through an AP.

Although multihop transmission takes more transmission capacity than a directed frame from the sender to the receiver, it has two major advantages:

1. An infrastructure BSS is defined by the distance between it and the AP. All mobile stations are required to be within reach of the AP radius, but no restriction is placed on the distance between individual mobile stations. The transmission capacity can be saved if direct communication between mobile stations is provided. On the contrary, the cost of increased physical layer complexity rises because mobile stations would need to maintain neighboring relationships with all other mobile stations within the service area.
2. APs in infrastructure networks are installed to assist stations that try to save power. APs are aware when a station enters power-saving mode and buffer frames for it. Battery-operated stations can turn the wireless transceiver off and turn it on to transmit and retrieve buffered frames from the AP only. In an infrastructure network, stations must associate with an AP to obtain network services. This association is the process whereby mobile station joins an 802.11 network; this association is logically equivalent to connecting to the network cable on an Ethernet, although this is not a symmetric process. Mobile stations always initiate this association process, where the APs may choose to grant or deny access based on the contents of an association request. Associations are exclusive on the part of the mobile station: a mobile station can be associated with only one AP [3]. There is no limit to the number of mobile stations that an AP may serve in 802.11. Implementation considerations may, of course, limit the number of mobile stations an AP might serve. Practically, the relatively low throughput of wireless networks is less possible in limiting the number of stations placed on a wireless network.

4.2.1.3 IEEE 802.11 Services

The 802.11 standard states that each conformant WLAN must provide nine services of two categories: five distributive and four stationary. The first category is related to managing cell membership and interacting with stations outside the cell. By contrast, the second category is related to activities served within a single cell.

The five distributive services are provided by the base stations and deal with station mobility as they enter and leave cells, by attachment and detachment. They are presented as follows [13]:

- **Association**: Every station must initially invoke the association service with an AP before it can send information through a distribution system. The base station may accept or reject the mobile station. If the mobile station is accepted, it must then authenticate itself. Each station can associate with only a single AP.
- **Disassociation**: Either the station or the base station may become disassociated, thus breaking the relationship.
- **Re-association**: A station may change its preferred base station using this service. This service is applicable in mobile stations when moving from one cell to another. If applied correctly, no data will be lost as a consequence of the handover.
- **Distribution**: This service determines how to route frames sent to the base station. If the destination is local to the base station, the frames can be sent out directly through the aerial relay; otherwise, they will have to be forwarded through the wired network.
- **Integration**: If a frame should be transmitted through a non-802.11 network having a different addressing scheme or frame format, the integration service handles the translation from the 802.11 format to the format required by the destination network.

The remaining four services are intracell—relating to actions within a single cell, used when the association has taken place. They are presented as follows:

- **Authentication**: Because wireless communication can easily be sent or received by unauthorized stations, a station must authenticate itself before it is allowed to send data. After a mobile station is associated by the base station (i.e., accepted into its cell), the base station sends a special challenge frame to it to check whether the mobile station knows the code (password) assigned to it. This service proves its knowledge of the secret key by encrypting the challenge frame and sending it back to the base station. If the result is correct, the mobile is fully enrolled in the cell. In the initial standard, the base station does not have to prove its identity to the mobile station. Therefore, researchers in this field are seeking the means to resolve this problem in the upcoming standards.
- **Deauthentication**: When a previously authenticated station is to leave the network, it becomes deauthenticated. After deauthentication, it may no longer use the network.
- **Privacy**: For the information sent over a WLAN to be kept confidential, it must be encrypted. This service manages the encryption and decryption of the information. The encryption algorithm specified is RC4, which was introduced by Ronald Rivest of MIT.

- **Data delivery**: In the field of interest, data transmission is the essence. IEEE 802.11 naturally provides a way to transmit and receive data. Because 802.11 is modeled on the Ethernet, where data transmission is not guaranteed to be 100% reliable, transmission over 802.11 is not guaranteed to be reliable either. In this context, higher layers must deal with detecting and correcting errors.

An 802.11 cell has some parameters that can be inspected and adjusted, in some cases. They relate to encryption, time-out intervals, data rates, beacon frequency, and so on.

4.2.1.4 Handover in 802.11

In 802.11 networks that use infrastructure mode [13], every mobile station is associated with an AP, which provides access to the fixed network infrastructure (e.g., the Internet). When a mobile station moves, it may need to change the AP it is associated with, because each AP covers only a limited geographical area. This process is called *handover*. Figure 4.5 shows a sample situation where handover is required. The more Wi-Fi network applications in telephony and multimedia, the more important such handovers. In particular, there is a need to speed up the handover process in a manner that does not interrupt application-level sessions [16].

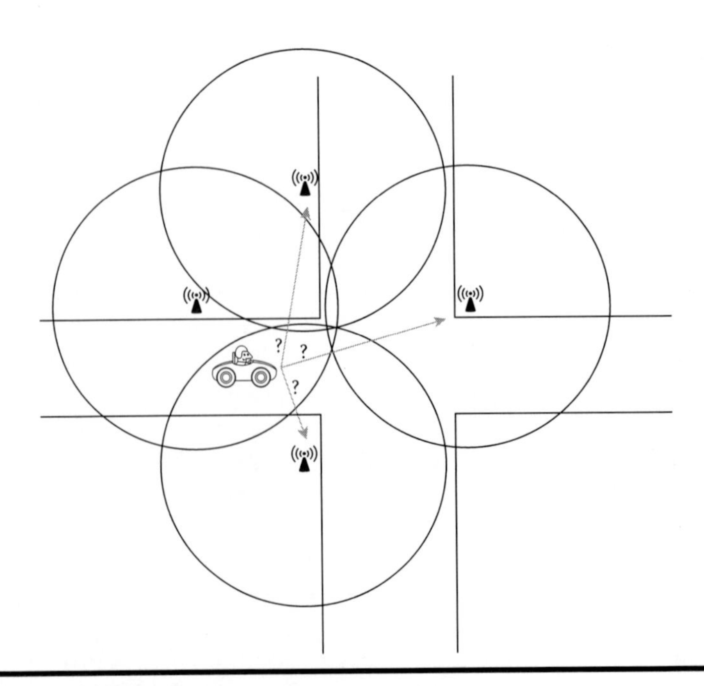

Figure 4.5 A sample situation where handover is required.

The handover process is composed of four main phases: (i) detecting the possible set of next APs to which the handover could be aimed at (probing phase); (ii) choosing the destination AP; (iii) associating with that AP; and (iv) reauthenticating the mobile station to the network.

4.2.2 IEEE 802.11 for VANET

The proliferation of IEEE 802.11 APs in residential and commercial building in addition to public and private transportation units in recent years has tremendously increased the coverage of wireless Internet in densely populated urban areas. Researchers have started looking at the type of performance possible for such access and whether this will suffice for certain applications [3]. The IEEE 802.11 is not suitable for vehicular applications. The fact that WLANs are not meant to support outdoor communications further increases the challenges associated with 802.11-based vehicular communications [16]. Wireless connectivity among moving vehicles can be provided by available 802.11a-compliant devices, with data rates of up to 54 Mbps being achieved through 802.11a hardware. However, vehicular traffic scenarios caused by varying driving speeds, traffic patterns, and driving environs have greater challenges than fixed wireless networks. Traditional IEEE 802.11 MAC operations require significant overhead when used in vehicular scenarios [17]. For instance, to ensure timely vehicular safety communications, fast data exchanges are required. Under these circumstances, the scanning of channels for beacons from an AP along with multiple handshakes required to establish communication are associated with high complexity and overhead (for example, when two vehicles approach each other from opposite directions, the duration for possible communication between them is very short [3], making it difficult to establish communications).

Here, in general, four major difficulties are observed in IEEE 802.11 applications with respect to opportunistic vehicular communication [17]:

First, connection establishment latency is usually very high due to continuous probing delays for AP discovery (a typical client probes on all 11 channels before connecting), high loss rates (during connection establishment, control packets may get lost and require retransmission), and delay in acquiring an IP address via dynamic host configuration protocol (DHCP). Several methods have been proposed to decrease the components of this type of delay, such as using selective scanning, reducing time-out periods in case of losses, and using static IPs instead of DHCP [18,19]. A recent work proposes improving link layer performance by applying macro-diversity and opportunistic reception [20].

Second, a good handoff strategy is critical. Several APs may express their presence at any location along the drive, and it is important to choose the best AP to connect to. Regular Wi-Fi clients initiate handoff only when disconnected and choose the AP with the strongest signal strength. Although this works well in home or office scenarios where clients are rarely mobile, it is not a good option for vehicular mobility. The opportunity to connect to a strong AP (both in terms of signal strength and available backhaul bandwidth) may be lost. A method of active scanning in operation is proposed to address this issue [21].

Third, current end-to-end data transfer and congestion control protocols do not work well when connectivity is only a few seconds long and intermittent and wireless loss rates are high (due to moving cars).

Fourth, the default wireless bit-rate selection algorithms are tuned to nonmoving users, resulting in suboptimal performance in vehicular applications.

Because of these difficulties, applying IEEE 802.11 in VANETs is a challenging issue. However, 802.11 APs are randomly deployed across most urban cities, which suggests a great deal of motivation to apply them in VANETs. To fulfill this motivation some modifications and improvements are necessary to prepare IEEE 802.11 for application in VANETs. In this respect, some of the studies conducted in this regard are presented in the following subsections.

4.2.2.1 Adapted 802.11 for VANETs

Here, some real-world VANET projects are reviewed where 802.11 is applied as the communication technology.

- **CarTel Project** [18,22]: The researchers on the CarTel project investigated general architectures for vehicular sensor networks, and characterized the extent to which wireless APs are deployed in cities that can be used as an uplink network for moving cars. The CarTel architecture is presented in Figure 4.6. For over a year, the CarTel project studied six cars running on a small scale in several metropolitan areas in the United States. This experiment did not address network performance issues, like optimized association, scanning, data transport protocols, or rate selection. Their measurements demonstrate that at urban vehicular speeds, a regular mobile client can gain connectivity for several seconds (median of 13 seconds per AP) and transfer a large amount of data (median of 216 KB using Transmission Control Protocol [TCP]) [18]. This study used stock implementations of network protocols like 802.11 connection establishment and TCP. Various improvements are possible using custom protocols as demonstrated by several related studies [20,21].
- **Cabernet** [19]: Cabernet is a system for delivering data to and from moving vehicles using open 802.11 (Wi-Fi) APs encountered opportunistically during travel. The architecture of Cabernet is shown in Figure 4.7. Network connectivity in Cabernet is both fleeting (APs are typically within range for a few seconds) and intermittent (because the APs do not provide continuous coverage) and causes high packet loss rates over the wireless channel.

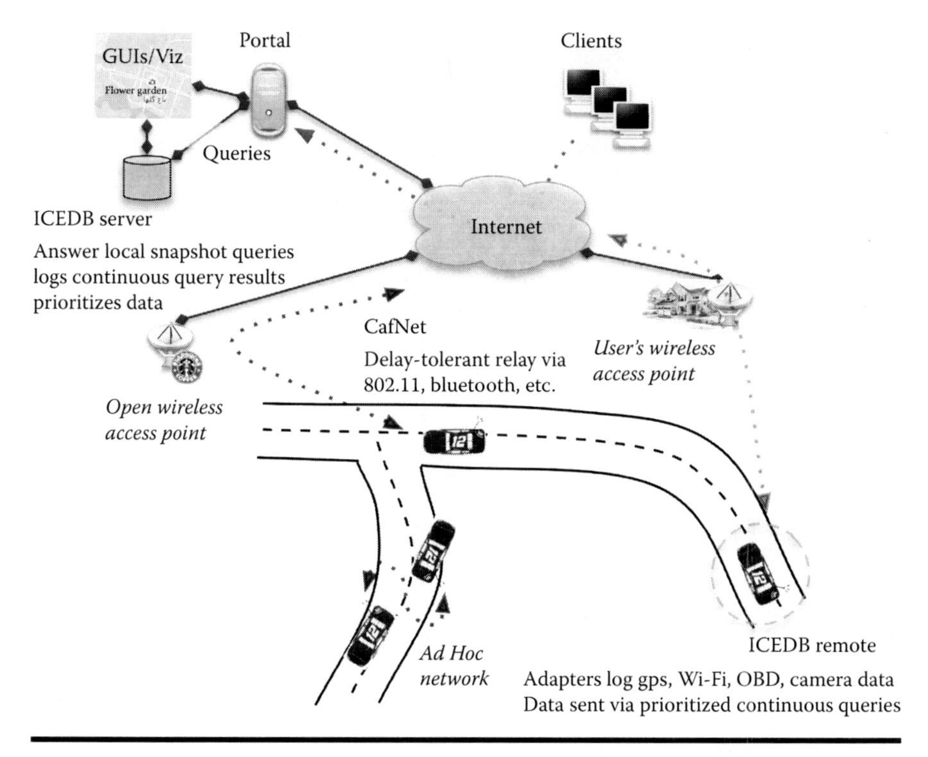

Figure 4.6 The CarTel system architecture. (From Bychkovsky, V., et al., A measurement study of vehicular internet access using in situ Wi-Fi networks, in *Proceedings of ACM MobiCom Conference*, 2006.)

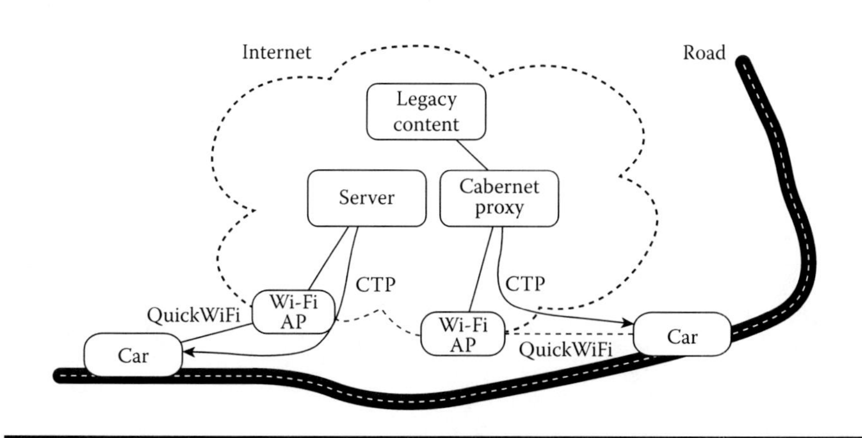

Figure 4.7 The architecture of Cabernet. (From Eriksson, J., et al., *Proceedings of ACM MobiCom Conference*, 2008.) CTP, Cabernet Transport Protocol; CAR, Context-Aware Routing.

Association, scanning, data transport protocols, and rate selection are optimized through Cabernet. Cabernet incorporates the following three techniques to solve the abovementioned problems.

The first issue was to reduce the time span between when the wireless channel is connected to an AP and in use, and when Internet connectivity through the AP is actually achieved. Here, QuickWiFi was developed, a streamlined process combining all the different protocols involved in obtaining connectivity (across all layers) into a single process. This process includes a new optimal channel scanning policy. QuickWiFi reduces the mean connection time to less than 400 ms, from over 10 seconds when standard wireless networking software is applied.

To improve end-to-end throughput over lossy wireless links, the Cabernet Transport Protocol (CTP) was developed, which outperforms TCP over opportunistic Wi-Fi networks by not confusing Wi-Fi losses in network congestion. Unlike previous work on efficient wireless transport protocols, CTP does not require AP modifications (which are not under the user's control); instead, it uses a lightweight probing scheme to determine the loss rate from Internet hosts to an AP.

To improve link rates, the impact of bit-rate selection was studied in vehicular Wi-Fi. Based on the obtained results, a static 11 Mbit/s Wi-Fi bit rate was obtained from data transfers from the car. This optimization could not be applied for downloads from Internet hosts, as the AP controls the bit rate in this direction. The researchers at this point deployed CTP on a fleet of 10 taxis in the Boston area. The long-term average transfer rate achieved was approximately 38 Mbytes/hour per car (86 kbit/s), making Cabernet a viable system for a number of noninteractive applications.

- **Drive-Thru Internet** [23]: In the Drive-Thru Internet project, the usability of providing network connectivity and, ultimately, Internet access to mobile users in vehicles is being investigated. The idea of Drive-Thru Internet is to provide hot spots along the road—within a city, on a highway, or even on high-speed freeways such as autobahns. Drive-Thru Internet must be placed in such a manner that a vehicle driving by would obtain WLAN access for some relatively short period of time; if located in rest areas, the driver may exit and pass by slowly or even stop to prolong the connectivity period. One or more locally interconnected APs form a so-called connectivity island that may provide local services and Internet access. Several of these connectivity islands along a road or in the same geographic area may be interconnected and cooperate to provide network access with intermittent connectivity for a vast area.

APs may be provided at each street corner or co-located with traffic lights, emergency phones (which are placed every 2 km on a German autobahn), parking lots, rest areas, gas stations, or other public areas. Several APs may be grouped to extend the reach of a connectivity island. When driving, mobile devices may have free line of sight to the AP(s) on the roadside; or the AP(s) may be obscured by trees, fences, the user's own vehicle's bodywork, other vehicles, crash barriers, or even buildings. The sensing of AP(s) potentially depends on roadside obstacles, passing vehicles, and so on. WLAN connectivity will appear and disappear; short periods of connectivity will alternate with long periods of nonconnectivity. Here, even short connectivity periods may be further interrupted. Hence, a mobile device in a vehicle traveling along a road with usable APs occasionally located close to the road will (1) permanently scan for signals from available APs; (2) attempt to associate with the respective AP whenever such a signal is detected; (3) detect network access; (4) perform IP configuration (obtaining an IP address, performing neighbor discovery) for the respective link after the association succeeds in order to be able to send and receive data; (5) use the wireless network for general Internet access, for VPN tunneling, and so on and use regular Internet protocols after IP connectivity is established; (6) go through a series of handovers if a connectivity island is made up of several APs; (7) at some point, notice a weakened signal and eventually lose the signal when the vehicle has passed through the connectivity island and travel on, returning to Step 1. In this project, the idea of Drive-Thru Internet is to introduce the use of WLAN technology to provide network access for users traveling by car, particularly on highways or the autobahn.

By using three different measurement settings, reference parameters for the equipment were obtained. Proof-of-concept tests were carried out, and eventually the technical feasibility of this project was validated. The measurements indicate that the coverage obtained from a single AP is much greater than expected, providing more than 10 seconds of connectivity even at speeds of 180 km/h. Using several APs to extend the reach of a connectivity island has turned out to be more difficult and to require a greater distance between APs or different parameterization than would be used for stationary users. It was observed that the connectivity is—expectedly—poor at the edges of a connectivity island (entry and exit phase), with a negative impact on packet loss and transmission delay, while over a distance of more than 200 meters network performance is excellent. Transmission of a maximum of 9 Mbytes of data was managed in a single pass through a connectivity island with a single AP, which confirms the principal suitability of WLAN for Drive-Thru networking.

- **Multi-AP solutions**: Achieving ubiquitous connectivity or high aggregate throughput by using a series of Wi-Fi APs can lead to serious performance problems for highly mobile clients. Data transfer used with a single AP at a time is insufficient for supporting delay- and throughput-sensitive applications. Leveraging concurrent data transfers through multiple APs offers a plausible alternative. In this respect, a system was designed, implemented, and evaluated that establishes and maintains concurrent connections to 802.11 APs in a mobile environment. It was demonstrated that existing multi-AP solutions do not perform well in mobile settings due to the limitations imposed by the association and DHCP processes. Therefore an alternative design was proposed. The system was implemented and evaluated on a vehicular test bed. The obtained results indicate that this system provides manifold improvements in throughput and connectivity over stock Wi-Fi implementations [24].

- **Interactive Wi-Fi**: The Wi-Fi is used in the base station diversity to reduce the impact of disruption in the vehicular context [25]. Instead of associating with one base station at a time, the Wi-Fi allows a mobile node to communicate with a set of base stations. In this manner, if one base station fails to serve the mobile nodes, other auxiliary stations can offer network services instead [26].
- **Situ Wi-Fi** [18]: The impressive penetration of 802.11-based wireless networks in many metropolitan areas around the world, for the first time, provide the opportunity for a "grassroots" wireless Internet service where users would "open up" their 802.11 (Wi-Fi) APs in a controlled manner to mobile clients. Although there are many commercial, legal, and security issues to be ironed out for this vision to become reality, SITU Wi-Fi focuses on an important technical question surrounding such a system: can such an unplanned network service provide reasonable performance to network clients moving in cars at vehicular speeds? To answer this question, we present the results of a measurement study that was carried out over 290 "drive hours" using a few cars under typical driving conditions, in and around the Boston metropolitan area (some of the data come from a car in Seattle). With a simple caching optimization to speed up IP address acquisition, it was found that for driving patterns the median duration of link layer connectivity at vehicular speeds was 13 seconds, whereas the median connection upload bandwidth was 30 KB/s, and that the mean duration between successful associations to APs was 75 seconds. It was found that TCP connections were equally probable across a range of urban speeds (up to 60 km/h). The end-to-end TCP upload experiments had a median throughput of about 30 KB/s, which is consistent with typical uplink speeds of home broadband links in the United States. The median TCP connection is capable of uploading about 216 KB of data. The high-level conclusion is that grassroots Wi-Fi networks are viable for a variety of applications, particularly ones that can tolerate intermittent connectivity.

4.2.2.2 IEEE 802.11p: A New Amendment of 802.11 for VANETs

Since its development, the IEEE 802.11 has expanded continuously through its applicable amendments. These amendments vary in their technological architecture and related elements. Among these amendments, 802.11p was presented specifically for vehicular communications [27].

IEEE 802.11p is an amendment to IEEE 802.11, adding wireless access in vehicular environments (WAVE) to this standard. The dedicated short-range communications (DSRC) is a general purpose communications link between a vehicle and RSUs (or between vehicles) using the 802.11p protocol. A similar mechanism is used in MAC layer operation of IEEE 802.11p as it was used for the IEEE 802.11 legacy system. By using the IEEE 802.11p service classes, as defined in the IEEE 802.11e standards, service quality differentiation is provided. At the PHY layer, the use of an orthogonal frequency division multiplexing (OFDM) system is expected to allow V2V as well as V2I communications for distances of up to 1000 m and relative speeds of 200 km/hr. The protocol may operate on either the 10 or 20 MHz channels, allowing data rates from 3 to 56 Mbps.

When comparing IEEE 802.11p with the original IEEE 802.11, a fundamental difference is observed if the network has the ability to communicate outside the context of a basic service set in an ad hoc manner in a highly mobile network. The IEEE 802.11 authentication and association processes preceding a first frame exchange would last too long, for example, for communication between two vehicles with opposing driving directions. Consequently, authentication and association are not provided by the IEEE 802.11p PHY/MAC but have to be supported by the station management entity or a higher layer protocol. In the vehicle-to-X use case the protocols of the IEEE 1609 (IEEE 2006) standard family contain the necessary procedures. IEEE 802.11p adds to the mode of communication outside a BSS into the standard while this manner of operation is not predicted [28].

IEEE 802.11p supports a new mode of operation in addition to the ad hoc and infrastructure modes of 802.11a/b/g networks. This new mode of operation is termed *WAVE mode*. In emergency cases, vehicular nodes operating in WAVE mode can send and receive messages without associating with a BSS in the conventional manner. This allows quick commencement of information exchange with the added advantage of very low overhead. The information exchange using WAVE may be between vehicles or between vehicles and the roadside infrastructure.

4.3 Predictive Methods for Improving Opportunistic Vehicular Communication

Several challenges exist in implementation of vehicular communication based on 802.11, such as long latency to establish connection to an AP, lossy link performance, and frequent disconnections due to mobility [3]. Some methods have been developed to overcome these challenges. In one of the most successful classes of methods, information related to vehicle movement paths is applied to establish more stable routes. Studies indicate that people drive on familiar routes on frequent bases; thus the mobility- and connectivity-related information along their routes can be predicted with good accuracy using historical information such as GPS tracks with timestamps, RF fingerprints, and link and network-layer addresses of visible APs. This information can be exploited to develop new algorithms for overcoming existing challenges in opportunistic vehicular communication.

In the section "Vehicle Motion Path Prediction Methods," the presented methods for vehicle movement path prediction are reviewed. In the section "VANET Improvement Based on Vehicle Position Prediction," some algorithms are explained for performance improvement in opportunistic networks based on the predicted vehicle movements.

4.3.1 Vehicle Motion Path Prediction Methods

Most drivers' trips are repetitive and studies indicate that by taking into account the first part of a trip and the previous trips made by a driver, the rest of the trip can be predicted. This prediction is made by means of extracting the movement patterns of the vehicle [29]. In this regards, a variety of algorithms for predicting drivers' routes have been proposed [30]. In the following, we review some examples of the algorithms and methods that contribute to predicting drivers' routes.

A method was presented for predicting the driver's destination during the drive [30]. In this method, the probability that Route B would be covered to reach Destination A was extracted from GPS history data related to the driver. When the driver was placed on Route B during his or her trip, the probability of reaching Destination A was calculated by the Bayesian probability rule, where the probability of a new network of candidate destinations within the driver's trip can be estimated. As the driver chooses a certain route over other routes during the trip, the probability of reaching some destinations gradually becomes zero.

Simmons et al. [31] built a hidden Markov model (HMM) by using past driving experiences to predict a driver's intention for the next trip. They collected GPS data on every trip a driver took. Then, by using a digital map, they converted pure GPS locations to a street-level graph representation and updated the HMM model incrementally based on new experiences. They realized that this model can be adopted to make accurate predictions of a driver's routes and destinations through online observation of his or her positions during a trip.

Froehlic and Krumm [29] developed an algorithm for predicting the end-to-end route of a vehicle based on GPS observations made during the vehicle's past trips. This algorithm matches the initial portion of the current trip with a previously observed trip to predict the vehicle's future paths. This method includes three stages [29]:

1. Extracting trips from GPS raw data: In this stage, GPS data are continually stored in memory so they do not have an explicit indication of when a trip begins or ends.
2. Extracting the regular routes from the trips: This stage is used in order to identify and eliminate some false trips. These false trips would be generated by some vehicles that continued to power their GPS receivers even when the vehicle is turned off.
3. Predicting future paths using extracted regular trips: For this purpose, the Hausdorff distance between the current trip and regular routes are computed. The closest route is selected as the outcome.

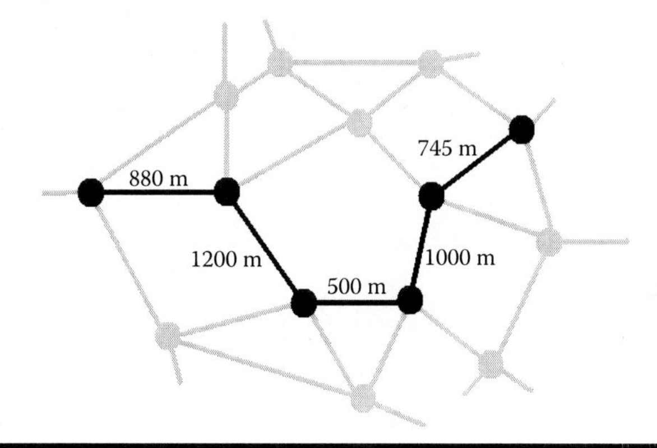

Figure 4.8 A sample trip in the road network. (From Bohlooli, A. and Jamshidi, K., *App. Intell.*, 36(3), 685–697, 2012.)

All methods reviewed in this section are GPS data dependent. Precision and efficiency are low when these GPS-based methods are applied in environments containing obstacles for GPS receivers. Bohlooli and Jamshidi [32] proposed a GPS-free method for future vehicle movement prediction with no loss of its precision in such environments. In this method, every trip is defined as movement between two long-term stops compared with temporary stops at the junctions. A sample trip is shown in Figure 4.8. A long-term stop is defined as a stop of more than 10 minutes. The drivers' trips are described by changes in vehicle direction and the distances between these directional changes. Every trip is characterized by the following features:

- Time-related, indicating the day and time of the week when the trip took place
- Movement, describing the changes in vehicle direction and the distance between these directional changes

This method applies a self-organizing map (SOM) neural network [33–38] for clustering the drivers' trips. After clustering the motion paths by the SOM, a representative vector for each movement pattern is extracted. These representative vectors symbolize general features of the movement patterns (i.e., time-related features and features related to the trip route). After extracting the representative vectors, the future movements of the vehicle can be predicted. To do this, when a portion of a trip is covered by that vehicle, the most similar pattern that could have been traveled on that route up to that time is used for detecting the future movement directions.

4.3.2 VANET Improvement Based on Vehicle Position Prediction

In the previous section, some presented methods for prediction of vehicle movement path were reviewed. The following two sections ("Applying Predicted Vehicle Position for Infrastructure-Based VANETs" and "Applying Predicted Vehicle Motions for Infrastructureless VANETs") consider the fact that communication in VANETs can be infrastructure-based or infrastructureless. These sections review some methods to apply the predicted vehicles' position for improving communication in infrastructure-based and infrastructureless VANETs, respectively.

4.3.2.1 Applying Predicted Vehicle Position for Infrastructure-Based VANETs

Several methods have been presented to improve communication in infrastructure-based VANETs based on the predicted position of vehicles. These methods are explained in the rest of this section.

4.3.2.1.1 Faster Connection Establishment

Driving on familiar routes provides the opportunity to learn and cache the relevant information about APs along a route, which can contribute to quick connection establishment. In this method, during periods of inactivity, clients listen for beacons and record the channel, network name (ESSID), and MAC address of the APs they can successfully associate with. This information is tagged with predicted locations. In addition, cooperation between APs and clients and/or use of auto configuration [39,40] can eliminate the need for IP address assignment through DHCP each time the client associates with an AP. All these accelerate the connection establishment process significantly.

4.3.2.1.2 Scripted Handoffs

In this method, the client builds a radio frequency (RF) fingerprint for the route by recording the signal strength from beacons and tagging them with the GPS location of the car where the beacon is heard. This data, when collected over a period of time, provides a rich estimate of the RF level connectivity of various APs along the route. This connectivity estimate combined with an estimate of the vehicle's mobility is considered input in an algorithm that computes the locations where the client needs to hand off to the designated AP. Thus, the handoffs are scripted. This computation is done offline; hence no bandwidth is wasted in scanning for better APs as in other online techniques. The algorithm ensures that the client is always connected to the best estimated AP. This is unlike most stock implementations, where a connection is maintained until it breaks.

4.3.2.1.3 Prefetching APs

Scripted handoffs and mobility estimates predict periods of future connectivity to various APs. This can provide further performance gains in downloading applications by having such prefetched APs as part of the content to be downloaded. In essence, the APs now collectively form a distributed cache for the mobile client and, in cooperation with the mobile client, prefetch predetermined portions of the content. This helps mask the long Internet delay on the Wide Area Network (WAN) side of the AP. Inaccurate estimation of mobility causes "cache misses," reducing the download performance. By contrast, duplicating prefetches (i.e., more than one AP prefetching the same bytes of content) increases the load on the WAN and the content server. These must be optimized carefully.

4.3.2.2 Applying Predicted Vehicle Motions for Infrastructureless VANETs

In infrastructureless networks, data are forwarded hop by hop. With due respect to the high speed of vehicles, selecting the vehicle for the next hop is very important because selecting nodes that can be established for a long time leads to link stability and improves network performance. In studies related to this issue, prediction of vehicle movement was applied to select more adequate hops to improve link stability.

In some studies frequent link breakages are overcome by choosing more stable links. In this context, the major attempts made to select stable links in VANET routing protocols are highlighted. In Ref. [41], Movement Prediction-Based Routing (MOPR) is presented for VANETs. The required duration of established connections is predetermined as an assumption in the MOPR algorithm. During route discovery, this algorithm selects nodes that will not move out of transmission range of each other during data transmission. The current speeds and movement directions of vehicles are used for computing their future locations and selecting proper nodes in this algorithm. By using obtained locations, it recognizes whether they will move out of each other's transmission ranges during the connection. The distributed Movement-Based Routing algorithm is presented in Ref. [42]. During the route discovery, this algorithm regards the current movement directions of nodes in addition to their locations. A node will not select any nodes that are moving opposite to its own movement direction as the next hop. This statement is justified through this algorithm; thus, establishing links with lifetimes that are too short is avoided. An algorithm called *Greedy Perimeter Stateless Routing with Movement Awareness* (GPSR-MA) was proposed for VANETs [43]. This algorithm is an extension of the Greedy Perimeter Stateless Routing (GPSR) algorithm [44]. GPSR is full of defaults when it comes to obtaining location of the nodes; therefore, the established link tends to break down quickly. GPSR-MA uses the current speed and the current angle of vehicle movements to acquire more exact locations of network nodes. Thus GPRS-MA leads to longer durations compared with the GPSR algorithm. A Prediction-Based Routing algorithm for VANET was introduced [45]. During route discovery, this algorithm approximates the duration of all possible routes between the source and destination using the current locations and current speeds of vehicles and seeks to choose the most stable route. If a vehicle discovers that a routing hole occurs on a link, then the proposed routing protocol does not seek to create a new route from the source vehicle but instead reroutes the data packets to a different block [46].

In Ref. [7], the Receive on Most Stable Group-Path routing algorithm was proposed for VANETs. It classifies vehicles according to their current speed vectors. During route discovery, it chooses the nodes with the same speed vectors. Bohlooli and Jamshidi [47] proposed a method for selecting more stable links based on the vehicle's trip history. In this method, each vehicle has a profile containing its movement patterns extracted from its trip history. The next direction to be chosen by each vehicle at the next junction is predicted through this profile and is sent to other vehicles. Afterwards each vehicle selects a node that has a future direction that is the same as its predicted direction.

4.4 Opportunistic Routing in Vehicular Communication

In this section, some routing solutions for an opportunistic vehicular ad hoc network are discussed. Opportunistic routing algorithms can be classified according to their dependency on roadside infrastructure. The taxonomy of opportunistic routing in VANETs is shown in Figure 4.9 [48]. In the first level, the routing algorithms are classified into algorithms designed for completely flat ad hoc networks

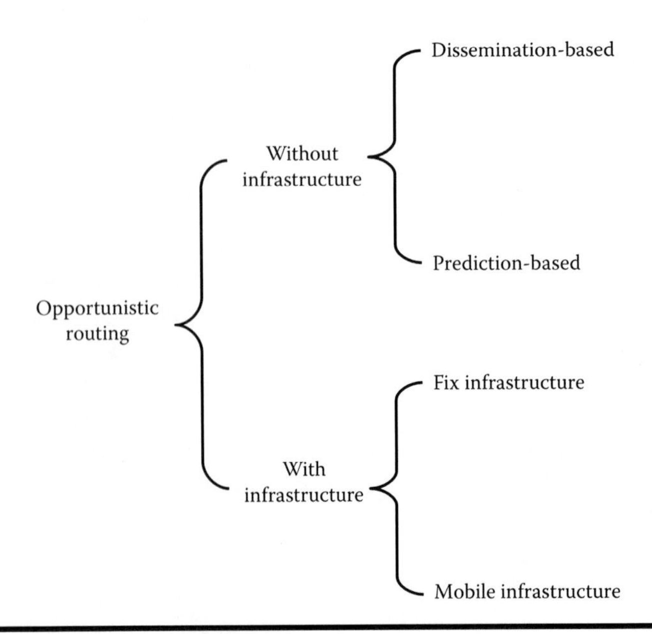

Figure 4.9 The taxonomy of opportunistic routing. (From Pelusi, L., et al., *IEEE Commun. Mag.*, 134–141, 2006.)

(without infrastructure) and algorithms where ad hoc networks exploit some form of infrastructure to opportunistically forward messages (with infrastructure). In the former case, approaches can be further classified into dissemination-based and context-based. Dissemination-based algorithms are essentially forms of controlled flooding and are distinguished by the policy used to limit flooding. Context-based approaches usually do not adopt flooding schemes but use knowledge of the context within which nodes are operating to identify the best next hop at each forwarding step. Algorithms that exploit some form of infrastructure can be classified (depending on the type of infrastructure they rely on) in fixed infrastructure and mobile infrastructure. In both cases, the infrastructure is composed of special nodes that are more powerful with respect to the nodes commonly present in the ad hoc network. They have high storage capacity; hence they can collect messages from many nodes passing by, even for a long period of time. They have high energy as well. The nodes in a fixed infrastructure are located at specific geographical points, whereas the nodes in a mobile infrastructure move around in the network following either predetermined, known paths or completely random paths.

4.4.1 Routing without Infrastructure

4.4.1.1 Dissemination-Based Routing

The objective of a data dissemination algorithm in VANETs is to spread information over a network of (mobile) nodes to the beneficiary nodes, for example, to avoid an accident with a parked vehicle, find a better route that bypasses a traffic jam, find an incident location on the node's future driving route, measure the duration of time since the incident, and identify its critical nature (type of incident) [49].

These routing techniques are involved in delivering a message to a destination by spreading it all over the network. These techniques work well in highly mobile networks where contact opportunities are common. They do limit the message delay, while being resource hungry. Dissemination-based techniques suffer from high contention and may potentially lead to network congestion because of the high number of transmissions. In order to increase the network capacity, the spreading radius of a message must typically be limited by imposing a maximum number of relay hops to each one of the messages, by limiting the total number of message copies present at the same time. Data dissemination is based on broadcasting, which can usually be classified into three main strategy categories according to the spreading of information packets in the network: single-hop broadcasting, multihop broadcasting, and epidemic broadcasting [50–63].

4.4.1.1.1 Single-Hop Broadcasting

Here, the information packets are not flooded by the vehicles but are kept in the vehicle's onboard database as received. Every vehicle selects some of the records stored in its database for broadcasting in a periodic manner; hence, in single-hop broadcasting, each vehicle must carry the traffic information with itself as it travels when this information is transferred to all others in its one-hop neighborhood in the next broadcast cycles. As a consequence, the vehicle's mobility is involved in spreading the information in the single-hop broadcasting protocol [64].

4.4.1.1.1.1 Fixed Interval-Based Single-Hop Broadcasting Protocols These protocols focus on the selection and aggregation of information only. TrafficInfo is an example of such a protocol, where every vehicle is equipped with a GPS and a digital road map through which the traffic information stored in its database is broadcast in a periodic manner. The road segment travel time is one of the special types of traffic information reported. During the broadcasting process, every vehicle stores its own travel time and the time taken by other vehicles during travelling into the database [65]. Figure 4.10 shows the principles of the TrafficInfo algorithm.

4.4.1.1.1.2 Adaptive Interval-Based Single-Hop Broadcasting Protocols Adjustment of the broadcast interval for adaptive broadcast interval protocols is of essence. The Collision Ratio Control Protocol applies adaptive broadcast intervals where each of the vehicles disseminates traffic information periodically [66]. Here, the traffic information consists of the location, speed, and road ID, which are measured per second. The adapted mechanism in the protocol for changing the broadcast interval dynamically is based on the number of packet collisions. In a sense, the main contribution of this protocol is to maintain the collision ratio at a targeted level with no vehicle density considered. Naturally, there exists a direct relation between the number of pocket collisions and an increase in network density.

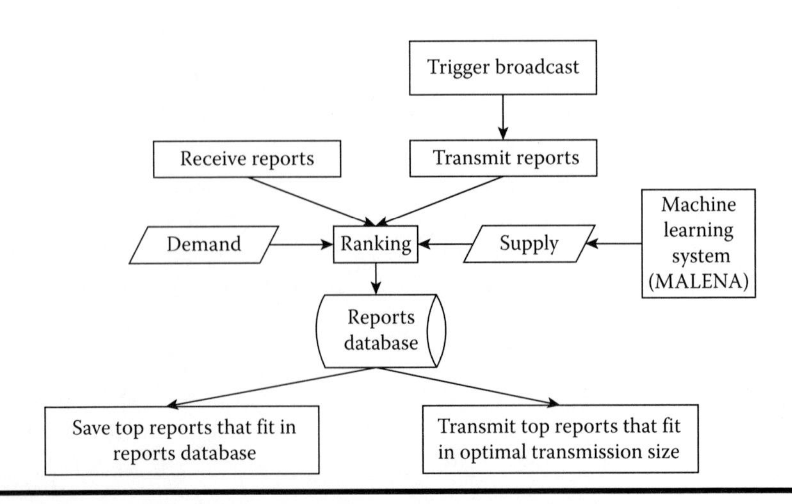

Figure 4.10 Principles of the TrafficInfo algorithm executed at each vehicle. (From Zhong, T., et al., *Proceedings of IEEE Intelligent Vehicles Symposium (IV)*, 2008.)

In addition to this protocol, there are three methods through which the selected data can be disseminated: Random Selection, Vicinity Priority Selection, and Vicinity Priority Selection with Queries.

The other example of an adaptive broadcast interval protocol is the abiding geocast protocol [67], which disseminates safety messages within a functional area where these messages are still relevant. In an emergency, a warning packet is transmitted through a vehicle. This packet indicates the area where the warning is still relevant. Upon receiving this packet, the other vehicle will act as a relay node and keep broadcasting the warning packet, provided that it is still traveling in the concerned area. A reduction in the number of redundant warning packets causes a vehicle to modify its rebroadcast dynamically. The rebroadcast interval is defined through the transmission range, the speed, and the relative distance between the emergency point and the vehicle.

The Segment-Oriented Data Abstraction and Dissemination protocol also adopts an adaptive broadcast interval [68]. As the name indicates, in this protocol the roads are divided into segments of predefined lengths (Figure 4.11). Each vehicle collects data by sensing the information and from reports received through other vehicles. Each vehicle adjusts its broadcast interval to reduce the redundancy

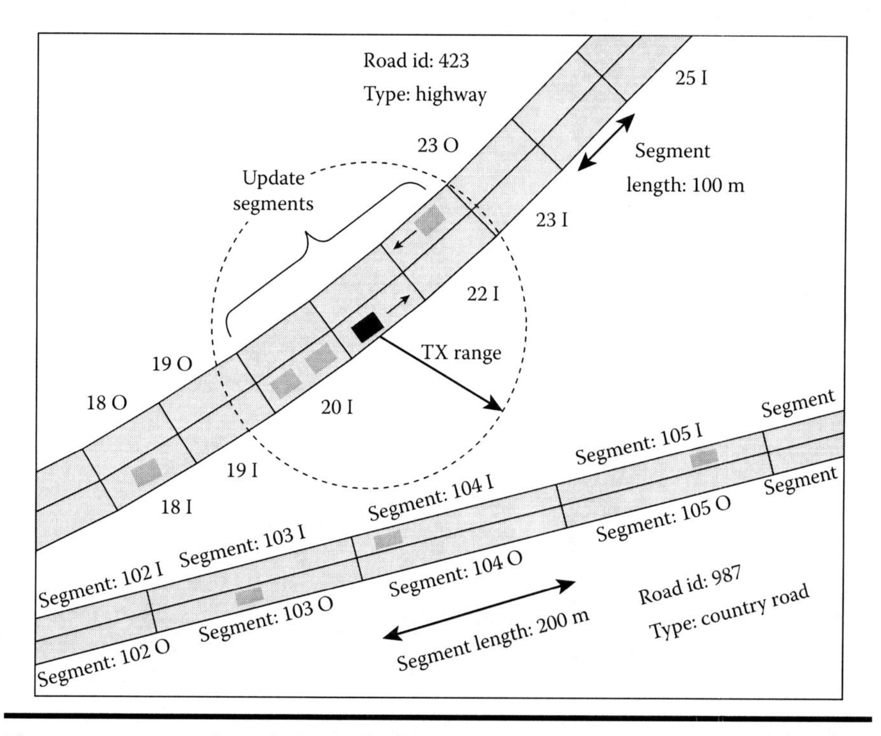

Figure 4.11 Map-based data abstraction by segmentation of roads in the Segment-Oriented Data Abstraction and Dissemination protocol. (From Wischhof, L., et al., *IEEE Trans. Intell. Transport. Syst.*, 6(1), 90–101, 2001.)

in an adaptive manner. Information received from other vehicles is characterized in two manners: (i) provocation and (ii) mollification. In provocation, the time until next broadcast is reduced, whereas in mollification the opposite is true. When a vehicle receives a packet, it determines whether it is a provocation or mollification event by assigning a weight to the received packet. This weight is calculated based on the discrepancy between the received data and those available in the vehicle's knowledge database. This weight can be higher if the received information is newer than the stored information. Based on this weight, the node determines whether a provocation or mollification event has occurred by comparing it with a threshold. The reduction or addition of the time for next broadcast depends on the weight value.

Another single-hop fixed interval broadcasting scheme is TrafficView, which is designed to enable traffic information exchange among vehicles [69]. The information exchanged among vehicles includes speed and position. Here, when a vehicle receives a packet, it stores the information in its database first; next, the information is rebroadcasted in the next broadcast cycle. However, after aggregating multiple records instead of broadcasting all stored records from the database, only a single record is broadcasted. The ratio-based and cost-based algorithms are the two applied in aggregation. In the ratio-based algorithm, a road is divided into small regions, where an aggregation ratio is assigned to each region according to its importance and the level of accuracy required for that region. In cost-based algorithm, the cost can be regarded as the loss of accuracy incurred from combining the records. According to the simulation, the cost-based algorithm yields better accuracy, but the ratio-based algorithm has more flexibility.

4.4.1.1.2 Multihop Broadcasting

In multihop broadcasting, a packet is spread in a network through flooding. Usually, when a sender vehicle broadcasts an information packet, a number of vehicles within its vicinity will become the next relay vehicles by rebroadcasting the packet further in the network. Similarly, after a relay vehicle (node) rebroadcasts the packet, some of the vehicles in their vicinity will become the next relay nodes and will perform the task of forwarding the packet further. As a result, the information packet is able to propagate from the sender to more distant vehicles.

4.4.1.1.2.1 Delay-Based Multihop Broadcasting Protocols

In a delay-based multihop broadcasting scheme, each vehicle is assigned a different waiting time before rebroadcasting the packet; here, the vehicle having the shortest waiting time gets the highest priority to rebroadcast the packet. In addition, when vehicles detect that the packet has already been rebroadcasted, redundancy is prevented by other vehicles through their waiting process.

Whereas different delays are assigned to each vehicle in delay-based broadcasting protocols, a different rebroadcast probability is assigned to each vehicle through a probabilistic-based protocol.

The Urban Multihop Broadcast (UMB) protocol is a delay-based multihop broadcasting protocol designed to control the broadcast storm, the hidden terminal, and the reliability problems in multihop broadcasting [70]. The UMB protocol divides a road inside the transmission range of a transmitter into smaller segments, and it gives rebroadcast priority to the vehicles that belong to the most distant segment. This protocol applies two types of packet forwarding: (i) direction broadcast and (ii) intersection broadcast. The UMB is inefficient because the next rebroadcast vehicle has to wait longer before being able to transmit the clear-to-broadcast (CTB) packet. This is due to the longest black burst duration of the next rebroadcast vehicle.

Smart Broadcast (SB) [71] is proposed to improve UMB protocol shortcomings. Here, when a source vehicle transmits a packet, a request-to-broadcast (RTB) packet containing its location and other information including packet propagation direction and contention window size should be transmitted by the same vehicle. All vehicles in the range of the source that receive the RTB packet determine the "sector" in which they belong by comparing their locations with that of the transmitting vehicle. Eventually, all vehicles that receive the RTB packet choose a contention delay based on the sector that they occupy.

The Efficient Directional Broadcast (EDB) protocol [72] is another delay-based multihop broadcast protocol somewhat similar to the UMB and SB protocols in practice. Here, the RTB and CTB control packets are not applied, but the use of directional antennas is exploited. In this proposed protocol, it is suggested that each vehicle be equipped with two directional antennas, with a 30-degree beam width. EDB uses two types of packet forwarding, just like UMB, directional broadcast on the road segment and directional broadcast at the intersection. In both types, collisions are decreased.

4.4.1.1.2.2 Probability-Based Multihop Broadcasting Protocols In this protocol, each vehicle rebroadcasts a packet according to the assigned probability. Because only a few vehicles will rebroadcast the packet, redundancy and packet collisions are reduced.

4.4.1.1.2.3 Network Coding-Based Multihop Broadcasting Protocols Network coding is a new manner of information dissemination, applicable in deterministic broadcast approaches, resulting in significant reductions in the number of transmissions in the network and yielding a much higher throughput than the traditional manners.

The COPE protocol [73] is based on the network coding principle. Although COPE is a unicast routing protocol, it provides the foundation for many multihop routing protocols. COPE is intended to realize the benefits of network coding beyond simple duplex flows. COPE is based on three key techniques: (i) opportunistic listening, (ii) opportunistic coding, and (iii) neighbor state learning. Opportunistic listening simply allows the nodes to take the advantage of the wireless broadcast medium by snooping all data packets. Each overheard packet will be stored in the node's buffer for a limited time. These overheard packets will later be used for network coding when the opportunity presents. Opportunistic coding defines some basic rules for a node to encode and transmit a packet. Basically, a node should ensure that its next-hop neighbor has enough information to decode the transmitted encoded packet. Usually, a node will be able to decode a packet i correctly from an encoded packet including p_1, p_2, …, p_n if it has $n - 1$ of these packets. Thus, learning what packets are in its neighbor's possession is crucial, and this is achieved through periodic broadcast of reception reports. Hence, each node announces the packets that are stored in its reception buffer to all its neighbors in a periodic manner. A simple scenario for improvement of throughput through network coding is shown in Figure 4.12.

CODEB is another network coding-based broadcasting protocol, introduced in Ref. [74]. It extends the concepts and techniques proposed in COPE in order to cover the broadcasting scenarios in wireless ad hoc networks. It uses opportunistic listening, where each and every node snoops all packets overheard by it. In addition, each node broadcasts the list of its one-hop neighbors in a periodic manner. This allows all nodes to build a list of their two-hop neighbors, which will further be applied in constructing a broadcasting backbone. CODEB relies on opportunistic coding, in which coding opportunities to transmit coded packets is specified. According to CODEB, opportunistic coding for broadcasting is slightly different from coding for unicasting. In broadcasting, all neighbors of the node

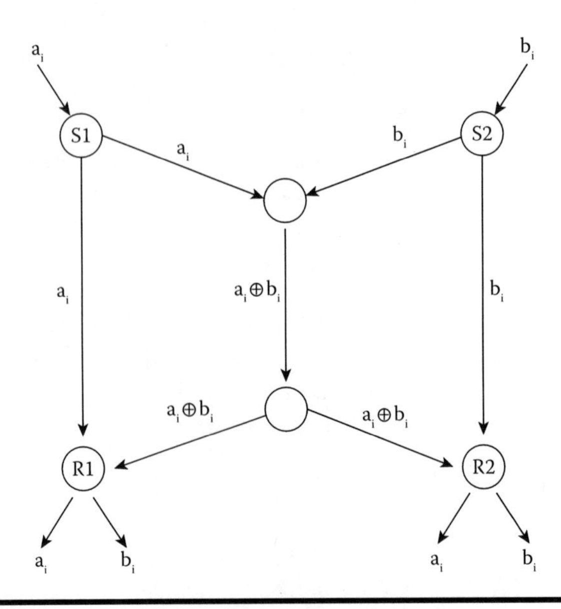

Figure 4.12 A simple scenario for improvement of throughput through network coding. (From Katti, S., et al., *IEEE ACM Trans. Netw.*, 16(3), 497–510, 2008.)

must receive the packet, whereas in unicasting only the intended next-hop node receives a given packet. Because broadcasting increases the level of complexity, all nodes that receive packets must be able to decode them.

DifCode is a network coding-based broadcasting protocol. Its objective is to reduce the number of transmissions required to flood packets in a wireless ad hoc network [75]. Similar to CODEB, DifCode chooses the next forwarding nodes in a deterministic manner. However, DifCode uses a selection algorithm based on multipoint relay (MPR) [76]. The MPR of a node is the list of its one-hop neighbors that covers its two-hop neighborhood. In DifCode, a node can encode and broadcast only packets received from nodes that select it as their MPR. DifCode and CODEB differ by their opportunistic coding techniques. In CODEB, all neighbors of a transmitter decode the received packets immediately, which limits the coding opportunities. In contrast, DifCode relaxes this constraint by allowing nodes to buffer packets that are not immediately decodable. Specifically, all nodes maintain buffers for keeping three different types of packets: (i) successfully decoded, (ii) not immediately decodable, and (iii) packets that need to be encoded and broadcasted further. Simulation results indicate that DifCode results in a lower redundancy rate than probabilistic broadcasting protocols.

4.4.1.1.3 Epidemic Routing

Epidemic routing is proposed as an approach for routing in sparse and/or highly mobile networks where there may not exist a contemporaneous path from source to destination [77]. It adopts a so-called store-carry-forward paradigm: a node receives a packet buffer and carries that packet as it moves, passing the packet on to new nodes that it encounters. Just like the spread of infectious diseases, each time a packet-carrying node encounters a node that does not have a copy of that packet, the carrier is said to infect this new node by passing on a packet copy; newly infected nodes, in turn, continue the trend. The destination receives the packet when it meets an infected node. When the traffic load is very low, epidemic routing is able to achieve minimal delivery delay at the expense of increased use of resources such as buffer space, bandwidth, and transmission power. This leads to link and/or storage congestion when the network is loaded. Variations of epidemic routing have recently been proposed that exploit the tradeoff between delivery delay and resource consumption, including k-hop schemes, probabilistic forwarding, and Spray and Wait [78]. These schemes differ in their "infection process," that is, the spreading of a packet in the network. They need to be combined with a so-called recovery process that deletes copies of a packet at infected nodes, following the successful delivery of the packet to the destination. Different recovery schemes are proposed: some are simply based on timers, and others actively spread the information that a copy has been delivered to the destination in the network.

It is clear that the efficacy of epidemic routing, as became apparent originally, is limited to a great extent by the scarcity of the storage capacity available at nodes.

Once a message is generated, large numbers of its copies are distributed throughout the network and would keep being stored even after the arrival of one of the replicas to the destination. Clearly, this would result in wastage of memory. A message is deleted only upon arrival of a new message to a node requiring storage if the buffer is full already. Therefore, in order to create the required space for the newly arrived message, an old message would be chosen from the buffer to be eliminated and substituted with the new one. Other highly effective techniques used for buffer replacement are studied in Ref. [48]. In particular, four alternative strategies are considered:

1. *Drop Random*: a packet is eliminated randomly. This strategy prioritizes the delivery of messages that are produced by the most silent sources (i.e., sources that produce packets with a slow rate). In fact, the higher the number of the packets produced by a source, the higher the chance of its packets being stored in the same buffer, and the higher the probability of one of them being eliminated.
2. *Drop Least Recently Received*: messages that have been kept in the buffer for much longer are eliminated first. In reality, the chance of these messages having already arrived at the destination is much higher.
3. *Drop Oldest*: messages that have been in the network for a much longer time are eliminated first. As before, these messages are highly expected to have already reached the destination.
4. *Drop Least Encountered*: the message with the least expected likelihood of delivery is eliminated from the buffer. The likelihood of delivery of a message means the probability of the node keeping that message being accessible to the destination node of that message or being accessible to a different node expected to encounter the destination. Every single node is expected to keep a list of likelihoods of delivery, one for every known node in the network. This list is brought up to date after pairwise contacts.

The results of simulations revealed that the Drop Oldest and the Drop Least Encountered policies perform better compared to other policies. Furthermore, when it comes to fixed buffer sizes at nodes and growing network load, the Drop Least Encountered algorithm performs much better than the Drop Oldest algorithm.

The Spray and Wait protocol [78] is attributed to flooding-based delivery arrangements, because it does not use any existing information on network topology or any knowledge about the past encounters of nodes, although it helps to considerably decrease the transmission load by putting a limit on the total number of copies that can be transmitted over the network for any single message. It is therefore more efficient energy-wise than flooding-based protocols. Furthermore, any delay experienced by messages very much resembles that of the optimal case of epidemic routing. Lastly, the Spray and Wait protocol is remarkably tough and scalable. The way this protocol performs is explained in detail as follows. Message delivery is subdivided into two stages: Stage 1, the spray phase, and Stage 2, the wait phase. During the spray phase, a number of copies of the message are distributed over the network by the source node as well as those nodes that have received the message from the source node itself in the first place. This phase would be terminated when a specified number of copies, denoted L, have been spread over the network. Then, in the wait phase, every node that keeps a copy of the message (the so-called relay nodes) would only store its copy and, at the end, deliver it to the destination when or if it becomes accessible. There are a number of ways in which the spray phase can be carried out. Based on the Source Spray and Wait heuristic, the source node sends all L copies of the message to the first L unique nodes with which it comes into contact. Spray and Wait is very flexible. In reality, with an increase in the number of nodes in the network, the percentage of nodes that need to be transformed into relays to accomplish the same level of performance drops. In comparison, many of the other multicopy arrangements carry out a rapidly increasing number of transmissions as the density of the node rises. The performance of the Spray and Wait protocol has been put to the test theoretically as well as in a simulation study.

4.4.1.2 Context-Based Routing

Most methods that are dissemination-based put a limit on message flooding by taking advantage of knowledge about direct contact with destination nodes. Context-based routing takes advantage of more information on the context that nodes are performing in to pinpoint suitable next jumps in the direction of the eventual destination; for example, the home address of a user is a critical piece of context knowledge when deciding the next jump. The degree to which a host is suitable as the next jump for a message is referred to as the *utility* of that host from this point forward. Context-based routing methods in general can greatly decrease message duplication when it comes to dissemination-based techniques. Context-based techniques, however, have a habit of reducing the delay experienced by each message while being delivered. This is because of the possible errors and inaccuracies in choosing the best relay candidates. Furthermore, utility-based techniques have greater costs associated with computation compared with dissemination-based techniques. Nodes are required to preserve a state to keep track of the utility values that are related to the rest of the nodes existing in the network (i.e., all the possible destination nodes) and therefore require storage capacity for state as well as messages. Last, the cost associated with keeping and updating the state and location of every node should be taken into account in the overall load imposed by the protocol.

The Context-Aware Routing (CAR) protocol presented by Musolesi et al. [79,80] provides asynchronous transmission for the delivery of a message. In a

DTN, due to the fact that the receiver is not normally in the same interconnected network, synchronous delivery of a message is not generally possible. In CAR, if synchronous delivery of a message cannot be carried out, the message is forwarded to the host that has the greatest probability of delivering it successfully, which acts as a carrier for the message. The process of delivering based on probability is built on evaluating and predicting context information through the use of Kalman filters. The process of prediction is utilized during temporary disconnection and the process goes on until it is possible to guarantee a specific accuracy level. Furthermore, epidemic routing can be taken as the best case when it comes to the delivery ratio due to the fact that every message is distributed across all reachable nodes that have large enough buffers to store the messages. Musolesi et al. showed in their simulations that in the case of a small buffer size, the CAR ratio of packet delivery is better than the ratio of packet delivery of epidemic routing because of the fact that CAR only generates a single copy for every message.

LeBrun [81] suggested a technique using the motion vector (MoVe) of mobile nodes to foresee their location in the future. The MoVe arrangement utilizes the information pertinent to relative velocities of a node and its surrounding nodes to predict the shortest distance between two nodes. Once the future locations of the nodes are computed, messages are forwarded to nodes that are approaching the destination. When compared to epidemic routing, this method carries less load associated with controlling the packet and utilization of buffer.

Leguay et al. [82] provided a technique that utilizes virtual coordinate routing named *mobility pattern spaces* (MobySpace). In this strategy, the node coordinates are made up of a set of dimensions, in which every dimension signifies the probability of a node being found in a particular location that is a virtual manifestation of the mobility pattern and does not provide the geographic coordinates of the node. A number of destination functions are calculated on this vector. The authors concluded that this strategy takes fewer resources than epidemic routing when delivering large loads of bundles.

4.4.2 Routing with Infrastructure

4.4.2.1 Routing Based on Fixed Infrastructure

When it comes to infrastructure-based routing, a source node intending to deliver a message normally holds it until it becomes accessible to a base station that is part of the infrastructure, then sends the message to it. Base stations are normally gateways used to access less-challenged networks, for example, they can facilitate Internet access or get connected to a LAN. Therefore, the aim of an opportunistic routing algorithm is the delivery of messages to the gateways, which are assumed to be capable of finding the final destination more easily.

Overall, there are two possible variations of the protocol. The first one performs exactly as already described above and the only communications allowed are node-to-base-station ones. Therefore, quite large delays are experienced by messages. The *Infostation* model is a classic example of this approach [83].

The second version of the protocol allows node-to-base-station as well as node-to-node communications. In other words, a node tending to send a message to a destination node provides the message to the base station directly, if it is within communication range; otherwise, it delivers the message opportunistically to a neighboring node that will forward it to the base station once encountered (routing arrangements presented in the section "Vehicle Motion Path Prediction Methods" can be used in this stage). In fact, such a protocol has been suggested in the Shared Wireless Infostation Model [84]. Because they simply act as information sinks, historically, fixed-base stations play a passive role in the opportunistic forwarding strategy (e.g., Infostations [83]). Through running an opportunistic routing algorithm also at base stations, many benefits can be realized. Base stations, for instance, can simply gather the messages forwarded by visiting nodes and after that wait for the destination nodes to become accessible so the stored messages can be forwarded to them. Base stations of a mobile infrastructure (described in the following section) normally have this type of active role.

4.4.2.2 Routing Based on Mobile Infrastructure (Carrier-Based Routing)

When it comes to carrier-based routing, nodes of the infrastructure are mobile data gatherers. They move around in the area of the network, following either predefined or random routes, and collect messages from nodes as they come into contact with them. These particular nodes are called *carriers*, *supports*, *forwarders*, mules, or even *ferries*. In a situation where only node-to-carrier communications are allowed, these nodes can be the only bodies responsible for the delivery of the messages, or they can provide help with increasing connectivity in low-competence networks and providing certainty that isolated nodes can also be accessed. In the second case, message delivery is achieved by carriers as well as ordinary nodes, and both node-to-node and node-to-carrier communication is allowed.

The focus of the Data-MULE system [85] is on the recovery of data from low-competence wireless sensor networks. It is made up of a three-tier architecture (see Figure 4.13):

1. The lower level is occupied by the sensor nodes, which regularly carry out data sampling from the neighboring environment.
2. The middle level consists of mobile agents, called MULEs, which, in order to collect their information, move around in the area covered by sensors.
3. The upper level is made up of a collection of wired APs and sources of data, which are the receivers of information from the MULEs. They are connected to a centrally located source of data where all the received data is stored and processed.

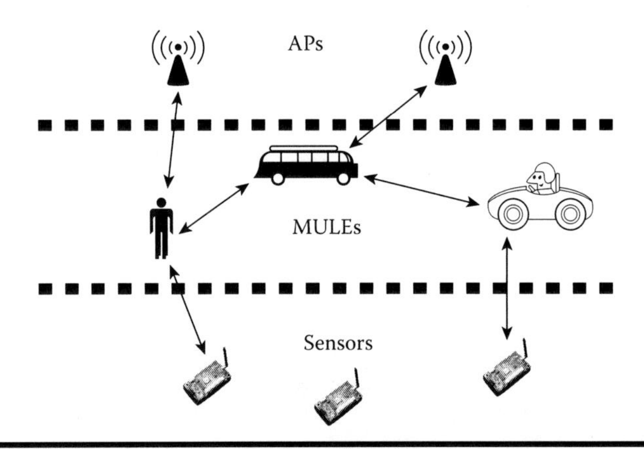

Figure 4.13 The three-tier architecture of MULE. (From Jain, S., et al., *ACM Kluwer Mobile Netw. Appl.*, 11(3), 327–339, 2006.)

In the Message Ferrying Approach [86], additional mobile nodes are opportunistically taken advantage of so they can offer a message relaying service. Such nodes are called *message ferries*, and they move around in the network in order to collect messages from source nodes. Message collection can be carried out in two possible ways:

1. *Node-Initiated Message Ferrying:* the ferry node moves around following a predetermined path. Each node in the network carries information pertaining to the paths followed by active ferries and moves to track ferries when it holds data that need to be delivered.
2. *Ferry-Initiated Message Ferrying:* the ferry node, once again, moves around following a predetermined path. Any source node tending to carry out message delivery sends a service request to the ferry (through a long-range radio signal) that also carries information about its current position. Once the request is received from the source node, the ferry adjusts its path to encounter the source node.

4.5 Summary

The opportunistic VANET is a kind of opportunistic network presented for maintaining high speed and reliable communication for vehicles in an urban environment. IEEE 802.11 (Wi-Fi) is one of the most important technologies used in VANETs. However, major difficulties have been observed in IEEE 802.11 application with respect to opportunistic vehicular communication. To address these challenges, some modifications and improvements are necessary to prepare IEEE 802.11 for application in VANETs. For this purpose, in some real-world projects, 802.11 has been adapted for VANETs. In addition, IEEE 802.11p is an amendment to IEEE 802.11 presented specifically for vehicular communications. Several routing algorithms have been presented so far for opportunistic VANETs. These opportunistic routing algorithms can be classified into two groups according to their dependency on roadside infrastructures: routing without infrastructure and routing with infrastructure. Several challenges such as long latency to establish connection, lossy link performance, and frequent disconnections exist in implementation of VANET routing. In one of the successful methods developed to overcome these challenges, information related to vehicle movement paths is predicted and applied to establish more stable routes. In this chapter, all of the issues mentioned regarding opportunistic VANETs were discussed.

References

1. L. Andreone and C. Ricerche, Activities and applications of the vehicle to vehicle and vehicle to infrastructure communication to enhance road safety, in *5th European Congress and Exhibition on ITS*, 2005.
2. H. Hartenstein and P. Laberteaux, A tutorial survey on vehicular ad hoc networks, *IEEE Commun Mag*, 46(6), pp. 164–171, 2006.
3. S. F. Hasan, N. Siddique, and S. Chakraborty, *Intelligent Transport Systems: 802.11-Based Roadside-to-Vehicle Communications*, Springer, New York, 2013.
4. H. Hartenstein and K. P. Laberteaux, *VANET: Vehicular Applications and Inter-Networking Technologies*, Wiley, United Kingdom, 2010.
5. A. Sebastian, M. Tang, Y. Feng, and M. Looi, A Multicast routing scheme for efficient safety message dissemination in VANET, in *Proceedings of IEEE Wireless Communications and Networking Conference* (*WCNC*), 2010.
6. M. Taha and Y. Hasan, VANET-DSRC protocol for reliable broadcasting of life safety messages, in *Proceedings of IEEE Int'l Symp. on Signal Processing and Information Technology*, 2007.
7. T. Taleb, E. Sakhaee, A. Jamalipour, K. Hashimoto, N. Kato, and Y. Nemoto, A stable routing protocol to support ITS services in VANET networks, *IEEE Trans Veh Technol*, 56(6), pp. 3337–3347, 2007.
8. M. K. Denko, *Mobile Opportunistic Networks: Architectures, Protocols and Applications*, Taylor & Francis, 2011.
9. A. I. Filali, Intelligent strategies of access point selection for vehicle to infrastructure opportunistic communications, in *IEEE Vehicular Networking Conference*, Tokyo, Japan, 2009.
10. J. Sushant, K. Fall, and R. Patra, Routing in a delay tolerant network, in *Proceedings of SIGCOMM*, 2004.
11. I. F. Akyildiz, D. M. Gutierrez-Estevez, and E. C. Reyes, The evolution to 4G cellular systems: LTE advanced. *Phys Commun*, 3, pp. 217–244, 2010.
12. A. S. Tanenbaum, *Computer Networks* (Third Edition), Prentice Hall, 1996.
13. P. Roshan and J. Leary, *802.11 Wireless LAN Fundamentals*, Cisco Press, 2003.
14. M. S. Gast, *802.11® Wireless Networks: The Definitive Guide*, O'Reilly, 2005.

15. M. Piao, C. Nam, and D. Shin, WVaMode: A Wi-Fi vehicular ad hoc mode for the unmanned vehicle, in *Advanced Communication Technology* (*ICACT'13*), 2011.
16. P. Deshpande, X. Hou, and S. R. Das, Performance comparison of 3G and metro-scale WiFi for vehicular network access, in *Proceedings of the 10th ACM SIGCOMM Conference on Internet Measurement* (*IMC '10*). ACM, New York, 2010.
17. P. Deshpande, A. Kashyap, C. Sung, and S. R. Das, Predictive methods for improved vehicular WiFi access, in *Proceedings of ACM MobiSys*, 2009.
18. V. Bychkovsky, B. Hull, A. Miu, H. Balakrishnan, and S. Madden, A measurement study of vehicular internet access using in situ Wi-Fi networks, in *Proceedings of ACM MobiCom Conference*, 2006.
19. J. Eriksson, H. Balakrishnan, and S. Madden, Cabernet: A WiFi-based vehicular content delivery network, in *Proceedings of ACM MobiCom Conference*, 2008.
20. A. Balasubramanian, R. Mahajan, A. Venkataramani, B. N. Levine, and J. Zahorjan, Interactive WIFI connectivity for moving vehicles, *ACM SIGCOMM Comput Commun Rev*, 38(4), pp. 427–438, 2008.
21. A. Giannoulis, M. Fiore, and E. W. Knightly, Supporting vehicular mobility in urban multi-hop wireless networks, in *Proceedings of ACM MobiSys Conference*, 2008.
22. B. Hull, V. Bychkovsky, Y. Zhang, et al., CarTel: A distributed mobile sensor computing system, in *Proceedings of the 4th International Conference on Embedded Networked Sensor Systems* (*SenSys '06*), ACM, New York, 2006.
23. J. Ott and D. Kutscher, Drive-thru Internet: IEEE 802.11b for "Automobile" users, in *Proceedings of IEEE INFOCOM*, 2004.
24. H. Soroush, P. Gilbert, N. Banerjee, B. Levine, M. Corner, and L. Cox, Improving mobile networking with concurrent Wi-Fi connections, in *ACM SIGCOMM, Poster Session*, Toronto, ON, 2011.
25. A. Subramanian, R. Mahajan, A. Venkataramani, B. N. Levine, and J. Zahorjan, Interactive WIFI connectivity for moving vehicles, *ACM SIGCOMM*, 38(4), pp. 427–438, 2008.
26. S. F. Hasan, N. H. Siddique, and S. Chakraborty, Stochastic analysis of variation in disruption due to changing traffic patterns, *J Inf Sci Eng*, 2014 (Accepted for publication) [SCIE indexed].
27. H. Guo, *Automotive Informatics and Communicative Systems: Principles in Vehicular Networks and Data Exchange*, IGI Global, 2009.
28. M. Amadeo, C. Campolo, and A. Molinaro, Enhancing IEEE 802.11p/WAVE to provide infotainment applications in VANETs, *Ad Hoc Netw*, 10(2), pp. 253–269, 2010.
29. J. Froehlic and J. Krumm, Route prediction from trip observations, in *Society of Automotive Engineers (SAE) 2008 World Congress*, 2008.
30. J. Krumm, Real time destination prediction based on efficient routes, in *Society of Automotive Engineers (SAE) 2006 World Congress*, 2006.
31. R. Simmons, B. Browning, Y. Zhang, and V. Sadekar, Learning to predict driver route and destination intent, in *2006 IEEE Intelligent Transportation Systems Conference*, Toronto, Canada, 2006.
32. A. Bohlooli and K. Jamshidi, A GPS-free method for vehicle future movement directions prediction using SOM for VANET, *App Intell*, 36(3), pp. 685–697, 2012.
33. T. Kohonen, The self-organizing map, *Proc IEEE*, 78(9), pp. 45–52, 1990.
34. J. Freeman and D. Skapura, *Neural Network, Algorithms, Applications and Programming Techniques*, Wesley, 1991.

35. P. Sankar, D. Biswarup, and M. Pabitra, Rough self organizing map, *Appl Intell*, 21(3), pp. 289–299, 2004.
36. H. Ye and B. Lo, Feature competitive algorithm for dimension reduction of the self-organizing map input space, *Appl Intell*, 13(3), pp. 215–230, 2000.
37. J. Vesanto and E. Alhoniemi, Clustering of self organized map, in *IEEE Transactions on Neural Networks*, 2000.
38. R. Schalkoff, *Artificial Neural Networks*, McGraw-Hill, 1997.
39. S. Cheshire, B. Aboba, and E. Guttman, *Dynamic Configuration of IPv4 Link-Local Addresses*, 2005. RFC 3927.
40. E. Guttman. Autoconfiguration for IP networking: Enabling local communication, *IEEE Internet Comput*, 5(3), pp. 81–86, 2001.
41. H. Menouar, M. Lenardi, and F. Filali, A movement prediction based routing protocol for vehicle-to vehicle communications, in *Proceedings of V2VCOM 2005*, 2005.
42. F. Granelli, G. Boato, and D. Kliazovich, MORA: A movement-based routing algorithm for vehicle ad hoc networks, in *Proceedings of 1st IEEE Workshop on Automotive Networking and Applications* (*AutoNet 2006*), San Francisco, CA, 2006, pp. 1–10.
43. F. Granelli, G. Boato, D. Kliazovic, et al. Enhanced GPSR routing in multi-hop vehicular communications through movement awareness, *IEEE Commun Lett*, 11(10), pp. 781–783, 2007.
44. B. Karp and H.T. Kung, GPSR: Greedy perimeter stateless routing for wireless networks, in *Proceedings of ACM/IEEE International Conference on Mobile Computing and Networking*, 2000.
45. V. Namboodiri and L. Gao, Prediction-based routing for vehicular ad hoc networks, *IEEE Trans Veh Technol*, 56(4), pp. 2332–2345, 2007.
46. J. Kim and S. Lee, Reliable routing protocol for vehicular ad hoc networks, *Int J Elec Comm*, 65(3), pp. 268–271, 2011.
47. A. Bohlooli and K. Jamshidi, Profile based routing in vehicular ad-hoc networks, *Sci China Inf Sci*, 57(6), pp. 685–697, 2014.
48. L. Pelusi, A. Passarella, and M. Conti, Opportunistic networking: Data forwarding in disconnected mobile ad hoc networks, *IEEE Commun Mag*, pp. 134–141, 2006.
49. A. Leal, Information-centric opportunistic data dissemination in vehicular ad hoc networks, in *2010 13th International IEEE Conference on Intelligent Transportation Systems* (*ITSC*), 2010.
50. F. Li and Y. Wang, Routing in vehicular ad hoc networks: A survey, *IEEE Vehicular Technol Mag*, 2(2), pp. 12–22, 2007.
51. T. Willke, P. Tientrakool, and N. Maxemchuk, A survey of inter-vehicle communication protocols and their applications, *IEEE Comm Surv Tutorials*, 11(2), pp. 3–20, 2009.
52. S. Panichpapiboon and W. Pattara-atikom, A review of information dissemination protocols for vehicular ad hoc networks, *IEEE Commun Surv Tutorials*, pp. 1–15, 2011.
53. A. Festag, P. Papadimitratos, and T. Tielert, Design and performance of secure geocast for vehicular communication, *IEEE Trans on Vehicular Technol*, 59(5), pp. 2456–2471, 2010.
54. R. J. Hall, An improved geocast for mobile ad hoc networks, *IEEE Trans Mobile Comput*, 10(2), pp. 254–266, 2011.
55. L. Junhai, Y. Danxia, X. Liu, and F. Mingyu, A survey of multicast routing protocols for mobile ad-hoc networks, *IEEE Commun Surv Tutorials*, 11(1), pp. 78–91, 2009.

56. O. Badarneh and M. Kadoch, Multicast routing protocols in mobile ad hoc networks: A comparative survey and taxonomy, *EURASIP J Wirel Comm Netw*, 2009(1), pp. 1–42, 2009.
57. A. Sebastian, M. Tang, Y. Feng, and M. Looi, A multicast routing scheme for efficient safety message dissemination in VANET, in *Proceedings of IEEE Wireless Communications and Networking Conference* (*WCNC*), 2010.
58. E. K. Lua, J. Crowcroft, M. Pias, R. Sharma, and S. Lim, A survey and comparison of peer-to-peer overlay network schemes, *IEEE Comm Surv Tutorials*, 7(2), pp. 72–93, 2005.
59. P. Ruiz and P. Bouvry, Survey on broadcast algorithms for mobile ad hoc networks, *ACM Comput Surv*, 48(1), pp. 121–137, 2015.
60. L. Zhou, Y. Zhang, K. Song, W. Jing, and A. V. Vasilakos, Distributed media services in P2P based vehicular networks, *IEEE Trans Veh Technol*, 60(2), pp. 692–703, 2011.
61. U. Shevade, Y. Chen, L. Qiu, et al., Enabling high bandwidth vehicular content distribution, in *Proceeding of ACM Int'l Conf. on Emerging Networking Experiments and Technologies* (*CoNEXT*), 2010.
62. M. Guo, M. Ammar, and E. W. Zegura, V3: A vehicle-to-vehicle live video streaming architecture, *Pervasive and Mobile Comput*, 1(4), pp. 404–424, 2005.
63. Y. Chu and N. Huang, Delivering of live video streaming for vehicular communication using peer-to- peer approach, in *Proceedings of IEEE Workshop on Mobile Network for Vehicular Environments*, pp. 1–6, 2007.
64. R. Kumar and M. Dave, A review of various VANET data dissemination protocols, *Int J. Serv Sci Technol*, 5(3), pp. 27–44, 2012.
65. T. Zhong, B. Xu, and O. Wolfson, Disseminating real-time traffic information in vehicular ad-hoc networks, in *Proceedings of IEEE Intelligent Vehicles Symposium* (*IV*), 2008.
66. T. Fujiki, M. Kirimura, T. Umedu, and T. Higashino, Efficient acquisition of local traffic information using inter-vehicle communication with queries, in *Proceedings of IEEE Intelligent Transportation Systems Conference* (*ITSC*), 2007.
67. Q. Yu and G. Heijenk, Abiding geocast for warning message dissemination in vehicular ad hoc networks, in *Proceedings of IEEE Int'l Conference on Communications* (*ICC*), 2008.
68. L. Wischhof, A. Ebner, and H. Rohling, Information dissemination in self-organizing inter-vehicle networks, *IEEE Trans Intell Transport Syst*, 6(1), pp. 90–101, 2001.
69. T. Nadeem, S. Dashtiezhad, C. Liao, and L. Iftode. Trafficview: Traffic data dissemination using car-to-car communication, in *Proceedings of the MC2R*, 2004.
70. G. Korkmaz, E. Ekici, and F. Ozguner, An efficient fully ad-hoc multi-hop broadcast protocol for inter- vehicular communication systems, in *Proceedings of IEEE Int'l Conference on Communications* (*ICC*), 2006.
71. E. Fasolo, A. Zanella, and M. Zorzi, An effective broadcast scheme for alert message propagation in vehicular ad hoc networks, in *Proceedings of IEEE Int'l Conference on Communications* (*ICC*), 2006.
72. D. Li, H. Huang, X. Li, M. Li, and F. Tang, A distance-based directional broadcast protocol for urban vehicular ad hoc network, in *Proc. of IEEE Int'l Conf. on Wireless Communication, Networking and Mobile Computing* (*WiCom*), 2007.
73. S. Katti, H. Rahul, W. Hu, D. Katabi, M. Medrad, and J. Crowcroft, XORs in the air: Practical wireless network coding, *IEEE/ACM Trans on Networking*, 16(3), pp. 497–510, 2008.

74. L. Li, R. Ramjee, M. Buddhikot, and S. Miller, Network coding-based broadcast in mobile ad-hoc networks, in *Proceedings of IEEE Conference on Information Computing* (*INFOCOM*), Ad Hoc Networks, pp. 1739–1747, 2007.
75. N. Kadi and K. Agha, MPR-based flooding with distributed fountain network coding, in *Proceedings of IFIP Annual Mediterranean Ad Hoc Networking Workshop*, pp. 1–5, 2010.
76. A. Qayyum, L. Viennot, and A. Laouiti, Multipoint relaying for flooding broadcast messages in mobile wireless networks, in *Proceedings of IEEE Hawaii Int'l Conference on Systems Sciences* (*HICSS*), pp. 3866–3875, 2002.
77. X. Zhang, G. Neglia, J. Kurose, and D. Towsley, Performance modeling of epidemic routing, *Comput Netw*, 51, pp. 2867–2891, 2007.
78. T. Spyropoulos, K. Psounis, and C. S. Raghavendra, Spray and wait: An efficient routing scheme for intermittently connected mobile networks, in *Proceedings of the ACM SIGCOMM 2005 Workshop on Delay Tolerant Networks*, Philadelphia, PA, 2005.
79. M. Musolesi, Context-aware Adaptive Routing for Delay Tolerant Networking, Phd dissertation, Department of Computer Science, University of London, London, UK, 2007.
80. M. Musolesi, S. Hailes, and C. Mascolo, Adaptive routing for intermittently connected mobile ad hoc networks, Proceedings of 6th IEEE International Symposium on a World of Wireless, Mobile and Multimedia Networks (WoWMoM'05). Taormina, Italy, 2005.
81. J. Anda, J. Le Brun, C.-N. Chuah, D. Ghosal, and H.M. Zhang, VGrid: Vehicular ad hoc networking and computing grid for intelligent traffic control, in *Proceedings of IEEE VTC*, 2005.
82. J. Leguay, T. Friedman, and V. Conan, DTN routing in a mobility pattern space, in *Proceedings of the ACM SIGCOMM 2005 Workshop on Delay Tolerant Networks*, Philadelphia, PA, 2005.
83. D. Goodman, J. Borras, N. Mandayam, and R. Yates, INFOSTATIONS: A new system model for data and messaging services, in *IEEE VTC'97*, 1997.
84. T. Small and Z. J. Haas, The shared wireless infostation model—A new ad hoc networking paradigm (or Where there is a Whale, there is a Way), in *Proceedings of the Fourth ACM International Symposium on Mobile Ad Hoc Networking and Computing* (*MobiHoc 2003*), Annapolis, MD, 2003.
85. S. Jain, R. Shah, W. Brunette, G. Borriello, and S. Roy, Exploiting mobility for energy efficient data collection in wireless sensor networks, *ACM Kluwer Mobile Netw Appl*, 11(3), pp. 327–339, 2006.
86. W. Zhao, M. Ammar, and E. Zegura, A message ferrying approach for data delivery in sparse mobile ad hoc networks, in *Proceedings of the 5th ACM International Symposium on Mobile Ad Hoc Networking and Computing* (*Mobihoc*), ACM Press, 2004.

Chapter 5

Routing Protocols in Opportunistic Networks

Anshul Verma and K. K. Pattanaik

Contents

5.1 Introduction

Opportunistic networking is an evolution of the classic mobile ad hoc network (MANET). Chapter 3 discusses MANET-related issues in greater detail. A MANET is characterized by its lack of infrastructure, autonomous nature, and mobile nodes. Nodes communicate directly when they are within communication range of each other. Each node can play two roles: end node (source node or destination node) and intermediate node (relay node or router). In a MANET, end-to-end connection between source and destination is necessary for eventual transmission of any message. However, when nodes are highly mobile, the connection opportunity between nodes may become intermittent. As a result, traditional MANET routing protocols are not able to perform eventual transmission due to lack of an end-to-end path between source and destination. Therefore, several properties of traditional MANETs, such as disconnection of nodes, mobility of users, network partitions, and link instability, are treated as drawbacks. This makes the design of MANET routing protocols significantly more difficult (Borgia et al. 2005). The aim of opportunistic networking is to provide routing functionality in this intermittently connected environment.

Opportunistic networks (Pelusi et al. 2006) are formed out of portable mobile devices carried by people without the assumption of any preexisting network infrastructure. They provide connection opportunity between mobile devices by exploiting their mobility while removing the physical end-to-end connection requirement. Disconnections, partitions, mobility, and so on are treated as features instead of drawbacks (Conti and Kumar 2010). Eventual transmission is achieved by using the store-carry-forward approach (Fall 2003). In this approach, intermediate nodes are used to store messages when there is no forwarding opportunity toward the destination, and any future contact opportunity with other mobile devices is exploited to bring messages more close to the destination.

Traditional routing protocols for wireless networks and MANETs, in which paths are decided on the basis of topological network information, cannot be directly used for opportunistic networks. Therefore, routing has become the most demanding issue in opportunistic networks. In this scenario, flooding-based routing protocols provide a good delivery ratio but generate more traffic overhead and unnecessary resource consumption. By contrast, context-based routing protocols reduce traffic overhead and unnecessary resource consumption by optimizing the forwarding task with the help of context information (Jain et al. 2004). Context information plays a significant role in designing efficient routing protocols (Boldrini et al. 2008). It is basically used to make routing protocols capable of learning the network state, automatically adjusting to its dynamic nature, and thus improving their operations (Conti and Giordano 2007a, 2007b). This chapter classifies the important existing routing protocols proposed in the literature based on the amount of context information they exploit. It identifies three main classes: context-oblivious, partially context-aware, and fully context-aware protocols. Further, it provides a comparative study of the routing protocols of all these classes, followed by detailed description of each. Finally, it presents a few cases of opportunistic networking.

The rest of this chapter is organized as follows. The section "Context Information" presents a general definition of the term *context information* for any context-sensitive network communication application. The section "Routing in Opportunistic Networks" describes routing in opportunistic networks and presents a comparative study of protocols. The section "Context-Oblivious Routing" describes the context-oblivious class of routing protocols, followed by partially and fully context-aware category protocols in the sections titled "Partially Context-Aware Routing" and "Fully Context-Aware Routing," respectively. These sections are followed by a discussion of case studies and applications for opportunistic networks. The chapter is summarized in the last section.

5.1.1 Context Information

With the evolution of pervasive and ubiquitous computing, new techniques to compute or provide services are emerging. The main advantage of these new techniques is that they do not provide the same services for each user in the same way. Rather, they provide services on the basis of the interests, preferences, or abilities of single users or specific user groups. Therefore, they are known as *context-aware services*. To make utilization of these services, the context information about a user, his or her device, and environment must be described, collected, and analyzed properly. However, before the use of context information, its meaning must properly be defined according to the suitability to an application. A general definition of *context information* based on a widespread literature study is as follows: Context information is classified into three main categories—user context, terminal context, and communication network context. For different applications or services, these classes can be combined or interpreted suitably.

Brown et al. (1997) describe *context* as the identities of the user's neighbors, position, time of day, temperature, and season. However, these items represent more specialized context information and are not included in the general definition. Ryan et al. (1997) define this concept similarly but temperature and season are not considered context. Schmidt et al. (1999) define *context* as applications, environment, status, surroundings, and situation. Furthermore, in the authors' opinion the definition given by Dey and Abowd (2000) is widely acceptable, as it describes the complete range of context information in a common way that can be used in mostly real-life applications: "Context is any relevant information that describes the environment of several objects (i.e., a person, object, or location) that can be used to the interaction between an application and a user. Context may be place, person, groups, and physical and computational objects."

Three different entities—people, places, and things—were identified by Dey et al. (2001). *Places* describe geographical spaces like homes, schools, buildings, and so on. *People* are categorized as groups or individuals. *Things* are identified as physical objects or software components. These entities can be grouped into the following four categories:

- *Identity*—each entity of the application is characterized by a unique identifier.
- *Location*—includes location-related information of identities such as positioning data and orientation as well as information about regional relations to neighboring entities.

- *Status*—represents the nature, condition, and environment of entities. For example, the status of a place can be described as the current temperature, the weather conditions, or the noise level.
- *Time*—refers to both date and time.

With respect to opportunistic networking, a key piece of knowledge to design efficient routing protocols is information about the context in which the users communicate. This information, such as the users' working address and institution or the probability of meeting with other users or visiting particular places, is also a type of context information and can be exploited to identify suitable forwarders based on context information about the destination. In the following, we classify the main routing protocols proposed in the literature on the basis of context information they exploit.

5.2 Routing in Opportunistic Networks

Opportunistic networks have several similarities with delay-tolerant networks (DTNs). DTNs consist of several independent network clusters; each cluster uses an Internet-like protocol for providing connectivity within the cluster. These clusters are interconnected through DTN overlay. DTN overlay exploits occasional communication opportunities among the clusters to provide end-to-end connectivity. The communication opportunity between clusters might be either scheduled over time or completely random. In general, in conventional DTNs the points of possible disconnection are known. Opportunistic networks are similar to DTNs except in some aspects; specifically, in opportunistic networks the possible points of disconnection are not known in advance and independent Internet-like clusters do not exist. Opportunistic networks are formed by individual nodes that are possibly disconnected for long periods and that opportunistically exploit any contact with other nodes to forward messages. The routing approach is therefore quite different for conventional DTNs compared to opportunistic networks. Because the points of disconnection (and, sometime, the duration of disconnections) are known in DTNs, routing can be performed along the same lines used for conventional Internet protocols, by simply considering the duration of disconnections as an additional cost of the links. Because opportunistic networks do not assume the same knowledge about network evolution, routes are computed dynamically while the messages are being forwarded toward the destination. Each intermediate node evaluates the suitability of encountered nodes to be a good next hop toward the destination. For example (see Figure 5.1), the user at the desktop opportunistically transfers a message intended for a destination through a user passing nearby, via a wireless link, expecting that this user will carry the information closer to the destination. Later this user finds a traveler in a train going to the same

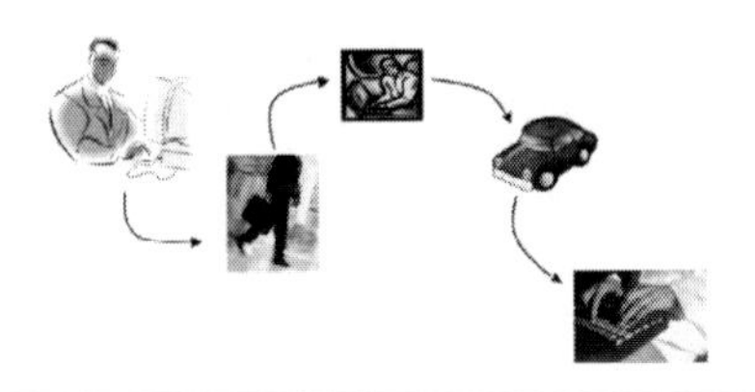

Figure 5.1 Message forwarding in an opportunistic network. (From Boldrini, C., et al., *Int. J. Auton. Adapt. Commun. Syst.*, 1, 122–147, 2008.)

city where the destination user works and forwards the message to the traveler. At the destination station, the traveler discovers that a car driver is going to the same neighborhood of the destination user's workplace and hands over the message. The car driver meets the destination user on his way, and the message is finally delivered to the destination user (Boldrini et al. 2008).

Opportunistic networks do not require prior establishment of the complete path from source to destination. This phenomenon drastically reduces the complexity of routing protocols in opportunistic networks. However, new routing challenges arise that are different from traditional networks. The routing protocols of opportunistic networks are capable of transmitting data with some reliability even during frequent network disconnections or when establishment of end-to-end paths is not possible. Moreover, because connectivity among nodes is intermittent, other conventional wireless network routing approaches can't be applied and, in such cases, flooding-based routing protocols suit well, although they generate extra traffic overhead and increase energy consumption (Nguyen and Giordano 2009).

The performance of routing protocols improves when context information of the network is exploited—knowledge about the network topology, users' behavior, information about the users themselves, and so on. Each context-aware routing protocol exploits all or some specific types of context information according to the design specification. The context information could be the home address, office, school, profession, phone number, the mobility pattern of users, the frequency of visiting a particular place, the communities that users belong to, and so on. All such information is helpful in making the decision to forward messages. For example, to find out the best forwarder for a communication toward the final destination, the home address of an intermediate node is a precious piece of context information.

In the following, the main existing routing protocols are classified into three main classes—context-oblivious, partially context-aware, and fully context-aware protocols—on the basis of the amount of context information they exploit (Conti et al. 2008). Basically, all the routing protocols of the context-oblivious class use some form of flooding. The heuristic behind this policy is when there is knowledge neither of a possible path toward the destination nor of an appropriate next-hop node, a message should be disseminated as widely as possible. Protocols of this class might be the only solution when no context information is available. Clearly, they generate a high overhead, may suffer high contention, and potentially lead to network congestion. A comparison of the main routing protocols of the context-oblivious class is given in Table 5.1. Partially context-aware protocols exploit a particular piece of context information to optimize the forwarding task. In contrast, fully context-aware protocols not only exploit several types of context information to optimize routing but also provide general mechanisms to handle and use context information. The main difference between partially and fully context-aware protocols is the fact that the latter usually provide a full-fledged set of algorithms to gather and manage any type of context information, whereas the former are customized for a specific type of context information. A comparison of the main routing protocols of the partially context-aware and fully context-aware classes is depicted in Tables 5.2 and 5.3, respectively.

A comparison of routing protocols shows which information is used as the context information, what routing approach is followed, and the consequences. A comparison in general indicates that context-based routing protocols are efficient mechanisms for forwarding messages in opportunistic networks. They are stronger alternatives to classical flooding-based protocols and can be seen as an effective means of controlling congestion in this network. With respect to context-based routing protocols, fully context-aware routing is better than partially context-aware routing in terms of delivery probability and congestion control. A comparative study of several routing protocols is given, and a detailed description of each routing protocol is presented in the next sections. The use of context information in decision-making for the routing is also emphasized in the subsequent sections.

5.3 Context-Oblivious Routing

The context-oblivious routing category contains flooding-based routing protocols. These routing protocols either follow blind flooding or controlled flooding techniques. The routing protocols of this category are only a solution when information about a possible path toward the destination or suitable next-hop node is not available. This means that they don't exploit any form of context information. The message is transmitted to the destination by disseminating it as widely as possible. However, flooding-based techniques disseminate a huge number of packets into the network, which generates a network congestion problem and is also very costly in terms of memory and energy consumption (Jindal and Psounis 2007). Some common approaches to solving this problem is to control flooding by limiting the maximum number of packets a network can have at a time,

Table 5.1 Comparison of the Context-Oblivious Class of Routing Protocols

Protocol	*Context Information*	*Approach*	*Remarks*
Direct Transmission Routing	No	Uses a single copy of message; source directly transmits to destination.	Single transmission per message; unlimited transmission delay
Epidemic Routing	No	Exchanges copy of missing messages with neighbors.	Best delivery ratio; misuse of resources
Spray and Wait Routing	No	Spreads copies of message, uses maximum two hops.	Reduces number of transmissions and delay; not efficient in IID[a] mobility model
Binary Spray and Wait Routing	No	Spreads copies of message in controlled way; uses multiple hops.	Gives more optimal results in case of IID mobility model
Network Coding	No	Merges several messages into one to reduce number of transmissions.	Performs better in case of poor connectivity; increased overhead in case of good connectivity
Erasure Coding (EC)	No	Converts message into several code blocks; original message is reconstructed by a few code blocks.	Provides best worst-case delay performance; does not utilize full contact duration
Aggressive-Erasure Coding (A-EC)	No	Same as EC; fully utilizes contact duration by packet transmission.	Provides best very small delay performance; poor worst-delay performance in case of black-hole nodes
Hybrid-Erasure Coding (H-EC)	No	Combination of EC and A-EC; first copy of blocks is sent using EC and second copy of blocks is sent by A-EC.	Good worst-delay and very small delay performance

[a] Independently and identically distributed.

Table 5.2 Comparison of the Partially Context-Aware Class of Routing Protocols

Protocol	*Context Information*	*Approach*	*Remarks*
Randomized Routing	Last encountered time	Uses single copy of message, forwards message to relay with higher delivery probability.	Shows progress at each forwarding step; position information calculation of nodes is not absolute.
Utility-Based Routing	Last encountered time with speed and mobility pattern	Uses single copy of message, forwards message to relay with higher delivery probability.	Position information calculation of nodes is absolute; suffers from slow-start initial phase problem.
Seek and Focus Routing	Last encountered time with speed and mobility pattern	Uses single copy of message, forwards message to relay with higher delivery probability.	Solves slow-start initial phase problem; combination of randomized and utility-based routing.
Prioritized Epidemic Routing	Last encountered time	Estimates routing cost, assigns deletion and transmission priority to each packet.	Reduces burden on resources without overly affecting delivery ratio.
PROPHET Routing	Frequency of meetings between nodes	Forwards copies of message to neighbors with higher delivery probability.	Extension of Epidemic Routing with the concept of delivery probability.
Meetings and Visits (MV) Forwarding Algorithm	Frequency of meetings and visits to physical locations	Learns patterns in the movement of participants and uses them to enable informed message passing.	Performs significantly better, reaching 84% of maximum possible delivery rate.

(Continued)

Table 5.2 *(Continued)* Comparison of the Partially Context-Aware Class of Routing Protocols

Protocol	*Context Information*	*Approach*	*Remarks*
MaxProp Protocol	Frequency of meetings and visits to physical locations	Based on prioritizing both packets to be transmitted to other peers and packets to be dropped.	Uses several other techniques with core to increase delivery rate and reduce latency.
MobySpace	Mobility pattern of nodes	The best forwarder is the node that is closest to the destination in Euclidean space.	Performs controlled flooding to achieve better performance in terms of delivery and delay.
Spray and Focus	Last encountered time	The source sends multiple copies to neighbors; they can forward it using a single-copy utility-based scheme.	Reduces delay of Spray and Wait Routing up to 20 times in some scenarios.
BUBBLE Rap Routing	Social community information	Message is forwarded to nodes belonging to the destination node's community or to increasingly sociable nodes.	Better forwarding efficiency compared to context-oblivious routing and PROPHET routing.
Integrated Routing Protocol	Frequency of meetings between nodes	Same as PROPHET when context information is available; falls back to Epidemic when context information is missing.	Provides better results in terms of delivery probability and delay in both cases regardless of whether the context information is available.

Table 5.3 Comparison of the Fully Context-Aware Class of Routing Protocols

Protocol	*Context Information*	*Approach*	*Remarks*
Context-Aware Routing Protocol	Residual battery life, rate of connectivity change, probability of meeting between nodes	Sender sends message to node in its cloud having highest delivery probability. This node sends message to destination or to node of other cloud having higher delivery probability.	Even without message replication, it gives better message delivery with minimum overheads than Epidemic Routing.
History-Based Opportunistic Routing Protocol	Node profile, social relationships, encounter history information	Nodes are able to learn current context and remember past context. Such context data are used to decide good next hops toward destinations.	Drastically reduces resource consumption, message loss rate, and message delay compared to Epidemic and PROPHET Routing.

by limiting the maximum number of intermediate nodes a packet can travel, or by setting a time to live (TTL) for each packet. The main routing protocols of this class are as follows.

5.3.1 Direct Transmission Routing

This is a single-copy-based routing protocol where a single copy of a message is maintained in the entire network. This is the simplest forwarding approach: a source forwards a message when it directly encounters the destination. Thus, it performs a single transmission per message with unlimited transmission delay (Spyropoulos et al. 2004).

5.3.2 Epidemic Routing Protocol

Epidemic Routing (Vahdat and Becker 2000) transmits a message to an arbitrary destination by exploiting periodic pairwise connectivity between nodes. For eventual transmission of messages, each node contains a buffer in which to store messages, including messages for which it is serving as an intermediate node.

Each node maintains a hash table to index the list of messages on the basis of the unique identifier associated with each message. In addition, each node also has a summary vector to indicate which entries are set in their local hash tables. When two nodes come within communication range of one other, they exchange summary vectors to identify missing messages on each node. A summary vector (SV) is a compact representation of all the messages stored on a node's buffer. This process is called an *anti-entropy session*, and it is always initiated by the node with a smaller identifier. Each node then requests a copy of the missing messages from the other node.

Figure 5.2 describes the process of the Epidemic Routing protocol. Nodes A and B are within communication range of each other and start an anti-entropy session. First, Node A transmits its summary vector (SV_A) to Node B. Node B uses the negation of its summary vector ($\overline{SV_B}$) to represent the missing messages and performs a logical AND operation between the negation of its summary vector, $\overline{SV_B}$ and SV_A. Node B sends the results of the logical AND to Node A, representing the messages missing from Node B and buffered at Node A. Finally, Node A transmits the requested messages to Node B. This process is performed each time two nodes come within communication range of each other. To avoid redundant connections, each node stores a list of nodes that it has encountered within a predefined time period. The anti-entropy process is not re-initiated between these nodes. In this way, given sufficient buffer space and time, Epidemic Routing guarantees eventual message delivery.

For eventual message delivery, Epidemic Routing associates a unique message identifier, a hop count, and an optional acknowledgment request with each message. The message identifier is a unique 32-bit number that consists of the node's ID and the message ID (16 bits each). The hop count represents the maximum number of intermediate nodes that a message can travel. It is similar to the TTL field in IP packets. The optional acknowledgment request field indicates the destination of a message to provide an acknowledgment of message delivery to its sender (this facility is required by certain applications).

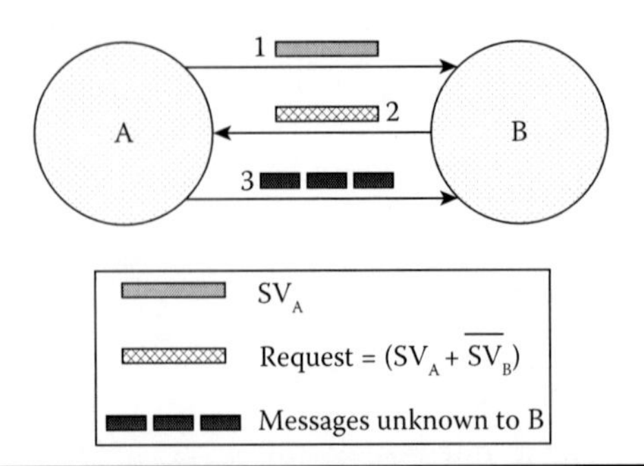

Figure 5.2 Message transmission between nodes in Epidemic Routing. (From Vahdat, A. and Becker, D., *Epidemic routing for partially connected ad hoc networks*, Computer Science Department, Duke University, 2000.)

5.3.3 Spray and Wait Routing

Spray and Wait Routing (Spyropoulos et al. 2005) operates in two phases: spray and wait. In the spray phase, the source transmits *N* copies of a message to *N* relays. Further, in the wait phase, if the destination is not found within *N* relays during the spray phase, each of the *N* relays directly transmits a copy of the message to its destination. If the *N* relays do not have contact with the destination, they wait for contact in the future.

Spray and Wait Routing combines the features of Epidemic and Direct Transmission Routing. Initially, it starts flooding of messages similar to Epidemic Routing, and when sufficient copies of the message have been flooded in the network to guarantee eventual transmission, it stops flooding and each relay carrying a copy of the message directly transmits it to the destination, similar to Direct Transmission Routing. In other words, Spray and Wait is a combination of single- and multicopy-based routing protocols. Results show that it performs better than other single- and multicopy-based routing protocols in terms of number of transmissions and delay.

An extension of this routing protocol is the Binary Spray and Wait Routing. In this protocol, the source initially transmits *N* copies of the message. Further, any node (source or relay) having more than one copy hands over half the number of copies to an encountered node, provided it does not have any copies of this message, and keeps the rest for itself. When a node is left with only one copy, it switches to direct transmission. Binary Spray and Wait Routing gives more optimal results in case of the independently and identically distributed mobility model.

5.3.4 Network Coding Protocol

The Network Coding Routing protocol (Widmer and Boudec 2005) is similar to flooding-based routing or probabilistic routing, except it uses network coding to limit message flooding. In this routing protocol, a message is converted into another format prior to transmission. The routing protocol inserts additional information into the coded blocks such that the original message can be successfully reconstructed with only a few coded blocks. In replication-based routing, the transmission of each individual data block is necessary to eventual delivery of a message. In contrast, in Network Coding Routing a message is eventually delivered when the necessary number of data blocks are transmitted and are sufficient to reconstruct the original message, which can be a small subset of the total data blocks transmitted. Therefore, Network Coding Routing performs better than replication-based routing when network connectivity is extremely poor. However, due to additional information inserted in the code blocks Network Coding Routing is less efficient than replication-based routing when network connectivity is stable.

Figure 5.3 describes an example in which A, I, and B are the three nodes of a network, where message transmission between Nodes A and B has to relayed by Node I. Node A generates a message x addressed to Node B, and Node B

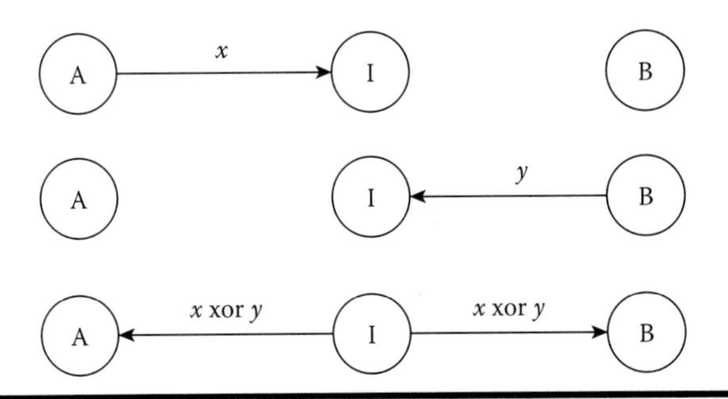

Figure 5.3 Message transmission between nodes in Network Coding Routing. (From Widmer, J. and Boudec, J. L., *ACM SIGCOMM 2005 Workshop on Delay Tolerant Networking*, 284–291, 2005.)

generates a message y addressed to Node A. In the traditional forwarding approach, Node I relays message x to B and y to A. In contrast, in Network Coding Routing, Node I broadcasts a single packet consisting of x xor y. Once having received x xor y, both Nodes A and B can decode and extract the message intended for them. A step-by-step description of Network Coding Routing is given by Widmer and Boudec (2005).

5.3.5 Erasure Coding-Based Data Forwarding

Erasure Coding (EC)-based approaches (Wang et al. 2005) provide better fault tolerance in comparison to replication-based approaches, by replicating the code blocks of the original message without the overhead of strict replication. They convert a message into a number of code blocks such that only a few relevant code blocks are required to reconstruct the original message. The main examples of this approach are Reed–Solomon coding and low-density parity-check-based coding. These algorithms have different coding/decoding efficiency, replication factors, and minimum numbers of code blocks needed to reconstruct a message. The selection of algorithms depends on the tradeoffs between these parameters.

EC takes an input message of size M and a replication factor r, and the message is divided into several code blocks of equal size b. The total number of coded blocks are represented as $\frac{M \times r}{b}$. Moreover, the original message is reconstructed by a minimum of $\frac{N}{r}$ coded blocks, which is equal to $\frac{M}{b}$. Figure 5.4 shows an example of EC-based data forwarding where erasure coded blocks are equally split among $n = 4$ relays. These relays transmit coded blocks directly to the destination. An equal number of code blocks are forwarded by each relay obtained by $\frac{N}{n} = \frac{Mr}{bn}$.

EC provides the best worst-case delay performance with a fixed amount of overhead. However, it does not provide good very small delay performance in comparison with other replication-based approaches.

5.3.6 Aggressive Erasure Coding-Based Data Forwarding

As discussed earlier, the EC approach provides better worst-case delay performance. However, it results in prolonged overall delivery latency in most non-worst cases. This main drawback of EC is due to its block allocation method. EC transmits a fixed number of blocks $\frac{Mr}{bn}$ during each contact without considering the length of each contact duration. Therefore, it gives better results when the contact duration is not much longer than the required time for sending the relayed data. In case of good network connectivity, this approach wastes the rest of the contact duration and results in ineffectiveness (see Figure 5.4).

The Aggressive Erasure Coding (A-EC) (Chen et al. 2006) approach overcomes this problem by completely utilizing each contact duration (see Figure 5.5). The source transmits as many coded blocks as possible during each contact (total of $\frac{Mr}{bn}$ blocks). Figure 5.5 shows that during Contacts 1, 2, 3, and 4 the source transmits six, one, three, and five coded blocks, respectively. In this way, the A-EC approach fully utilizes the network contact duration for very small delay performance cases, and as a result it performs better than the EC scheme.

However, A-EC gives a poor delivery ratio and/or a very large delivery delay for worst-delay performance cases when most employed relays are either unreliable or

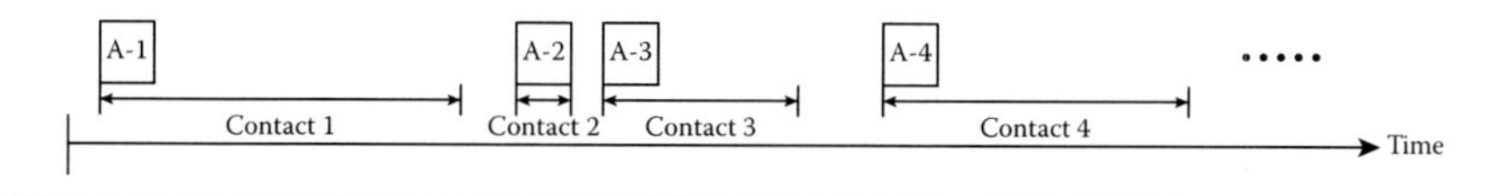

Figure 5.4 One erasure code block transmitted through relays (n = 4) in Erasure Coding routing. (From Chen, L., et al., *SIGCOMM'06 Workshops*, Pisa, Italy, 213–220, 2006.)

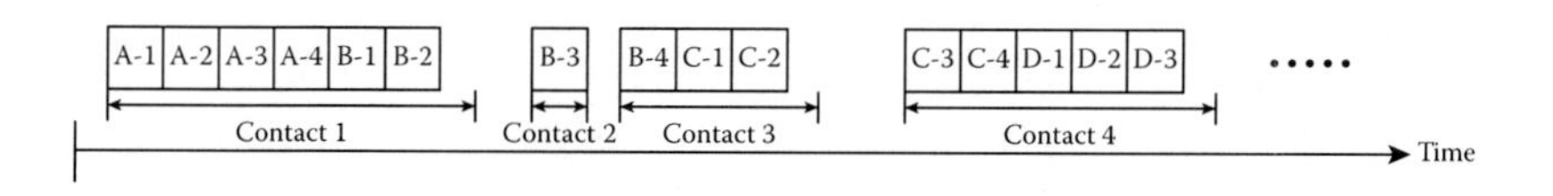

Figure 5.5 Four erasure coded blocks transmitted among four relays in A-EC routing. (From Chen, L., et al., *SIGCOMM'06 Workshops*, Pisa, Italy, 213–220, 2006.)

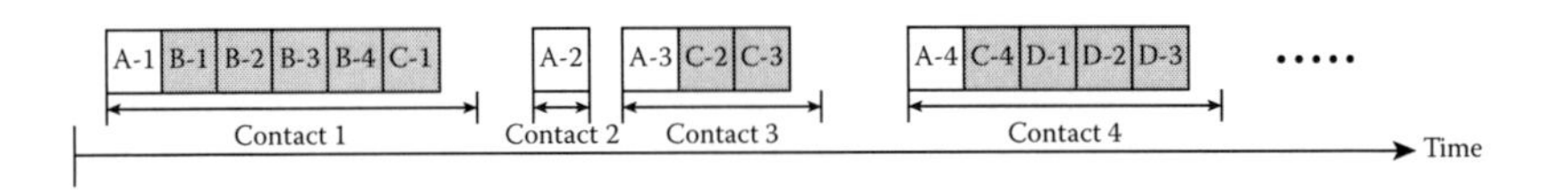

Figure 5.6 Two copies of four erasure coded blocks transmitted among four relays in H-EC. (From Chen, L., et al., *SIGCOMM'06 Workshops*, Pisa, Italy, 213–220, 2006.)

hardly moving closer towards the destination. All these relays are called *black-hole nodes*. In case of black-hole nodes, A-EC degrades the overall performance of the message delivery.

5.3.7 Hybrid Erasure Coding-Based Data Forwarding

Hybrid Erasure Coding (H-EC) (Chen et al. 2006) based data forwarding incorporates advantages of both the EC and A-EC approaches to provide better message delivery performance in both the worst-delay performance and very small delay performance cases. In this approach, two copies of erasure-coded blocks are sent to the relays. The first copy of erasure-coded blocks is transmitted according to the EC approach to provide each coded block to a single relay, and the second copy of erasure-coded blocks is transmitted according to the A-EC approach to utilize the rest contact duration. In Figure 5.6, the white blocks show transmission of the first copy through the EC approach, and the gray blocks show transmission of the second copy through A-EC. In the normal scenario, the contact duration is better utilized in H-EC when the network does not have black-hole nodes. However, when the network contains black-hole nodes, the H-EC approach delivers results similar to EC. This way, H-EC gives better results in the worst-delay performance case and in the very small delay performance case.

The advantage of context-oblivious-based routing protocols is due to their simplicity in finding a route—in particular, a minimum delay ratio for a connection request, as it does not require any global information about network topology or any context information. However, flooding causes a huge number of control packets in control channels, which can result in network congestion. Moreover, such techniques are very energy- and memory-intensive. Network performance is particularly important in opportunistic networks because of device constraints. One way to address this problem is to use users' context information to estimate the best forwarders and allow only these forwarders to forward the message toward the destination. Routing protocols based on this phenomena are discussed in the next section.

5.4 Partially Context-Aware Routing

Partially context-aware routing protocols use a specific type of context information, on the basis of which they make forwarding decisions. The main difference between partially and fully context-aware routing protocols is that the former use a particular type of context information, whereas the latter exploit and manage any type of context information (Verma et al. 2013). A description of main routing protocols of this class follows.

5.4.1 Randomized Routing

Randomized Routing (Spyropoulos et al. 2004) is a single-copy-based routing protocol that uses relays to transmit a message from source to destination. It is an extension of Direct Transmission Routing. In this routing protocol, Node A forwards a message when it encounters Node B only if B has a higher probability of delivering the message to its destination than A. The time of last encounter for other nodes is used as contact information to calculate the delivery probability. This routing protocol shows progress at each forwarding step. To calculate the probability of delivering a message to its destination, each node maintains timers containing the time since last encounter with every other node.

5.4.2 Utility-Based Routing

In Randomized Routing, each node maintains timers representing the last encountered time of every other node. This information indirectly represents the position information of the encountered nodes and is not absolute. A node with a low timer value for an encountered node indicates it is expected to be nearby. The current position of an encountered node depends not only on the meeting time but also on the speed and mobility pattern of that node. Utility-Based Routing (Spyropoulos et al. 2004) works similarly to Randomized Routing, except it also considers the speed and mobility pattern of the nodes with last-encountered time as context information to calculate the probability of message transmission to the destination. In this, each node uses a utility function $U_A(B)$ for every other node that indicates the probability of Node A to deliver a message to Node B. For example, Node X forwards a message to Node Y destined for Node Z, if $U_Y(Z) > U_X(Z)$.

5.4.3 Seek and Focus Routing

Utility-Based Routing suffers from the slow-start initial phase problem. Specifically, in a large network, where the distance between source and destination is large, intermediate nodes may not have any information regarding destination. Thus the source takes a long time at the beginning to find a node with a higher utility value. Seek and Focus routing (Spyropoulos et al. 2004) overcomes this problem. It combines the Randomized and Utility-Based Routing protocols. Seek and Focus routing operates in two phases: during the seek phase, if the utility around the source node regarding destination is low, the source node forwards the message through

Randomized Routing to quickly search for a better relay. Thereafter, during the focus phase, when a node finds a node with a utility value higher than a predefined value, it switches to Utility-Based Routing.

5.4.4 Prioritized Epidemic Routing

Epidemic Routing performs better in terms of its message delivery ratio when the load on resources is low or network connectivity is highly intermittent. In contrast, in a highly connected network where there is already a large number of messages prevailing in the network, protocols based on intelligent route discovery perform better than Epidemic Routing. An important drawback of Epidemic Routing is that it blindly disseminates copies of messages, causing an increased burden on resources. Redundant messages can be dropped to lessen the burden on resources without overly affecting the delivery ratio. Prioritized Epidemic (PREP) (Ramanathan et al. 2007) follows a similar approach through assigning a relative priority to packets for dropping and transmitting when required. PREP always maximally utilizes the full capacity of resources and drops packets only when necessary. The working of PREP is divided into a topology awareness scheme for estimating the routing cost from a node to the destination, and assigning deletion and transmission priority to each packet.

Topology awareness: Each node uses a neighbor discovery algorithm to discover and maintain bidirectional links with its current neighbors. Each link is assigned an average availability (AA) that describes the average time a link will be available for use in the near future. When a node observes a change in value of AA for one of its links, it generates a link state advertisement (LSA) message, spread in the network by the epidemic technique, containing a list of all current links with their AA value, and assigns it an incremented version number. This way, each node possesses knowledge of the other nodes from which it has received an LSA within the recent time period. This topology awareness is used to compute and assign a routing cost to each link as a function of the average availability: (1–AA) + 0.001. The least-cost route is estimated using Dijkstra's technique.

Packet drop and transmit priority: Each packet is assigned a drop priority p_d and a transmit priority p_t, which can be any real number. A lower value indicates higher priority. For example, if a contact comes up, packets with a lower value of p_t are transmitted before packets with higher p_t. On buffer fill, packets with higher p_d are dropped before packets with lower p_d. Packets have a hop count field that decrements on each successful hop toward the destination. Each node maintains a low watermark and a high watermark to control buffer occupancy. When buffer occupancy exceeds the higher watermark, the packet drop procedure is initiated; it is stopped when buffer occupancy falls below the low watermark. Packets are deleted according to their drop priority. A transmission procedure is initiated for an encountered neighbor only when no communication has happened in the past within a predefined time. Packets with higher transmit priority are sent first according to transmission procedure (Ramanathan et al. 2007).

5.4.5 Probabilistic Routing Protocol Using History of Encounters and Transitivity

One of the most admired examples of this class is the Probabilistic Routing Protocol using History of Encounters and Transitivity (PROPHET) (Lindgren et al. 2003). It is an extension of the Epidemic Routing protocol (Spyropoulos et al. 2004) with the concept of delivery probability. The delivery probability is the predictability for a node to achieve successful message delivery to a specific destination. In real life, users mostly move in a predictable manner according to their repeating behavioral patterns—for example, if a node has visited a place many times before, there is a higher probability that it will visit that place again. On the basis of these observations, the protocol uses the frequency of meetings between nodes as context information to improve the routing performance. According to the protocol, each node maintains a probabilistic metric called *delivery predictability* $P_{(a,b)} \in [0,1]$, for each known destination. Like Epidemic Routing, when two nodes come within communication range of each other, they exchange summary vectors along with their delivery predictability information. The protocol uses this information to update the internal delivery predictability. Moreover, the summary vector information is used to request the desired messages from other nodes according to the forwarding strategy.

5.4.5.1 Delivery Predictability Calculation

The delivery predictability calculation (Equation 5.1) follows a three-step process. First, when nodes meet they update their probabilistic metrics, so the nodes that often meet have high delivery probabilities.

$$P_{(a,b)} = P_{(a,b)old} + (1 - P_{(a,b)old}) \times P_{init} \tag{5.1}$$

where $P_{init} \in [0,1]$ represents the initialization constant. If two nodes don't encounter each other within a predefined time, they are not good forwarders of messages to each other, so the delivery probability values are reduced. Equation 5.2 represents aging, where $\gamma \in [0,1]$ represents the aging constant, and k represents the number of time units lapsed since the last time the nodes met. The time units may differ according to application and network requirements.

$$P_{(a,b)} = P_{(a,b)old} \times \gamma^{k} \tag{5.2}$$

The delivery predictability follows the transitive property as well. According to this, if Node A frequently meets Node B, and Node B frequently meets Node C, then Nodes A and C also have a good message delivery probability with respect to each other. The relation between transitivity and delivery predictability is expressed in Equation 5.3, where $\beta \in [0,1]$ represents the scaling constant, which decides the impact of transitivity on delivery predictability.

$$P_{(a,c)} = P_{(a,c)old} + (1 - P_{(a,c)old}) \times P_{(a,b)} \times P_{(b,c)} \times \beta \quad (5.3)$$

5.4.5.2 Forwarding Strategies

In traditional routing protocols, selection of a forwarder is relatively simple: the neighboring node having the shortest path or lowest cost to the destination is selected. However, the scenario is different in opportunistic networks. When a node receives a message, it may store the message in the buffer due to non-availability of a forwarder. Then upon each encounter with other nodes, the decision whether to forward the message depends on the estimated delivery probabilities above a predefined threshold.

5.4.6 Meetings and Visits (MV) Forwarding Algorithm

The Meetings and Visits (MV) routing protocol (Burns et al. 2005) was designed for efficient message delivery in opportunistic networks. It exploits the frequency of meetings between nodes and also exploits information about the frequency of visits to particular physical locations. Historical data are used to rank each message in a node's buffer according to the delivery probability. MV recognizes nodes' mobility patterns and exploits them to improve the routing performance.

5.4.6.1 Assumption

MV routing only supports networks that follow three assumptions.

- Nodes have an unlimited size buffer for their own messages and only a limited size buffer for the messages they receive from others.
- When nodes have a chance to transfer, they transmit with a fully reliable and unlimited bandwidth link layer. MV breaks this limitation and isolates routing protocols independent of the limits of the data link layer.
- Messages are eventually transmitted to stable destinations.

5.4.6.2 The MV Algorithm

In the MV routing algorithm, the encountered nodes exchange messages in a number of steps. Initially, Node A provides a list of the messages along with their destinations to Node B. Each message explained by Node A is associated with a delivery probability. Similarly, Node B supplies the same list to A and A estimates the delivery probability for the messages of B. The merged list is sorted according to the delivery probabilities in decreasing order and messages with relatively lower delivery probabilities are deleted. Node A then chooses the remaining top n messages and requests from B all the messages that have not already been received. This is an enhancement to the work of Davis et al. (2001), in a way where the technique follows a probabilistic approach to obtain the delivery probabilities $P_n^k(i)$, where k is the current node that can eventually deliver a message to destination i in n attempts.

5.4.6.3 Probability of Delivery

The expression $P_0^k(i)$ describes the probability of transmitting a message to some node k in a single attempt. In this situation, the message delivery probability will be equal to the node's probability of visiting the destination area. MV assumes the probability of visiting an area in the future is highly correlated with the peer's history of visiting a region. Accordingly, for each node k, MV has a vector P_0^k with one entry for each area. Each entry i of $P_0^k(i)$ depends on the recorded movement of the node during the last t round, where a round is defined as a fixed length of time (e.g., 1 day or 1 hour, depending on the velocity of the node): $P_0^k(i) = t_i^k / t$, where t_i^k represents the number of times node k visited cell i during the previous t visits. Second, MV assumes messages can be transmitted to maximum one node before finally reaching their destination. Both the current node k and the intermediate node j have a copy of the message, and any or both are capable of delivering it. Let $P_1^k(i)$ be the probability of eventually delivering a message to area i starting with node k and by using a maximum of one intermediate node. This is described by the following:

$$P_1^k(i) = 1 - \prod_{j=1}^{N} \left(1 - m_{jk} P_0^j(i)\right) \tag{5.4}$$

where N represents the number of nodes in the system, and m_{jk} is the probability of nodes k and j visiting the same area together. Like movement probability, MV has meeting probability on the basis of meetings during the last t visits: $m_{jk} = t_{j,k} / t$, where $t_{j,k}$ represents the number of times nodes j and k have been in the same area. Equation 5.4 describes the probability that neither node k nor any other node k will visit the destination directly. Finally, MV assumes that messages can be forwarded to no more than n other nodes:

$$P_n^k(i) = 1 - \prod_{j=1}^{N} \left(1 - m_{jk} P_{n-1}^j(i)\right) \tag{5.5}$$

Equation 5.5 is unable to scale with the number of intermediate hops or nodes in the system. To compute the probability, the meeting patterns of all other nodes must be known. Through evaluations, it is found that $P_1^k(i)$ can be approximated to $P_n^k(i)$.

5.4.7 The MaxProp Protocol

The MaxProp protocol (Burgess et al. 2006) is based on prioritizing both the schedule of packets transmitted to other peers and the schedule of packets to be dropped. These priorities are based on the likelihood of paths to peers, based on historical data and also on several complementary mechanisms, including acknowledgments, a head start for new packets, and lists of previous intermediaries. The working of the MaxProp protocol is described in the following.

5.4.7.1 Model

The protocol assumes each node has an unlimited size buffer to store its own messages but a fixed size buffer to store messages generated by others. Duration and bandwidth are the constraints for the transfer opportunity. The protocol assumes nodes don't have any prior knowledge about their environment—network connectivity, node movement patterns, nodes' geographic location, and so on. Opportunistic network routing performs three major operations.

- *Neighbor discovery*: Nodes must discover their neighbor nodes before a transmission starts.
- *Data transfer*: When two nodes come within communication range of each other, they can exchange data. Nodes do not know the duration of meeting.
- *Storage management*: Each node must manage its fixed size buffer by deleting stored packets using any buffer management technique.

Each node carries stored messages until it meets other nodes. A node will continuously forward a message to any number of encountered nodes until the message times out, delivery of the message is notified by an acknowledgment, or the message is dropped due to buffer overflow.

5.4.7.2 Protocol Definition

The MaxProp protocol uses various approaches to increase the delivery rate and reduce the latency of delivered packets. MaxProp exploits several approaches to describe the order of packet transmission and deletion. The heart of the protocol is a ranked list of packets, stored in a node, according to the cost assigned to each destination. The cost is an approximation of the delivery possibilities. Moreover, MaxProb uses acknowledgments to confirm packet delivery. MaxProb assigns a higher priority to new packets and also prevents reception of duplicate packets. In the following, the mechanisms of destination cost estimation and buffer management are described.

Estimating delivery likelihood: In the literature, it has been demonstrated that the optimal path for delivery in opportunistic networks can be generated by creating a directed graph of nodes (Jain et al. 2004). It uses a variant of Dijkstra's algorithm to determine the optimum shortest path. MaxProp assigns link weights according to the following rules.

Let s be a set of nodes in the network. Each node i calculates and maintains a probability of meeting with node j, where $i \in s$ and $j \in s$. The probability f_j^i describes the possibility that the next encountered node will be j. Initially, f_j^i is set to $1/(|s|-1)$ for all nodes. The value of f_j^i is incremented by 1 on each meeting with node j, and all values of f are also recalculated. This method is called *incremental averaging*; the nodes that are encountered less frequently are assigned lower values over time. These values are exchanged each time two nodes meet. With the help of other nodes' values, the node computes cost, $c(i, i+1, \ldots, d)$, for each possible path to the destination d by using up to n intermediate hops. The total cost for a path is the sum of the probabilities, where the probability represents the likelihood of not having a connection. It is calculated as one minus the probability of link availability:

$$c\left(i, i+1, \ldots, d\right) = \sum_{x=1}^{d-1}\left[1-\left(f_{x+1}^{x}\right)\right] \tag{5.6}$$

The lowest cost path is selected as the cost for a destination. In MaxProp, the highest priority rank packets are transmitted first during a transfer opportunity. The lowest priority rank packets are deleted to make space for new incoming packets. In a situation when two packets have the same cost to their destinations, packets that have traveled via fewer intermediate nodes is given priority.

Buffer management: There are three mutually exclusive situations when node n can drop a packet p, without affecting the overall delivery rate of the network.

- A copy of packet p has already been delivered to its destination.
- Non-existence of a route with adequate bandwidth during the lifetime of packet p.
- No copy of packet p has been delivered, but at least one copy will be delivered in the future even if node n deletes its copy.

MaxProp follows packet deletion priority by removing acknowledged packets instantly, followed by packets that have crossed the predefined threshold of intermediate hops with low scores, followed by packets with maximum hops below the same threshold.

5.4.8 MobySpace: A Mobility Pattern Space

In this routing protocol (Leguay and Friedman 2006), the mobility pattern of nodes is used as context information to make routing decisions. The protocol uses high-dimensional Euclidean space, called *MobySpace*. Each axis of MobySpace represents the possibility of two nodes meeting and the gap between them represents the probability of that meeting occurring. Two nodes that contact a similar set of nodes with similar frequencies are very close in MobySpace. The node closest to the destination is treated as the best forwarder. Packets are forwarded toward nodes that have similar mobility patterns to the destination node. In MobySpace, the mobility pattern of a node indicates its coordinates, named MobyPoint; eventually delivery is performed by giving packets to the nodes that have MobyPoints closest to the MobyPoint of the destination. Mobility patterns and ways of managing these patterns are explained in the following.

5.4.8.1 Mobility Pattern Characterization

Mobility patterns are characterized on the basis of the number and dimension of MobySpace. MobySpace doesn't represent the geographical location of nodes, but it does describe some features of the node's mobility pattern. Mobility patterns should be very simple to measure in order to reduce the computational cost and to reduce the overhead associated with spreading them within the network. A recent study of mobility patterns was conducted by Hui et al. (2005). To transmit a packet from one node to another, a relay that frequently meets the destination is selected. Each permissible contact is indicated by an axis, and the distance along that axis represents the probability of meeting. Two nodes that meet a similar set of nodes with similar frequency are very close in MobySpace. In contrast, nodes that meet with different sets of nodes or meet with the same set of nodes but with different frequencies, have a larger distance in Euclidean space.

5.4.8.2 Mobility Pattern Acquisition

A node can recognize its own mobility pattern by several methods. A node can observe its mobility pattern by learning its contacts or its frequency of visits to various places. The tags are fixed to each location, so that nodes can identify their current positions. Alternatively, nodes may be able to investigate an existing infrastructure to construct these patterns. In the same manner, a node can obtain the mobility patterns of others. First, mobility patterns can be spread through a dissemination-based approach. Second, nodes can spread only efficient information of mobility patterns to minimize the buffer and network resource consumption. Finally, nodes can store their mobility patterns at predefined storage locations, and they update their information with the data stored at these storage locations.

5.4.8.3 Mobility Pattern Usage

Packets are given to nodes having similar mobility patterns as the destination. Let U and L represent the set of all nodes and their locations, respectively. The MobyPoint for a node $k \in U$ indicates a point in an n-dimensional space, where $n = |L|$. $mk = (c1k, \ldots, cnk)$ represents the MobyPoint of node k. Moreover, $d(mi,mj)$ represents the distance between two MobyPoints i and j. The directly connected neighbors of node k at any time t are represented by $W_k(t) \subseteq U$. The group of neighbors with k as one of the nodes is represented by $W_k^+(t) = W_k(t) \cup \{k\}$. MobySpace routing chooses only one of these neighbors for receiving or keeping packets. The routing function f exploits the neighbor that is nearest to the destination b. The decision to transmit packets from k to b is made as follows:

$$f\left(W_k^+(t), b\right) = \begin{cases} b \text{ } if \text{ } b \subset W_k(t), \text{ else} \\ i \in W_k^+(t): d(mi, mb) = min_{j \in W_k^+(t)} d(mj, mb) \end{cases} \tag{5.7}$$

The distance function d is very important for the routing process and some other distance functions can be found in the study by Leguay et al. (2005).

5.4.9 Spray and Focus

The Spray and Focus routing protocol (Spyropoulos et al. 2007b) uses the time since two nodes last met as context information to calculate the delivery probability of messages. The spraying approaches available in the literature (Spyropoulos et al. 2005, 2007) create and spread (spray) copies of packets to nodes within the communication range. Each intermediate node then stores and carries the packet until it meets the destination or the TTL for the packet expires. These schemes exploit multiple hops simultaneously and independently for seeking the destination; therefore, they are more capable of exploring the maximum possible routes to destination and can find the shortest path.

To gain high performance from such schemes, the network nodes should be highly mobile. However, in many real-life scenarios, most of the time node mobility is restricted to a small area, for example, node movement on a university campus. Operation of this scheme is divided into two phases: in the first phase it transmits some copies of a packet to all nodes within communication range, and in the second phase these neighbor nodes directly transmit copies of the packet to the destination when they encounter it. This way an intermediate node waits for transmission until it comes within communication range of the destination. The waiting problem is solved in Spray and Focus. In the second phase (the focus phase) the node doesn't wait for direct communication with the destination; it forwards the packet to a more relevant forwarder with a better probability of transmission. The protocol is described in the following.

5.4.9.1 Spray Phase

When the source generates a new message for transmission, it also creates some copies of the message. These copies of the message are called *forwarding tokens.* Only the owner of the message can generate and forward forwarding tokens of the message, according to the following rules:

- Each node has a "summary vector," which contains a compact representation of the messages currently stored in the node's buffer. When two nodes meet, they exchange summary vectors.
- Upon exchanging summary vectors, nodes exchange a copy of messages that the encountered nodes do not possess and also hand over $n/2$ forwarding tokens.
- When a node has a single copy of a message with only one forwarding token for this message, it proceeds with the focus phase.

5.4.9.2 Focus Phase

When an intermediate node has a message and a single corresponding forwarding token, it transmits the message using the focus phase. In this phase, the message is sent to different intermediate nodes according to the forwarding policies. These forwarding policies are constructed based on last encounter timers that record the time lag (or age) from the last meeting of two nodes.

Age of Last Encounter Timers with Transitivity: In this protocol, each node i has a timer $T_i(j)$ to record the time elapsed after the last meeting with every other node j in the network. The protocol initially sets $T_i(j) = 0$ and $T_i(j) = \infty$. Whenever i and j meet, both timers are initialized to zero, that is, set $T_i(j) = T_i(j)=0$, and increment their timers by one at every clock tick. On the basis of these timers, a utility function is constructed to identify the usefulness of a node to transmit a message to another. Intuitively, messages are transmitted to the nodes that have smaller timer values for the destination.

Single-Copy Utility-Based Routing: Let each node i have a utility value $U_i(j)$ for every other node j in the network to make the routing more efficient. If Node A wants to transmit a message to Node D via Node B, then Node A forwards the message to Node B only if $U_B(D) > U_A(D) + U_{th}$, where U_{th} is a utility threshold parameter in the algorithm. The timers become a poor identifier as their values increase; therefore to improve the efficiency of routing it is essential to minimize the uncertainty for greater timer values. The protocol removes this problem by introducing "transitivity" to update the utility function. When Node A meets Node B frequently and Node B meets Node C frequently, due to the transitivity rule A can transmit a message to C through B with a high probability, even if A rarely meets C. Therefore, when Node A meets with Node B, it also updates its utility values for all nodes for which B has greater utility values. The protocol uses the following transitivity function.

Timer transitivity: Let Node A meets Node B at distance d_{AB} and let $t_m(d)$ represent the time that a node takes to travel a distance d while following a given mobility model. Then $\forall j \neq B$: $T_B(j) < T_A(j) - t_m(d_{AB})$, set $T_A(j) = T_B(j) + t_m(d_{AB})$.

For example, the transitivity functions for the random waypoint and random walk mobility models are as follows:

$$T_A(D) < T_B(D) - d_{AB}, \text{ set } T_B(D) = T_A(D) + d_{AB}, \text{ (waypoint)} \tag{5.8}$$

$$T_A(D) < T_B(D) - d_{AB}^2, \text{ set } T_B(D) = T_A(D) + d_{AB}^2 \text{ (walk)} \tag{5.9}$$

These transitivity functions are capable of fast dissemination of utility information in the network and minimizing uncertainty related to the location of a particular node.

5.4.10 BUBBLE Rap Routing

The BUBBLE Rap routing protocol (Hui and Crowcroft 2007) introduces the concept of community in opportunistic networking to improve the efficiency of the forwarding task. The protocol exploits the community information of nodes to make forwarding decisions. The protocol follows two major perceptions. First, nodes have varying popularities in a community, so the routing protocol prefers forwarding messages through nodes that are more popular than the current node. Second, there are more chances of interaction between nodes of the same community, so the routing protocol always appreciates the selection of a node as forwarder that belongs to the destination community. The following are the two main assumptions of the routing protocol.

- Each node must be related to at least one community. A community can also have only a single node. A node can also belong to several distinct communities.
- Each node has two types of rankings: global ranking and local ranking. Global ranking represents the popularity of the node across the network, whereas local ranking indicates only the popularity of the node within the community to which it belongs. If a node belongs to several communities, it has several local rankings.

When a source wants to transmit a message to the destination, to forward the message, it constructs a hierarchical ranking tree on the basis of global ranking, until it reaches a node of the destination community. Thereafter, the message is forwarded through a local ranking tree, which is constructed by local ranking rather than global ranking, until it reaches the destination or the expiry of TTL. In this method, it is not compulsory that every node know the ranking of all other nodes in the network, but they must be capable of comparing rankings with neighbor nodes and forwarding the

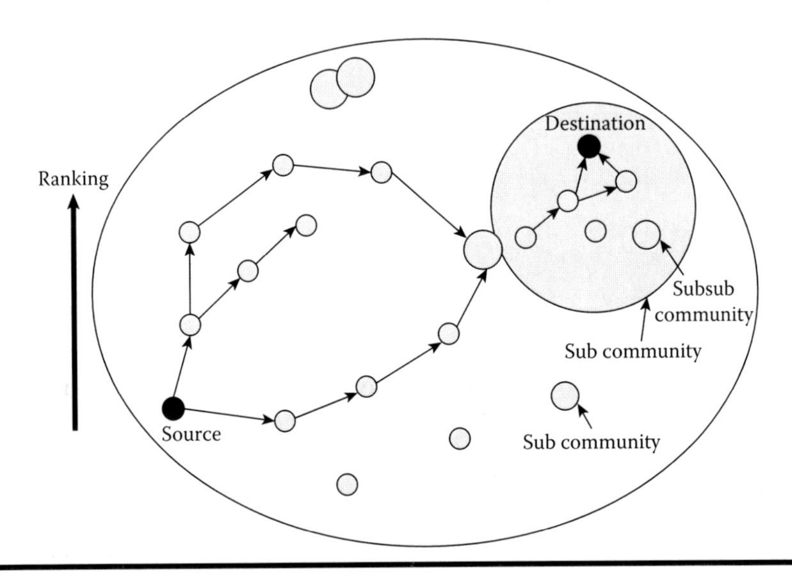

Figure 5.7 Illustration of the BUBBLE routing protocol.

message to the more popular node. In order to minimize the cost, when a message is transmitted to a node that belongs to the destination community, the source node of the message can delete this message from its buffer to make space and to prevent further propagation. The forwarding procedure of this routing protocol is illustrated with a real-life example. For example, if you want to transmit a message to another person, first of all you try to forward the message through a neighboring person more popular than you, and then the neighbor forwards it to a more popular person belonging to the wider community, for example, a postal worker. When the postal worker encounters a person of the destination community, the message is forwarded to that person. The person who receives the message tries to locate more popular persons within the community. In this way the message is forwarded within the destination community until it reaches the destination or the TTL expires. Figure 5.7 illustrates the BUBBLE Rap routing protocol.

5.4.11 Integrated Routing Protocol

This routing protocol (Verma and Srivastava 2011) is an extension of PROPHET routing (Lindgren et al. 2003) and uses the frequency of meeting between nodes as context information to make forwarding decisions. The integrated routing protocol exploits the mobility patterns of the user to bring the message closer to its destination. Real users move in either a predictable or random fashion, which means that a user may visit a place that was visited previously or may move to a new location that was never visited. Therefore, users' mobility behavior may be predictable or unpredictable. The integrated routing protocol exploits this observation to remove the limitations of PROPHET. It combines the features of flooding-based routing with probabilistic routing and creates a new integrated routing protocol that has the features of both. To achieve this task, each node maintains contact probabilities for all other nodes currently available in the network, in a probability matrix as described by Lindgren, Doria, and Schelen (2003), in which each cell represents the contact probability between two nodes. Each node also maintains a time attribute that represents the time the matrix was last updated. Whenever two nodes encounter each other, they update their contact probabilities and time attributes. Nodes exchange their contact probability matrices and compare them; the matrix is updated with another matrix that has a more recent time attribute. Therefore, after each interaction between two nodes, each node will have the same probability matrix.

5.4.11.1 Probability Calculation

A complete evaluation of delivery probabilities is given by Lindgren et al. (2003). The probability values are updated when two nodes meet, so that the nodes frequently encountered have high message delivery probabilities. When node x encounters node y, the delivery probability of node x for y is updated by Equation 5.10.

$$P'_{xy} = P_{xy}(1 - P_{xy})P_0 \quad (5.10)$$

where P_0 represents the initial probability. When two nodes don't meet for a long time, the delivery probability between them must be decreased over time. For example, if nodes x and y do not meet in a predefined time, the delivery probability is decreased by Equation 5.11.

$$P'_{xy} = \alpha^k P_{xy} \quad (5.11)$$

where α is the aging factor, and k represents the number of time units after the last update. The routing protocol also follows the transitivity rule to compute the delivery probability. According to this a node x can calculate the delivery probability of node z through node y.

$$P'_{xz} = P_{xz} + (1 - P_{xz})P_{xy}\,P_{yz}\,\beta \quad (5.12)$$

where β is a parameter representing the impact of transitivity.

5.4.11.2 Routing Strategies

When a node gets a message, it checks for availability of a path to the destination. If a path is not available, it stores the message in the buffer and upon each meeting with other nodes it decides whether to forward the message. Furthermore, a message should be forwarded to multiple nodes to increase the chances of eventual delivery to the destination. Whenever two nodes meet, they exchange probability matrices. If the receiver is the destination of the message, the message is eventually delivered. Otherwise, if the receiving node has a delivery probability greater than a predefined threshold, the message is transmitted to it; if not, it is discarded by the receiver. If none of the nodes within communication range of the sender have context information about the destination and the sender has waited for a predefined time, the sender disseminates the message to all its current neighbors, similar to the dissemination-based scheme. The same procedure is followed at each node until the message reaches its destination.

The partially context-aware routing protocols discussed above exploit some context information to make forwarding decisions, such as mobility information of nodes, and the frequency of visiting particular locations. This information impacts the effectiveness of routing in opportunistic networks and increases the performance of networks, especially in the routing of messages, when efficient routing techniques are deployed. The overhead carried by context-oblivious-based routing schemes can be further reduced by taking into account knowledge of a node's context information. The main difference between partially context-aware and fully context-aware protocols is that partially context-aware protocols exploit few types of context information such as the mobility patterns of the nodes, the information of the devices themselves, or the encountered history between nodes. Fully context-aware protocols not only exploit such context information but also take into account the social aspect of nodes as an important parameter for routing messages. Routing protocols that exploit several types of context information are described in the next section.

5.5 Fully Context-Aware Routing

Fully context-aware routing protocols provide both facilities while optimizing routing as well as techniques to manage and make efficient use of context information. Routing protocols in this category are more general than the protocols discussed in the section "Partially Context-Aware Routing." These protocols can work in every environment because they can use any set of context information to make routing decisions (Verma et al. 2013). The main routing protocols of this category are Context-Aware routing (Musolesi et al. 2005) and History-Based Opportunistic routing protocol (HiBOp) (Boldrini et al. 2007).

5.5.1 Context-Aware Routing

Communication between nodes in MANETs is only possible when nodes are simultaneously connected with the same network. In reality, this assumption is mostly not true, because MANETs consist of mobile devices that are highly mobile. When the network is highly disconnected, it creates a form of opportunistic network. Context-Aware routing (Musolesi et al. 2005) is a novel protocol to provide communication between nodes in such a partially connected MANET. The protocol provides communication between different disconnected MANET clouds and assumes an underlying existing MANET routing protocol to connect the nodes within the same cloud. To deliver a message outside the cloud, the sender selects a node from its current cloud that has highest probability of successfully delivering the message to its destination. This node buffers the message and waits to encounter either the destination or any node of the destination cloud that has a higher probability of encountering the destination. Therefore, nodes calculate the delivery probability proactively and spread them in their cloud. However, each node can calculate the delivery probability only for those destinations that it is aware of. The protocol is able to exploit different types of context information—residual battery life, the probability of nodes meeting, and the rate of connectivity change. Because of its proactive nature, each node periodically spreads the information about underlying synchronous routing and a list of delivery probabilities for other nodes. Each node updates its routing table when it receives this information.

5.5.1.1 Prediction and Evaluation of Context

The procedure for predicting and evaluating context information is outlined below.

- Each node computes its delivery probability to all other nodes of the cloud and periodically spreads in the network to update the routing information of other nodes.
- Each node maintains a routing table that contains information about the next hop and its delivery probability for all known destinations.
- Each node uses the local assumption of delivery probability during temporary disconnection and between information updates.
- If a node doesn't possess any information about the destination, it forwards the message to the node that has the highest mobility.
- If an intermediate node encounters a node with a higher delivery probability, it forwards the message to the higher probability node.

5.5.1.2 Local Evaluation of Context Information

There are various schemes for utilizing the multiple types of context information to identify the best forwarder for a particular message. The simplest method is assigning a static value to each type of context information on the basis of priority to create a static hierarchy. However, the assignment of static priorities is inefficient. To make the routing more realistic, the protocol simultaneously maximizes the utilization of several types of context attributes.

Significance-Based Evaluation of Context-Aware Information: The context information of a node can be represented by a set of attributes $(X_1, X_2, \ldots, X_n)$. The attributes in capital letters describe the set of all values an attribute can have, while small letter representation is related to a specific attribute value from this set. In the protocol, each node is capable of locally creating a utility function U(X_1, X_2, ..., X_n) representing the delivery probability for every other node. The goal of the protocol is to select the most suitable node for message delivery. The combined goal function is described as follows:

$$Maximize\left\{f\left(U\left(x_i\right)\right)=\sum_{i=1}^{n} w_i U_i\left(x_i\right)\right\} \tag{5.13}$$

where (W_1, W_2, ..., W_n) represents the significance weights that describe the importance of each goal.

Autonomic adaptation of the utility function: The protocol is able to accept the weights of each attribute dynamically as soon as the values of these attributes change. To do this, the protocol uses a runtime self-adaptation mechanism to adopt the weights being used for this evaluation process. The protocol modifies the previous formula by adding adaptive weights a_i, to make changes in utility function corresponding to the disparity of the context.

$$Maximize\left\{f\left(U\left(x_i\right)\right)=\sum_{i=1}^{n} a_i\left(x_i\right) w_i U_i\left(x_i\right)\right\} \tag{5.14}$$

where $a_i(x_i)$ may be a composite parameter. Following are the three important aspects that help to evaluate $a_i(x_i)$.

- Difficulty of specific ranges of values, $a_{range_i}\left(x_i\right)$
- Predictability of the context information, $a_{predictability_i}\left(x_i\right)$
- Availability of context information, $a_{availability_i}\left(x_i\right)$

The weights a_i can be calculated with the help of the above factors.

$$a_i\left(x_i\right)=a_{range_i}\left(x_i\right)\cdot a_{predictability_i}\left(x_i\right)\cdot a_{availability_i}\left(x_i\right) \tag{5.15}$$

The protocol models the adaptive weights corresponding to the possible ranges of values $a_{range_i}\left(x_i\right)$ as a function that can take 0 or 1. The protocol predicts the behavior of context information by exploiting several statistical attributes, for example, the autocorrelation function, to identify the degree of relation between values. Each context attribute has a different degree of availability, whose value may be predictable. Sometimes the values of some attributes are not predictable due to non-availability of sufficient information. The protocol solves this problem by assigning 0 to an adaptive weight a_i of missing context information.

$$a_{availability_i}\left(x_i\right)=\begin{cases}1, \text{ if the context information is currently available}\\0, \text{ if the context information is not currently available}\end{cases}$$

Automatic Adaptation of the Refresh Period of Routing Tables and Context Information: The protocol proposes the following function for the refresh time calculation based on the rate of context information dissemination by considering that such information is predictable and the prediction is mostly accurate.

$$t(x_1, x_2, \ldots, x_n) = c \sum_{i=1}^{n} | \rho_{k_i} | \tag{5.16}$$

where c is a constant of proportionality.

5.5.1.3 Prediction of the Information Attributes using Kalman Filters

Context-aware routing uses a Kalman filter (Kalman 1960) for two purposes, to predict the context information of a node more realistically and to optimize the use of bandwidth. Updating the routing tables that store delivery probabilities is a very expensive task, whereas such information can be easily predicted and used to update the routing tables even when fresh context information is unavailable. Such a prediction problem can be presented in the form of a state space model. A time series of observed values can be created from context information. It is helpful to create a prediction model with the help of inner state described by a set of vectors. An important feature of the Kalman filter is that it doesn't store the complete past history of the system, thus making it very suitable for mobile devices with limited storage.

5.5.2 History-Based Opportunistic Routing Protocol

5.5.2.1 Context Creation and Management

The context in which a user lives is made up of two components: the current context and the legacy of the evolution of the context over time. Information about the user itself is known as the *current context* (CC) of that user, which gets stored in an identity table (IT), as shown in Table 5.4. CC also includes information about the current neighbors of that node. Thus, CC is seen as a snapshot of the local environment of that user, and by using this information the effectiveness of the neighboring node is checked to decide whether it is a good forwarder. Making decisions based on instantaneous information alone does not seems to be a good decision, as context information does not describe users' past experiences and behavior. For example, a user can be called a good forwarder if he or she meets the destination at least once every day. To achieve this there is a need to maintain a history table (see Table 5.5). Other information in a history table helps HiBOp (Boldrini et al. 2007) calculate the probability of nodes meeting in the future. HiBOp stores the history of a greater number of encountered nodes compared to PROPHET. All attributes of encountered nodes provides some legacy in the HiBOp history. Information stored in the history table is refreshed periodically.

Table 5.4 Sample Identity Table

Personal Information	
Name	ABC
Surname	XYZ
E-Mail	abcxyz@iiitm.ac.in
Phone	+91-9826074XXX
NID	PXLBNGER06CC8Y
Residence	
Street	Morena Road
City	Gwalior
Work	
Street	Morena Road
City	Gwalior
Organization	ABV-IIITM
Work, Hobbies, and Fun	
Address	Sport Centre
City	Gwalior
Association	BH-2
System Information	
MAC-Bluetooth	01:23:45:67:89:AB
MAC-802.11	09:00:07:A9:B2:EB
IP address	192.168.1.103

Context Management Algorithms: IT (Table 5.4) consists of a set of personal information, behavioral information, and social information. The user decides about the information to be exposed in its IT, which should be unique in the entire network. Node identity (NID) is the hash of the IT, which is used to uniquely name the each node in the network. During the neighbor discovery phase, nodes exchange their ITs periodically and asynchronously to learn the environment around them. The time interval between two neighbor discovery phases is called the *signaling interval*. At the end of every signaling interval,

Table 5.5 History Table Structure

Aggregate	*Class*	P_c	*H*	*R*
Gwalior	City	The values of P_c, *H*, and *R* can be calculated by Equations 5.18, 5.19, and 5.20, respectively.		

Table 5.6 Repository Table Structure

Aggregate	*Class*	*Carriers*	*Cont Count*	*Het Count*	*Red Count*
Gwalior	City	A, B, C	2	1	2

each node sends its IT or NID to its neighbors. If it receives the IT or NID of any nodes that were previously present in its current context, then it marks their presence, and when a new node is introduced it broadcasts its IT. This way nodes exchange their ITs among the neighbors they encountered during the last signaling interval. During the last signaling interval, if a node is found to be absent, its entry is removed from the CC table. To tolerate transitory disconnections, the IT is removed from the CC table.

The second building block is the history table, which stores the values of nodes present in the ITs of neighbors that the node has met in the past. Three counters are tagged to each aggregate: continuity probability (P_c), redundancy (R), and heterogeneity (H). P_c is the probability of encountering the node, H is the average number of nodes that have been encountered, and H is the fault tolerance index. High H indicates several distinct chances of encountering aggregate on distinct nodes. R is the average number of occurrences of aggregates within the same IT. As an evolution to the CC, the history table is formed and dynamically updated using the contents of the repository table, which stores neighboring node information (see Table 5.6). After the signaling interval, the repository table is updated, whereas after flushing the interval history table is updated by merging the data of the repository table into the history table. After every signaling interval, HiBOp scans the CC and for attributes not having a row in the repository table, a new row is added. Such new attributes are initialized with a value of zero. As soon as these attributes are found in CC, their counter is incremented.

For each attribute, HiBOp follows the following steps:

- Increments the continuity counter (continuity counter shows how many times that attribute has been seen as a neighboring node during flushing interval).
- If the node whose IT stores the required attribute is not listed in the carrier list, the heterogeneity counter is incremented and NID is added to the carriers list. In addition, the redundancy counter is incremented by the number of times an attribute appears in the IT.

The flushing interval is an integer number of the signaling interval. For each attribute of the repository table, the continuity probability, heterogeneity, and redundancy are calculated as follows:

$$P_c^{(rep)} = \frac{ContCount}{M} \tag{5.17}$$

where M = the number of signaling intervals in a repository flushing interval. The continuity probability in the history table is updated as follows:

$$P_c \leftarrow \partial.P_C + (1-\partial)P_C^{(rep)} \tag{5.18}$$

where ∂ = classic smoothed average parameter ($0 < \partial < 1$). Similarly, heterogeneity and redundancy are calculated as follows:

$$H = \partial.H + (1-\partial)\,HetCount \tag{5.19}$$

$$R = \partial.R + (1-\partial)\frac{RedCount}{HetCount} \tag{5.20}$$

$\frac{RedCount}{HetCount}$ is a redundancy sample. Dividing RedCount by HetCount is computing the "average redundancy" of the attribute during the flushing interval—that is, the average number of times the attribute has been seen in a single IT during the flushing interval.

5.5.2.2 Forwarding Operations

Forwarding is done based on the probability of opportunity to reach destination. Messages are passed through nodes that have a higher probability of reading the destination. The improvement is in how the context can be used to compute delivery probabilities. The sender includes the destination IT, and this is matched with context information stored at each encountered node to calculate delivery probabilities. Nodes having a high probability indicate high similarity with destination; hence they are good forwarders. Replication of messages is done to maintain multiple copies in the network. HiBOp works in three phases (see Figure 5.8).

- *Emission*: Sender injects the message into the network and replicates it for reliability.
- *Forwarding*: Exploits nodes' mobility and contacts to forward messages towards destination.
- *Delivery*: When a node finds the destination, the process stops.

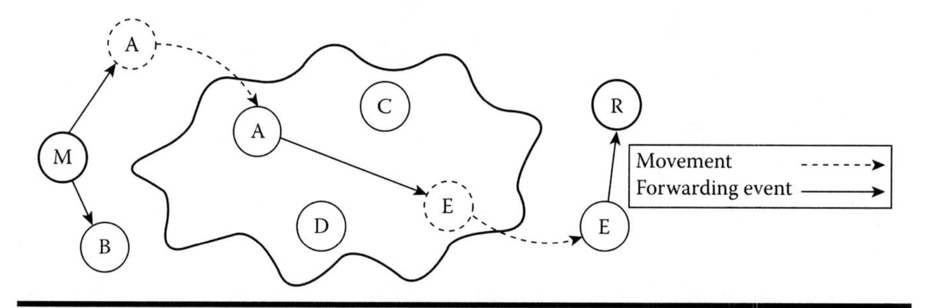

Figure 5.8 HiBOp forwarding process.

Emission phase: HiBOp maintains reliability using replication by replicating messages at the sender only. It assumes maximum tolerable message loss P_l^{max}. The sender node obtains the probabilities of successfully delivering the message $P_{succ}^{(i)}$ to its final destination from its neighbors, where *i* denotes the *i*th neighbor while the delivery probability of the sender is $P_{succ}^{(0)}$. Assuming these probabilities are independent, the number of neighbors to which the message is forwarded by the sender is as follows:

$$k = min\left\{ j \mid \prod_{i=0}^{j} \left(1 - P_{succ}^{i}\right) \leq P_l^{max} \right\} \tag{5.21}$$

To keep the joint loss probability below the maximum threshold specified by the application, the sender forwards the message to the minimum number of neighbors. In the event of non-availability of enough neighbors, the message is forwarded to available neighbors queued at the sender. As soon as new neighbors become available, they become ready for forwarding.

Forwarding phase: HiBOp calculates the delivery probabilities based on matches between the sender information in the message and the destination context information available on nodes (Jindal and Psounis 2007). Class weights represent the precision of that class in identifying the destination. Figure 5.9 represents the qualitative ranking of class precision (bigger circles represent lower precision).

The definition of the weights reflects this qualitative ranking. The weight should be monotonically increasing and the relative difference between classes should increase if the less significant one allows for a higher redundancy, because higher redundancy usually means lower significance. A node wishing to forward a message broadcasts context information about the destination by attaching its own delivery probability. Nodes that receive such a message evaluate their delivery probability and send it back to the questioning node if it is higher. At each node the delivery probability is calculated from the node's IT, CC, and history.

The advantage of fully context-aware routing protocols is that they are more general than partially context-based routing protocols. Indeed, these routing protocols can be used with any set of context information, and thus they can be easily

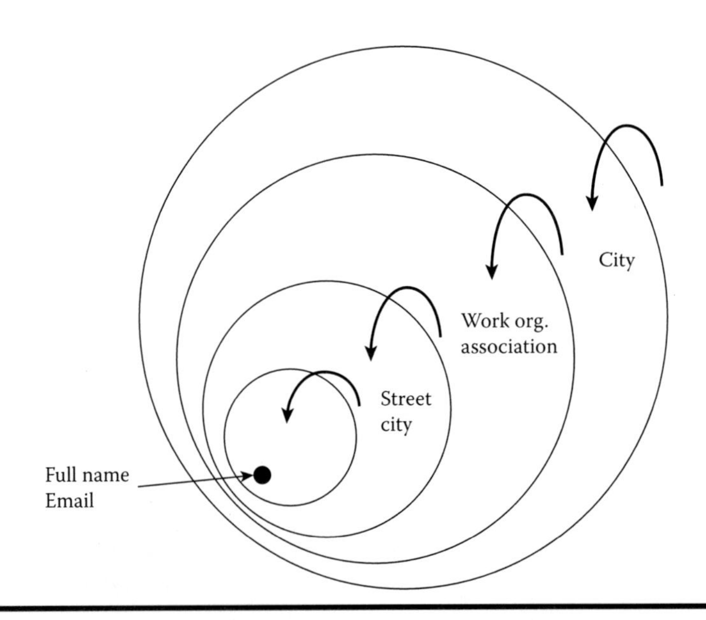

Figure 5.9 Precision of attribute classes.

customized to best suit an environment. In the fully context-aware routing protocols, as nodes are selected with social criteria, the delay ratio is higher than context-oblivious-based routing protocols and lower than partially context-aware protocols. In contrast, with the overhead parameter, fully context-aware routing protocols appear to be the best choice since only the nodes that match context information with a certain percentage are flooded. The next section describes some real-life case studies and applications of opportunistic networks. The above-described routing protocols are used in these applications according to the application's requirements.

5.6 Case Studies and Applications

Despite the fact that research on opportunistic networks dates back to just a few years ago, due to its simplicity and effectiveness concrete applications and real case studies are available. A few main applications of opportunistic networks are discussed in this section. The cases are structured around their objective(s), the approach/mechanism they followed, and their applications in different areas.

5.6.1 Haggle Project

The Haggle Project (Nordstrom et al. 2009) was funded by the European Commission in the framework of the Future and Emerging Technologies - Situated and Autonomic Communications (FET-SAC) initiative. It is a new autonomic networking architecture, designed to provide communication in an opportunistic network, especially where an increased probability of data transfer is required. It provides a simplified (no setup, no layers) means of direct communication between people/devices who are within communication range.

Haggle does not follow the TCP/IP protocol suite and eliminates layers above the data link layer. It uses application-driven message forwarding instead of relying on the network layer. It provides context-based message forwarding between mobile devices in an intermittently connected network. The project emphasizes measuring and modeling pairwise contacts between devices that are characterized by means of two parameters: contact duration and intercontact times. The statistical properties of these parameters are used to drive the design of forwarding policies (Chaintreau et al. 2007).

The application areas for Haggle includes transportation (car to car, or car to roadside), communicating in remote and developing regions, for emergency services, and so on. Furthermore, it is also the basis for the design of concrete applications. For example, the Haggle Project is working with epidemiologists to experimentally study the correlation between human contact patterns and the spread of diseases like the flu. The patterns of contact between people are also the basis for designing social-aware applications. An initial example of this approach is the design of a content distribution system in an urban setting (Leguay et al. 2006).

5.6.2 DakNet Project

DakNet (Pentland and Hasson 2004) is an example of an opportunistic network that provides asynchronous digital connectivity. It has been successfully deployed in the rural areas of several countries to provide villagers non-real-time Internet access at a cost much lower than that of traditional Internet infrastructure solutions. It exploits the existing communications and transportation infrastructure to distribute digital connectivity to areas where digital communications infrastructure is not available. The name *DakNet* is derived from the Hindi word *dakiya* (the postman). As a postman distributes letters collected from the central post office to recipients in rural areas, DakNet exploits existing transportation facilities and wireless communication techniques to distribute Internet connectivity from a central Internet hub point to rural areas.

DakNet transmits data over short point-to-point links between kiosks and portable storage devices, called *mobile access points* (MAPs), to preserve energy consumption. A MAP mounted on a bus, a motorcycle, or even a bicycle physically transports data among kiosks and a hub (Figure 5.10). It works in two steps: first, as the MAP-equipped vehicle comes within the range of a village Wi-Fi-enabled kiosk, it automatically senses the wireless connection and then uploads and downloads data. Second, when a MAP-equipped vehicle comes within the range of an Internet access point, it automatically synchronizes the data from all the rural kiosks, using the Internet.

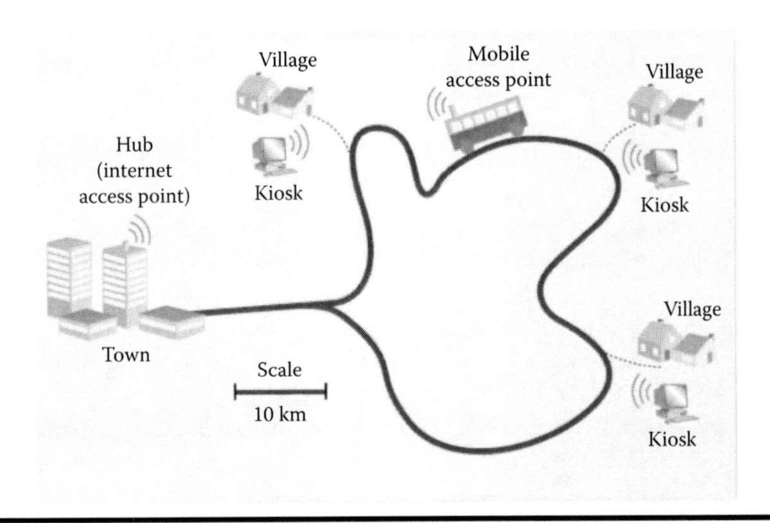

Figure 5.10 The DakNet concept. (From Pentland, R. F. and Hasson, A., *IEEE Computer*, 37, 78–83, 2004.)

DakNet has been successfully implemented in the villages of India and northern Cambodia. One of the earliest deployments was the Bhoomi e-governance project implemented in India by the state government of Karnataka to distribute a computerized land record facility from district headquarters to rural areas. DakNet was also implemented in remote areas of Cambodia, called the *Internet Village Motoman project*, to connect schools in rural areas with the capital of Ban Lung to provide Internet facility through asynchronous digital connectivity. The terrain in northern Cambodia is very difficult; therefore MAPs are installed on motorcycles and oxcarts instead of buses. Projects implementing these concepts are currently ongoing. For example, the KioskNet project (Guo et al. 2007) focuses on realizing a very low-cost asynchronous ICT infrastructure to provide connectivity to rural villages in India, while the Saami network connectivity project (Doria et al. 2002) provides connectivity to the inhabitants of Lapland.

5.6.3 ZebraNet Project

Opportunistic networks are also applied to interdisciplinary projects focusing on wildlife monitoring. The project goal was to use the least energy, storage, and other resources necessary to maintain a reliable system with a very high successful eventual data transmission rate. Usually, small monitoring devices are attached to animals, and an opportunistic network is formed to gather information and carry it to a few base stations possibly connected to the Internet. Contacts among animals are exploited to aggregate data and carry them closer and closer to the base station.

This is a reliable, cost-effective, and nonintrusive solution. Concrete applications implementing these ideas have been used in the ZebraNet project (Juang et al. 2002). It is an interdisciplinary project of Princeton University performing novel studies of animal migration and interspecies interactions, by deploying opportunistic networks on zebras in the vast savanna area of central Kenya under control of the Mpala Research Centre.

The ZebraNet system includes custom tracking collars (nodes) carried by the animals under study across a large wild area. The collars operate as a peer-to-peer network to deliver logged data back to base stations. Each collar contains a small wireless computing device and consists of a GPS, flash memory, wireless transceiver, and a small CPU. Because there is no cellular service or broadcast communication covering the region where animals are studied, ad hoc peer-to-peer routing is needed. The project 'Network on Whales' is similar to ZebraNet; the objective is to collect data under the ocean about whales and investigate their habits as well as how they react to human intervention in their living environment. The Shared Wireless Infostation Model is used to collect biological data associated with whales in the oceans (Small and Haas 2003).

Despite the fact that opportunistic networking is a new research area, its applications are being used to serve several diverse areas such as the environment, government, event/exhibition, transportation, distribution, health, life science, and agriculture due to its several vital properties such as autonomous behavior, self-organization abilities, adaptability to changing environments, and even self-healing when faced with component failures or malicious attacks.

5.7 Summary

Opportunistic networks are constructed by users' mobile devices based on social relationships among them that may be helpful in message forwarding. Developing a social relationship model to map users' social relationships and using them to construct routing protocols is an emerging research domain. This chapter discussed context-information-based routing protocols for opportunistic networks and how this information is used to automatically adapt to the dynamic environment resulting from the different mobility patterns of nodes. Further, a classification of the protocols was presented on the basis of the amount of context information they exploit to make forwarding decisions. A comparative study of the protocols was made to understand the issues they address and the approach followed by each. Context-based routing protocols can be stronger alternatives to classical flooding-based protocols and can be seen as an effective means of controlling congestion in opportunistic networks. Finally, a few application cases of opportunistic network were discussed. An interesting research direction is interoperability issues between completely infrastructureless opportunistic and infrastructure-based networks.

References

Boldrini, C., Conti, M., Jacopini, I. and Passarella, A. (2007) HiBOp: A history based routing protocol for opportunistic networks. *IEEE International Symposium on a World of Wireless, Mobile and Multimedia Networks, 2007* (*WoWMoM 2007*). Helsinki.

Boldrini, C., Conti, M. and Passarella, A. (2008) Autonomic behaviour of opportunistic network routing. *International Journal of Autonomous and Adaptive Communications Systems*. 1(1). pp. 122–147.

Borgia, E., Conti, M., Delmastro, F. and Pelusi, L. (2005) Lessons from an ad-hoc network test-bed: Middleware and routing issues. *Ad Hoc and Sensor Wireless Networks, an International Journal*. 1. pp. 125–157.

Brown, P., Bovey, J. and Chen, X. (1997) Context-aware applications: From the laboratory to the marketplace. *IEEE Personal Communications*. 4(5). pp. 58–64.

Burgess, J., Gallagher, B., Jensen, D. and Levine, B. (2006) MaxProp: Routing for vehicle-based disruption-tolerant networks. *25th IEEE Annual Joint Conference of the IEEE Computer and Communications Societies* (*INFOCOM 2006*).

Burns, B., Brock, O. and Levine, B. (2005) MV routing and capacity building in disruption tolerant networks. *24th IEEE Annual Joint Conference of the IEEE Computer and Communications Societies* (*INFOCOM 2005*).

Chaintreau, A. et al. (2007) Impact of human mobility on opportunistic forwarding algorithms. *IEEE Transactions on Mobile Computing*. 6. pp. 606–620.

Chen, L., Yu, C. and Sun, T. (2006) A hybrid routing approach for opportunistic networks. *SIGCOMM'06 Workshops*. Pisa, Italy. pp. 213–220.

Conti, M. and Giordano, S. (2007a) Multihop ad hoc networking: The reality. *IEEE Communications Magazine*. 45. pp. 88–95.

Conti, M. and Giordano, S. (2007b) Multihop ad hoc networking: The theory. *IEEE Communications Magazine*. 45. pp. 78–86.

Conti, M. and Kumar, M. (2010) Opportunities in opportunistic computing. *IEEE Computer Society*. pp. 42–50.

Conti, M. et al. (2008) Routing issues in opportunistic networks. In: Garbinato, B., Miranda, H. and Rodrigues, L. (eds.). *Middleware for Network Eccentric and Mobile Applications*. Springer, Berlin Heidelberg. pp. 121–147.

Davis, J. A., Fagg, A. H. and Levine, B. N. (2001) Wearable computers and packet transport mechanisms in highly-partitioned ad-hoc networks. *Inti. Synylusirrni on Wrcrrabk Cnnrpvters*.

Dey, A. K. and Abowd, G. (2000) Towards a better understanding of context and context-awareness. *Workshop on the What, Who, Where, When and How of Context-Awareness, affiliated with the CHI 2000 Conference on Human Factors in Computer Systems*. New York.

Dey, A. K., Salber, D. and Abowd, G. D. (2001) A conceptual framework and a toolkit for supporting the rapid prototyping of context-aware applications. *Special issue on Context-Aware Computing in the Human-Computer Interaction (HCI) Journal*. 16. pp. 97–166.

Doria, A., Uden, M. and Pandey, D. (2002) Providing connectivity to the Saami nomadic community. *2nd International Conference on Open Collaborative Design for Sustainable Innovation*. 1(2).

Fall, K. (2003) A delay-tolerant network architecture for challenged internets. *ACM Conference on Applications, Technologies, Architectures, and Protocols for Computer Communications* (*SIGCOMM*). pp. 27–34.

Guo, S. et al. (2007) Very low-cost internet access using kiosknet. *SIGCOMM Computer Communication Review*. 37. pp. 95–100.

Hui, P. and Crowcroft, J. (2007) *Bubble rap: Forwarding in small world DTNs in every decreasing circles*. Technical Report UCAM-CL-TR684. Cambridge, UK: University of Cambridge.

Hui, P. et al. (2005) Pocket switched networks and human mobility in conference environments. *WDTN '05 Proceedings of the 2005 ACM SIGCOMM Workshop on Delay-Tolerant Networking*. pp. 244–251.

Jain, S., Fall, K. and Patra, R. (2004) Routing in a delay tolerant network. *ACM Conference on Applications, Technologies, Architectures, and Protocols for Computer Communications* (*SIGCOMM*). pp. 145–158.

Jindal, A. and Psounis, K. (2007) Contention-aware analysis of routing schemes for mobile opportunistic networks. *1st International ACM MobiSys Workshop on Mobile Opportunistic Networking* (*MobiOpp 2007*). pp. 1–8.

Juang, P. et al. (2002) Energy-efficient computing for wildlife tracking: Design tradeoffs and early experiences with ZebraNet. *ACM SIGPLAN Notices*. 37. pp. 96–107.

Kalman, R. E. (1960) A new approach to linear filtering and prediction problems. *Transactions of the ASME Journal of Basic Engineering*. 82. pp. 35–45.

Leguay, J. and Friedman, T. C. V. (2006) Evaluating mobility pattern space routing for DTNs. *25th IEEE Annual Joint Conference of the IEEE Computer and Communications Societies* (*INFOCOM 2006*). pp. 1–10.

Leguay, J., Friedman, T. and Conan, V. (2005) DTN routing in a mobility pattern space. *WDTN '05 Proceedings of the 2005 ACM SIGCOMM Workshop on Delay-Tolerant Networking*. pp. 176–183.

Leguay, J. et al. (2006) Opportunistic content distribution in an urban setting. *SIGCOMM Workshop on Challenged Networks* (*CHANTS 2006*). pp. 205–212.

Lindgren, A., Doria, A. and Schelen, O. (2003) Probabilistic routing in intermittently connected networks. *ACM Mobile Computing and Communications Review*. 7. pp. 19–20.

Musolesi, M., Hailes, S. and Mascolo, C. (2005) Adaptive routing for intermittently connected mobile ad hoc networks. *IEEE International Symposium on a World of Wireless, Mobile and Multimedia Networks* (*WoWMoM 2005*). pp. 183–189.

Nguyen, H. A. and Giordano, S. (2009) Routing in opportunistic networks. *IJACI*. 1(3). pp. 19–38.

Nordstrom, E., Gunningberg, P. and Rohner, C. (2009) Haggle: A data-centric network architecture for mobile devices. *MobiHoc Workshop*. New York: ACM. pp. 37–40.

Pelusi, L., Passarella, A. and Conti, M. (2006) Opportunistic networking: Data forwarding in disconnected mobile ad hoc networks. *IEEE Communications Magazine*. 44. pp. 134–141.

Pentland, R. F. and Hasson, A. (2004) DakNet: Rethinking connectivity in developing nations. *IEEE Computer*. 37. pp. 78–83.

Ramanathan, R. et al. (2007) Prioritized epidemic routing for opportunistic networks. *1st International MobiSys Workshop on Mobile Opportunistic Networking*. pp. 62–66.

Ryan, N., Pascoe, J. and Morse, D. (1999) Enhanced reality fieldwork: the context-aware archaeological assistant. Bar International Series 750. pp. 269–274.

Schmidt, K. et al. (1999) Advanced interaction in context. *11th International Symposium on Handheld and Ubiquitous Computing (HUC99)*. Lecture notes in computer science, Springer, Berlin Heidelberg. 1707. pp. 89–101.

Small, T. and Haas, Z. J. (2003) The shared wireless infostation model—A new ad hoc networking paradigm (or Where there is a whale, there is a way). *4th ACM International Symposium on Mobile Ad Hoc Networking and Computing (MobiHoc 2003)*. Annapolis, MD. 1–3 June 2008. New York: ACM. pp. 233–244.

Spyropoulos, T., Psounis, K. and Raghavendra, C. S. (2004) Single-copy routing in intermittently connected mobile networks. *In Proceedings of IEEE Secon'04*. pp. 235–244.

Spyropoulos, T., Psounis, K. and Raghavendra, C. S. (2005) Spray and wait: Efficient routing in intermittently connected mobile networks. *ACMSIGCOMM Workshop on Delay Tolerant Networking (WDTN)*.

Spyropoulos, T., Psounis, K. and Raghavendra, C. S. (2007a) Efficient routing in intermittently connected mobile networks: The multiple-copy case. *ACM/IEEE Transactions on Networking*. 16. pp. 77–90.

Spyropoulos, T., Psounis, K. and Raghavendra, C. S. (2007b) Spray and focus: Efficient mobility-assisted routing for heterogeneous and correlated mobility. *IEEE PerCom Workshop on Intermittently Connected Mobile Ad Hoc Network*. New York.

Vahdat, A. and Becker, D. (2000) *Epidemic routing for partially connected ad hoc networks*. Technical Report CS-2000-06. Computer Science Department, Duke University.

Verma, A., Pattanaik, K. K. and Ingavale, A. (2013) Context-based routing protocols for OppNets. In: Woungang, I., Dhurandher, S.K., Anpalagan, A. and Vasilakos, A.V. (eds.). *Routing in Opportunistic Networks*. Chapter 3. New York: Springer. pp. 69–97.

Verma, A. and Srivastava, A. (2011) Integrated routing protocol for opportunistic networks. *(IJACSA) International Journal of Advanced Computer Science and Applications*. 2(3). pp. 85–92.

Wang, Y., Jain, S., Martonosi, M. and Fall, K. (2005) Erasure coding based routing for opportunistic networks. *ACM SIGCOMM Workshop on Delay Tolerant Networks*.

Widmer, J. and Boudec, J. L. (2005) Network coding for efficient communication in extreme networks. *ACM SIGCOMM 2005 Workshop on Delay Tolerant Networking*. pp. 284–291.

Chapter 6

Smart Environments: Exploiting Passive RFID Technology for Indoor Localization

Kevin Bouchard, Jean-Sébastien Bilodeau, Dany Fortin-Simard, Sebastien Gaboury, Bruno Bouchard, and Abdenour Bouzouane

Contents

6.1 Introduction

Smart environments have fascinated people and researchers for a few decades now (Weiser 1991). Originally, these environments were simply standard habitations or buildings equipped with automated systems such as lighting and heat control features (Robles and Kim 2010). By definition, a *smart environment* is an environment that possesses the qualities of one or many intelligent agents. It first needs to be enhanced with sensors to enable it to understand and see what activities are ongoing. It then requires effectors to modify the environment and interact with the other agents (usually the residents or employees). Finally, it requires one or many brains: computers or small independent devices equipped with microcontrollers. Smart environments nowadays appear in a range of applications from simple automation systems to assistive systems (Cook and Schmitter-Edgecombe 2009; Demerism et al. 2008; Phua et al. 2009) (for the cognitively/physically impaired, for the frail and elderly, etc.), including optimization of energy consumption (Ricquebourg et al. 2006) (thermostat, automatic windows, etc.). The progression, through the last decades, of such environments toward an offering of more complex and smarter services was made possible by the development of new technologies and the miniaturization of both sensors and effectors but also by the major breakthroughs that have been made in ambient intelligence and networking technologies. Research could advance a major step forward in these innovative areas if we could extract better knowledge from many of the sensing technologies that provide only imprecise and complex data. For instance, being able to precisely track everyday life objects in a house could lead to better assistive services in a smart home by improving the granularity of activity recognition (Patterson et al. 2005), which constitutes the main challenge of this field of research (Ramos et al. 2008). In the specific context of smart environments, the fundamental spatial information related to everyday objects (position, zone, movement, orientation, gestures, etc.) plays a significant role in the gathering of knowledge on the current context (Bouchard et al. 2011). Therefore, a key challenge of every smart environment is to implement a precise and effective localization system, allowing tracking of the most important objects in real time.

In the last few years, many researchers have tried to address this core issue by proposing positioning systems based on various approaches such as GPS (Zhou and Shi 2009), ultrasonic wave sensors (Hauschildt and Kirchhof 2011), video cameras (Hoey 2006), and radio frequency identification (RFID). However, RFID technology, because of its robustness, its low price, and its flexibility, seems to have imposed itself as one of the best solutions currently available for smart environments. Typically, an RFID system consists of three elements: radio frequency (RF) tags, at least one RF antenna, and a data collection module. This system works as follows. First, the RF antenna emits a wave of radiation. Then, if a tag is located within its coverage area, the tag intercepts the signal, and its internal chip retransmits a signal to the antenna. The transmitting antenna receives this new signal, and it returns the information to the collection module. These data can then be used by an algorithm to infer spatial information about the tagged objects, such as their positions. RF tags are subdivided into three families: active, semi-active, and passive. Active and semi-active tags are battery powered and often have an internal erasable memory. Therefore, an active system can transmit a low-power RF emission, and tags remain able to meet with high-level signals. On the other hand, passive RFID tags are in a dormant state and wake up when receiving power from an RF wave emitted from nearby antennas. They then use that same energy to power their inner chip and send an RF incorporating their unique ID (see Figure 6.1).

Because of that internal power, active tags achieve a much higher range and reliable accuracy. Moreover, the positioning systems for these technologies are considerably more advanced. For instance, Hekimian-Williams et al. (2010) obtained very precise results (millimeter range) by using software coupled with accurate clocks to estimate the phase difference from signals received from two separate antennas. Nevertheless, passive RFID seems more promising for smart environments over the long term because of many advantages. First, passive tags have a technically unlimited lifespan and do not need an external power supply. Therefore, a system relying

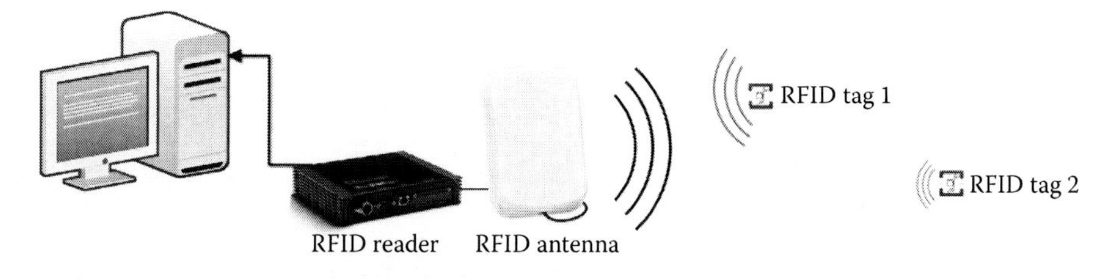

Figure 6.1 Passive RFID system. Farther tags receive less power and are then harder to detect.

on them will require little maintenance and will be more autonomous. Second, they are much smaller than their active counterparts and thus reduce intrusiveness. Consequently, it is possible to embed passive RFID tags in everyday life objects such as cups, plates, and so on. Finally, passive tags generally cost only a few pennies, whereas active and semi-active tags are in a much higher price bracket. In this chapter, RFID means passive RFID excepted when stated otherwise.

This chapter aims to explore the paradigms of passive RFID technology within smart environments. In particular, its goal is to analyze each element toward extraction of the full potential of RFID. This chapter is mainly based on our experiments through the years with this technology and covers the best solutions we found to keep the usage of RFID simple yet interesting. In particular, this chapter covers three localization solutions: proximity based, trilateration based, and learning based. For each category, we provide sample results we obtained at our smart home laboratory (the Laboratoire d'Intelligence Ambiante pour la Reconnaissance d'Activités (LIARA)), but we also provide an assessment of their advantages and disadvantages. We also wanted to discuss something that the literature often ignores, since it is less challenging: the format and manner of turning simple position data into more interesting knowledge.

The remainder of this chapter proceeds as follows. The section "Related Work" discusses the literature about indoor and outdoor localization systems. It first discusses non-RF and hybrid technologies. Then it provides an overview of the main active and passive RFID approaches. The section "Passive RFID Primers" gives an introduction to the idea of using RFID in smart environments and examines a few fields of application for RFID technology. It discusses two important challenges that dramatically decrease the accuracy of localization: false readings and the high standard deviation of received signal strength indication (RSSI). The section "Proximity-Based Localization" presents the proximity-based solution and explains how to configure an environment to use it. In particular, we show how it was exploited in LIARA's smart home. The section "Trilateration" discusses the trilateration method and explains how it can be adapted with ellipses instead of circles when dealing with directional antenna. The section "Learning-Based Localization" describes machine learning alternatives to complex localization algorithms and assesses the limitations of such methods. The section "Modeling and Inferring Information from Sensor Output" discusses what kind of knowledge can be extracted from basic RSSI and localization with RFID. In particular, it covers the recent exploration of gesture recognition from passive RFID technology and how fuzzy logic can enhance the information.

6.2 Related Work

Over the years, the question of localizing entities in noisy smart environments has attracted many researchers, resulting in hundreds of scientific publications covering diverse topics and technologies. Despite this fact, indoor localization of objects is still a challenging issue that could find application in several areas. In particular, techniques related to RFID technology have been blossoming in the last few years. However, due to the inherent imprecision, positioning and tracking with RFID are still very hard to achieve. That explains why many researchers have explored hybrid approaches based on ultrasonic sensors, accelerometers, cameras, and LEDs. These works are outside of the scope of this chapter, which concentrates on passive RFID technology, but we still briefly describe the most important while discussing why we think RFID is often more appropriate in smart environments. Thereafter, we present the most interesting RFID approaches.

6.2.1 Non-RFID and Hybrid Approaches

The most renowned localization system not based on RF is probably the one by Addlesee et al. (2001). It is one of the first successful systems that relies solely on ultrasonic sensors. In a controlled environment, it achieves a respectable precision (≈3 cm), which places it among the most precise indoor localization systems. However, it requires dense deployment of costly receivers installed throughout the ceiling of the targeted area. Choi and Lee (2009) followed their work by hybridizing ultrasonic sensors to passive RFID. They unfolded static and fixed tags at predefined positions (called *reference tags*) to infer the position of a mobile robot. They yield a much higher accuracy than any other RFID approaches (≈1–3 cm). Every sound-based system is limited by environmental conditions such as noise, obstruction of the line of sight, and so on. Hähnel et al. (2004) mixed an RFID reader with a laser range scanner onboard mobile robots. Their model starts by using machine learning to draw a virtual map of the environment with the laser. Despite its novelty, their system only achieves a high localization error of around 1–10 meters. Milella et al. (2009) combined vision sensors with the RFID technology to achieve a precision on the order of 20 cm. Sample et al. (2012) also exploited vision sensors. They enhanced the tags with LEDs to enable a robot equipped with a camera to precisely locate a tag in the environment. Recently, Parr et al. (2012) introduced a novel method for RFID tag tracking by fusing an inertial measurement unit with a handheld reader. Their technique uses acceleration data, without the knowledge of antenna position, to achieve accurate positioning at a reasonably low cost.

Although non-RFID and hybrid approaches usually give better performance than pure RFID localization, they are arguably less appropriate in many situations. First, they are more costly than RFID approaches. Second, they rely on technologies that suffer from high intrusiveness (cameras are particularly problematic in many cases). Third, they often impose line-of-sight constraints that RF avoids. Fourth, none of these systems offer robustness comparable to passive RFID tags. For example, they cannot be put into a dishwasher, and they often need batteries. They are also slower and harder to install than simple RFID methods. Finally, these technologies are too cumbersome for object tracking inside buildings.

6.2.2 RFID Localization

As we stated in the introduction, RFID technology can be subdivided into two families: active and passive. Both have been the object of study for localization. Active systems are generally much more precise on the same environmental scale than their passive counterparts. For instance, Hekimian-Williams et al. (2010) implemented phase difference to achieve millimeter accuracy in perfect conditions. Active tags suffer from many weaknesses that keep them from being incorporated into smart environments. First, they are much bigger and therefore impossible to install on everyday life objects with the goal of localizing them. Second, they are considerably more costly. Above all, they work with batteries, requiring regular maintenance, which is to be avoided in smart homes.

There are a large number of positioning approaches based upon the use of passive RFID tags. A substantial number of them arise directly or indirectly from the well-known LANDMARC system (Ni et al. 2004), which introduced the concept of localization from reference tags placed at strategic locations. Vorst et al. (2008) is one of them. Their model uses passive RFID tags and an onboard reader to localize mobile objects in an environment. A prerequisite learning step is required to define a probabilistic model. This model is exploited with a particle filter (PF) technique, which estimates position. It achieves a precision of 20–26 cm. The major drawback is the relatively high computational cost (at least for real-time tracking). Lei et al. (2012) addressed this issue by combining PF with weighted centroid localization. They switch between the two methods depending on the estimated velocity of the tracked object. In ideal conditions, they localize an antenna with an average error of 20 cm while greatly increasing the speed of the process. Another model, from Joho et al. (2009), uses reference tags in combination with different metrics. In particular, they are based on both the RSSI and the antenna's orientation to get an average localization error of 35 cm. Chawla and Robins (2011) developed a model based on the variation of antenna power to estimate the distance of nearby reference tags. They incrementally adjust the antenna decibel until the tag is in range. Thereafter, they use many tags' distance from the antenna to localize a mobile robot. Their approach yields an accuracy varying from 18 to 35 cm.

Some of these approaches provide very good results, more than enough to exploit them as support services for smart environments. However, they all rely on large deployment of tags of references. Although it is a fairly good solution for robot localization, it is not always feasible to deploy them in smart environment contexts. In fact, in many contexts, such as smart homes, the modifications to the infrastructure need to be minimal, and this installation of tags would be rather unwelcome to the residents. Finally, most of the previous techniques localize antennas with tags. Antennas are too big to be bundled on objects. Therefore, this is not an interesting solution for object tracking.

Trilateration has been largely ignored in the scientific literature despite its simplicity and potential applications. This is mainly because this technique is quite challenging to use with noisy and imprecise information. A recent instance of an RFID localization system based on this technique is the approach of Kim and Kim (2012). They performed a classical trilateration calculus from active tags by using the time of arrival of the signal to calculate the distance from each antenna. Another approach worth mentioning is that of Chen et al. (2010), who performed trilateration with a different RF technology (ZigBee). They developed a fuzzy inference engine with one variable that correlates the RSSI of an object transmitter to the distance separating it from a receiver. They achieved a precision of 119 cm.

6.2.3 Literature Assessment

As we have seen, there is a plethora of localization algorithms in the literature. Many possess their own advantages in their context of application over the methods presented in this chapter. Nevertheless, compared to other approaches, our models do not require reference tags, which are often not appropriate in smart environments (Chawla and Robins 2011; Hähnel et al. 2004; Joho et al. 2009; Vorst et al. 2008). Moreover, with these tags, it is necessary to perform calibration and precisely install them in a new environment. Speed is another very important criterion to evaluate the performance of a localization system. A fast system will be more likely to perform well when tracking moving objects. Reference tag–based systems require more computation compared to learning-based methods or even elliptical trilateration. Table 6.1 presents a summary of the main localization systems found in the literature.

6.3 Passive RFID Primers

RFID technology is used extensively in some industries such as retail business to track goods in big warehouses or in the shipment business to allow users to follow the delivery of their package in real time. However, in research, it is primarily robotics that has served to advance localization techniques (Choi and Lee 2009; Heesung and Kyuseo 2005). This technology is also increasingly used in smart homes (Fishkin et al. 2004), but most researchers in artificial intelligence consider the resulting data like any other, whether it is to perform recognition of activities or extract knowledge with data mining techniques. Whatever the field of application, to track people or objects, everyone would benefit from better exploitation of this technology for localization. This chapter concentrates on this topic and explores different ways to use it in smart environments.

Through this chapter, we discuss not only methods that we find interesting and promising for localization in smart environments but also our own experiments, which were all conducted in our prototyping smart home. Our laboratory infrastructures implement new cutting-edge technologies in a surface of about

Table 6.1 Summary of a Few Localization Systems

Authors	*Technology*	*Technique*	*Hardware*	*Precision (cm)*
Addlesee et al. (2001)	Ultrasonic	Reference receivers	750 receivers	3
Choi and Lee (2009)	Ultrasonic/ RFID[a]			1.6–2.4
Chen et al. (2010)	ZigBee	Trilateration/ RSSI[b]	1 device, 4 sensors	119
Jin et al. (2006)	Active RFID	Reference tags	20 tags, 4 antennas	72
Fu and Guenther (2009)	Active RFID	Trilateration/ RSSI		200–300
Hekimian-Williams et al. (2010)	Active RFID	Phase difference	1 tag, 2 antennas	≈1
Hähnel et al. (2004)	Passive RFID	Reference tags	100 tags, 2 antennas	100–140
Zhang et al. (2007)	Passive RFID	Direction of arrival	Simulation	100
Vorst et al. (2008)	Passive RFID	Reference tags	374 tags, 4 antennas	20–26
Joho et al. (2009)	Passive RFID	Reference tags	350 tags, 1 antenna	35.5
Chawla and Robins (2011)	Passive RFID	Reference tags	132 tags, 1 antenna	18–35
Lei et al. (2012)	Passive RFID	Reference tags	32 tags, 1 antenna	20

[a] Radio frequency identification.
[b] Received signal strength indication.

100 square meters and possess more than a hundred different sensors and effectors. More specifically, this laboratory possesses eight directional RFID antennas divided among two independent receivers in the apartment to track the position of the objects. These A-PATCH-0025 antennas from Poynting, Munich, Germany, which covers both EU and USA RFID bands, were designed to be waterproof and easy to install. In addition, they are circularly polarized to enhance reading in any orientation. Moreover, the application of the driver of these antennas is open

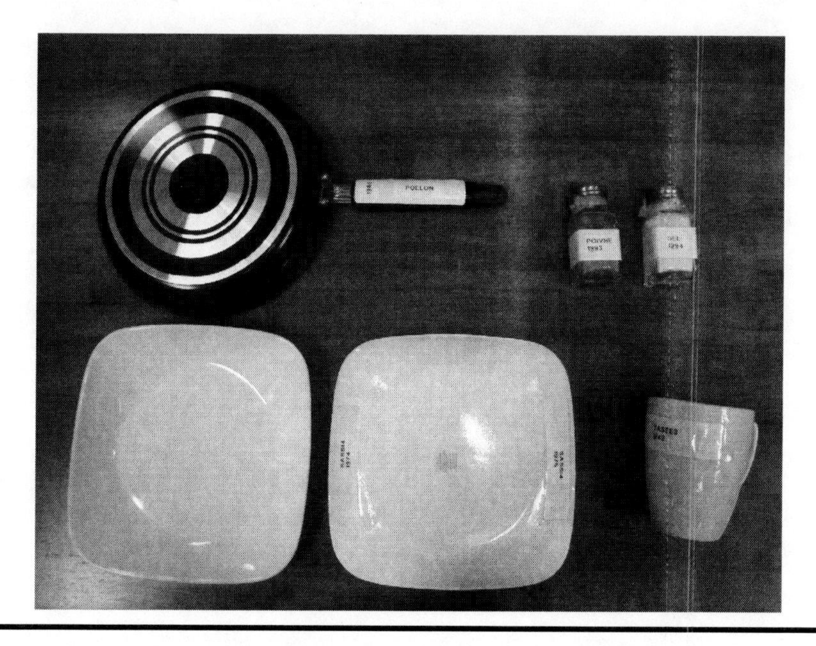

Figure 6.2 A few everyday objects with passive RFID tags.

source, and it is supplied by the company. Therefore, we are able to modify and adjust the system if required. We also acquired a small, inexpensive, and portative kit from Alien Technology, San Jose, CA that is especially useful when we want to try new prototypes in senior citizen centers. Figure 6.2 shows examples of tagged objects used in our laboratory.

The exploitation of RFID technology for localization offers many challenges and we will propose simple and more elaborate solutions for each of them through this chapter. For example, the physical configuration of such a system must be done carefully. The antenna's position might determine which techniques can be exploited for the localization. Moreover, it is important to try to avoid having many antennas broadcasting at the same time. Each receiver usually implements a time slice method to avoid interference between its antennas, but antennas from different receivers could be an issue. In the sections describing the different localization approaches, we discuss the placement of antennas. In addition to the physical configuration of antennas, it is important not only to choose good RFID technology but to select the tags carefully for good localization results. In order to minimize problems related to variation in receptivity, we prefer to use only one kind of tag for all objects. We suggest testing the various types of tags and retaining those that limit the variation of the signal strength. The type of tag is not the only thing that might decrease the precision or accuracy of localization. Even among tags that are technically identical, sometimes the sensitivity is very different (Chawla and Robins 2011). To address this problematic behavior, we propose testing every tag before installing it on an object and eliminating those that are too far from average sensitivity. In addition, if one wishes to use precise localization (such as trilateration), it might be interesting and profitable to put more than one tag on each object. In fact, the main issue of localization is bad angles or the arrival of radio wave on a tag. Consequently, covering more angles should ensure better quality information. Finally, there are two other very important challenges that are discussed in the subsections "False Reading Challenges" and "Variation of the RSSI."

6.3.1 False Reading Challenges

The first challenge encountered when trying to build a passive RFID localization framework is situated at the basic step of data collection. Due to the nature of the system operation, it is very common to obtain false-negative readings (FNRs). An FNR occurs when a tag is in the antenna coverage area but is not detected during a certain period of time. This type of problem happens in all the passive RFID systems we have tried through the years, considering that it is slightly more frequent on inexpensive systems. Brusey et al. (2003) identified three reasons to explain this situation:

- The reader can fail to see all tags for a certain time due to an unknown internal problem.
- RF emitted from more than one tag may collide.
- Interference might occur due to environmental emissions or due to surrounding metal shielding.

There is also the opposite situation. Although this is much less common, occasionally tags that are not in the normal area of the antenna are detected. These are called *false-positive readings* (FPRs). In a smart home environment, this translates as an object that might be stored somewhere (in the cabinet, for instance) and a signal that gets stronger because of an uncontrollable event for a period of time. For example, a person may use a metal object (such as a kettle) that enables the signal to rebound or travel farther than usual. This could cause a localization algorithm to decide that an object has moved if not handled correctly. Note that in many applications FPRs are not significant and are ignored.

One of the ways to solve the reading uncertainty issue would be to create enhanced tags with shielding properties. However, the tags would be bigger, and thus it would be harder to embed them in every object. We could also use the strategy of shielding part of the environment where the tag should not be detected. For instance, we could shield the kitchen cabinet to limit detection of objects when they are stored. Shielding is not necessarily a costly operation. It can be done with simple consumer aluminum paper sheets. However, it is best to reserve this method for rapid prototyping in laboratories, since it is not necessarily aesthetic.

6.3.1.1 *Iteration-Based Filter*

From our review of the literature on passive RFID localization systems, we found an interesting solution to FNRs (Brusey et al. 2003). The solution is a time filter based on the general rule that if an object is expected to be in antenna range but is not, it is considered as not present only after no detection has occurred for a period of time. That interval of time needs to be carefully tweaked. It must be as high as possible (for greater impact) but not too high because tags might become too hard to detect. To this end, the authors introduced a function named *top-hat* (Equation 6.1). This function excludes all readings that, from the current time (t_{now}), are separated by more than a certain time interval Δt_{hat}. The function $f_{hat}(t)$ returns true if there is presence or false otherwise. With this method, the tag is considered to have been detected as soon as there is more than one detection during the time interval.

In our smart home context, FPRs were also an issue and this function did not allow dealing with them. However, it is possible to slightly modify the function for that purpose. The function in Equation 6.2, denoted by $f_{ite}(i)$, is constructed by using iteration instead of time as a parameter. We find it preferable to use a fixed time interval because it is easier and more intuitive. We can decide whether a (boolean) tag's detection state (O_s) has changed by subtracting the first detection iteration (i_d) of a sequence of the opposite state from the current iteration (i_c) and comparing it with a Δi. The Δi is the minimum number of iterations the object's state needs to be stable before considering that the detected state has changed. The difference from the work of Brusey et al. (2003) is subtle, but it allows one to deal with both kinds of wrong readings (FNRs and FPRs).

$$f_{hat}(t) = \begin{cases} \text{True} & |t_{now} - t| < \Delta t_{hat} \\ \text{False} & \text{otherwise} \end{cases} \tag{6.1}$$

$$f_{ite}(i, O_s) = \begin{cases} !O_s & |i_c - i_d| \geq \Delta i \\ O_s & \text{otherwise.} \end{cases} \tag{6.2}$$

For example, suppose that Tag X is undetected at Iteration 1, then the next iterations would proceed as follows (for $\Delta i = 1$):

Iteration 2: X is *read*, $|2 - 2| \ngeq 1$, no change
Iteration 3: X is *not read*, $|3 - 3| \ngeq 1$, no change
Iteration 4: X is *read*, $|4 - 4| \ngeq 1$, no change
Iteration 5: X is *read*, $|5 - 4| \geq 1$, X is now detected
Iteration 6: X is *not read*, $|6 - 6| \ngeq 1$, no change
Iteration 7: X is *not read*, $|7 - 6| \geq 1$, X is now undetected

The performance of the function relies on the Δi, which depends greatly on the RFID configuration. The parameter can be automatically determined if one knows the rate of false readings (fr_{rate}). For example, suppose that fr_{rate} = 25%, which means that in roughly one of each four readings the tag should not be detected. If one aims to have a $fr_{rate} \approx 99.5\%$, the Δi would be 4 since

$$\left(\tfrac{1}{4}\right)^4 = \tfrac{1}{256} \approx 0.39\%$$

The false reading rate will be probably higher since the behavior is not random (the tags often disappear for few iterations), but if the calculation of Δi is not satisfactory it can be determined experimentally. It is important to note that the value of Δi will increase the response time when the tag state really changes ($\Delta i^* i_{length}$). Therefore, the value should be decided according to the need of the system.

We tested the $f_{ite}(i, O_s)$ function in the LIARA smart home with a proximity localization method and trilateration, for which only the antenna's configurations and the reading speed had a different impact on the false reading rate. We computed the rate in Figure 6.3.

The false reading problem could also be addressed with a simple average over a number of iterations. However, it is important to understand the implications, because the results would differ. The example below illustrates that difference (with Δi = 2).

$$O_s[\cdot] = \{T,T,T,T,F,T\} \quad \text{Average} = \text{True} \quad f_{ite}(i,O_s) = \text{True}$$

$$O_s[\cdot] = \{T,T,T,T,F,T,F,F,F\} \quad \text{Average} = \text{True} \quad f_{ite}(i,O_s) = \text{False}$$

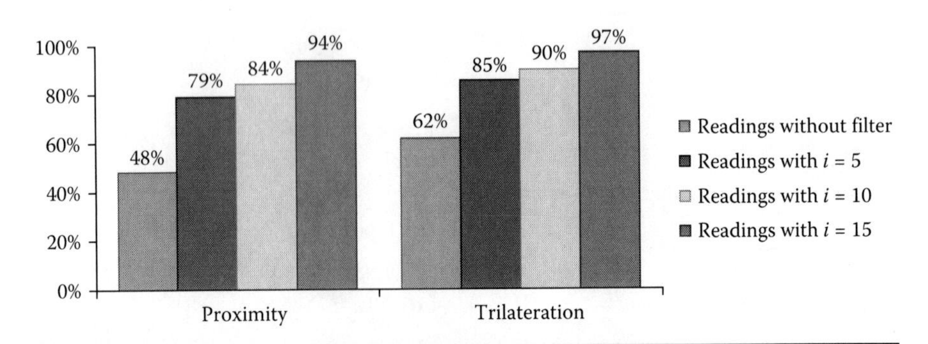

Figure 6.3 False reading rates with and without $f_{ite}(i,O_s)$ at a reading rate of five per second.

6.3.2 Variation of the RSSI

Another problem with passive technology is that the RSSI usually has high variation from one iteration to another even when the tag has not moved. At the step of localization, this high standard deviation results, on the tracked object, in a seemingly perpetual random movement. These variations cannot be completely eradicated because they are caused by an unchangeable fact. The RF signal is indeed greatly influenced by environmental variables (people, liquids, metal, etc.). The amount of flickering can easily be reduced with an average; however, it would give the same importance to old and late RSSIs. It would, consequently, delay the real movement of a tracked object. To overcome this issue, a Gaussian mean can be applied to the RSSI returned by the antennas (Equation 6.3). The bell-shaped curve of the distribution is centered on the current iteration number i_c. The parameter i is the iteration number associated with the RSSI record that we are weighting, and the constant σ is determined proportionally to the iteration length. Thereafter, the mean weighted RSSI of a tag is computed by making use of the next formula (Equation 6.4).

$$f_{Gaussian}(i) = e^{-\frac{1}{2}\left(\frac{i_c - i}{\sigma}\right)^2} \tag{6.3}$$

$$f_{\text{strength}}\left(t[i_c]\right) = \frac{\sum_{i=i_c-\Delta i}^{i_c} t[i]_{rssi} \times f_{Gaussian}(i)}{\sum_{i=i_c-\Delta i}^{i_c} f_{Gaussian}(i)} \tag{6.4}$$

where $t[i]_{rssi} * f_{Gaussian}(i)$ denotes the weighted RSSI for the ith iteration. This function receives as a parameter the RSSI of a tag to calculate the mean weighted RSSI. That parameter consists in an array ($t[\cdot]$) containing the RSSI readings for each iteration. Then, the sum of the weighted RSSI, for all iterations satisfying $i_c - i \leq \Delta i$, is divided by the total weight of the Δi readings. The constant Δi is the number of iterations considered for the RSSI mean calculation and is necessary only to limit the computation. Note that the history of RSSI values $t[\cdot]$ should be emptied when the object changes state from detected to undetected. Otherwise, the calculation could be done for very old values with new ones. For instance, if an object was undetected during Iterations 3–99, it would mean that at Iteration 101 (for $\Delta i \geq 3$) the computation would be performed on $t[\cdot] = \{i_1, i_2, i_{100}, i_{101}\}$, which obviously does not make sense.

The Δi and σ constants can be determined automatically corresponding to the reading speed (S) and the time one wants to weight. Suppose that at least one second of reading needs to be given importance. Then, when $S = 200$ ms it results in five iterations, at $S = 100$ ms ten iterations, and so on. The rule that was used in our various experiments (with different speed configurations) for sigma was $\sigma = number$

Table 6.2 Sample Weights of a Reading with Various Sigma Values

Iteration Rank	*σ = 2*	*σ = 5*	*σ = 10*	*σ = 25*	*σ = 100*
1	100.00%	100.00%	100.00%	100.00%	100.00%
2	60.65%	81.87%	90.48%	96.08%	99.00%
3	36.79%	67.03%	81.87%	92.31%	98.02%
4	22.31%	54.88%	74.08%	88.69%	97.04%
5	13.53%	44.93%	67.03%	85.21%	96.08%
–	–	–	–	–	–
11	0.67%	13.53%	36.79%	67.03%	90.48%
21	0.00%	1.83%	13.53%	44.93%	81.87%
–	–	–	–	–	–
51	0.00%	0.00%	0.67%	13.53%	60.65%
–	–	–	–	–	–
201	0.00%	0.00%	0.01%	1.91%	13.67%

of iterations/2. Table 6.2 gives sample weight values for an iteration with a few different sigma values with iterations ranked from the latest to the oldest.

6.3.2.1 Testing the Gaussian Filter

To validate that the Gaussian filter had an effect upon the accuracy of the position, we conducted a test with four antennas installed in the kitchen of our smart home. To do so, we used a basic trilateration algorithm (see the section "Trilateration") using the Friis equation (Friis 1946) without any other pretreatment filters or any aftertreatment. The experiments demonstrated that the Gaussian mean weighting filter greatly improves the accuracy. With this filter, the effect of the fluctuations on the signal strength is considerably reduced. After the application of this filter, the estimated positions are more accurate and grouped. The results can be seen in Figure 6.4.

6.4 Proximity-Based Localization

The proximity-based method uses an intuitive strategy to track the tags in a smart environment (Song et al. 2007). The idea behind this technique is to deploy a large number of antennas and to calibrate their range in order to reduce the overlapping as

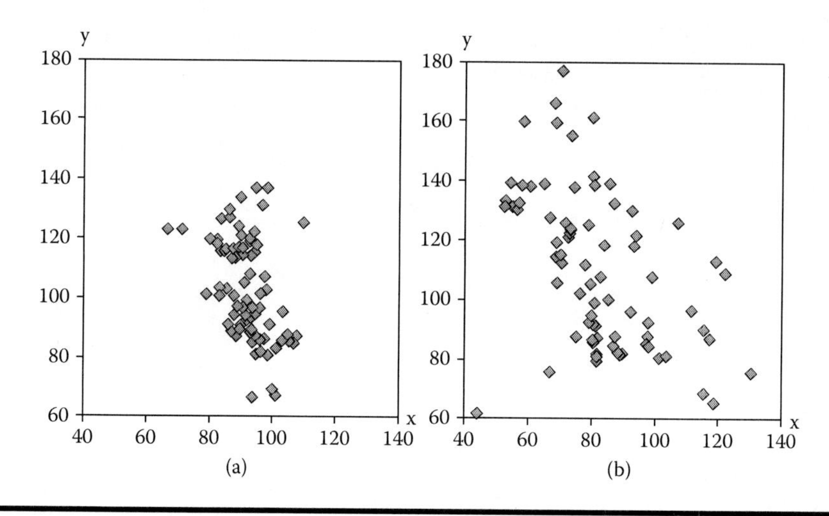

Figure 6.4 Concentration of the approximate positions of an object (a) with and (b) without the Gaussian filter.

much as possible. An object that enters an antenna detection zone is assumed to be at the same position as this antenna. When multiple antennas detect an object, the position is assumed to be at the position of the antenna receiving the most powerful signal. Therefore, the precision of the localization is proportional to the increase of the number of antennas and to the decrease of their reception range. Although apparently very basic, this technique works very well and is currently the most robust method to exploit with passive RFID tags. Another advantage of the method is the ease of installation and configuration. In fact, it was the default localization paradigm implemented at the LIARA smart home. To use it, we configured each antenna to make them cover a circular area approximately 1 meter in diameter. Their sensitivity was cut to half their full capacity, and they emitted at low power (33%). Figure 6.5 shows their normal coverage and the rough FPR range on an aerial map of the infrastructure. The reading time of the system was initially set to 750 ms, but with this localization technique time is generally not an issue. Therefore, this could be seen as another advantage if available computation power is limited.

This method was good enough to perform a few experiments on activity recognition (Bouchard et al. 2011) but also to give an assistance demonstration to health-care professionals. To this day, it is still the most accurate method that we use in our laboratory, especially with the combination of the $f_{ite}(i,O_s)$ filter. The shortcoming is the lack of precision of the information, which does not enable us to extract much knowledge (see the section "Modeling and Inferring Information from Sensor Output"). Moreover, as can be seen in Figure 6.5, eight antennas cover less than 10% of our smart home. Consequently, one would need to deploy an enormous number of antennas to cover the whole area of any big smart environment. The cost and difficulty of configuration would rise accordingly.

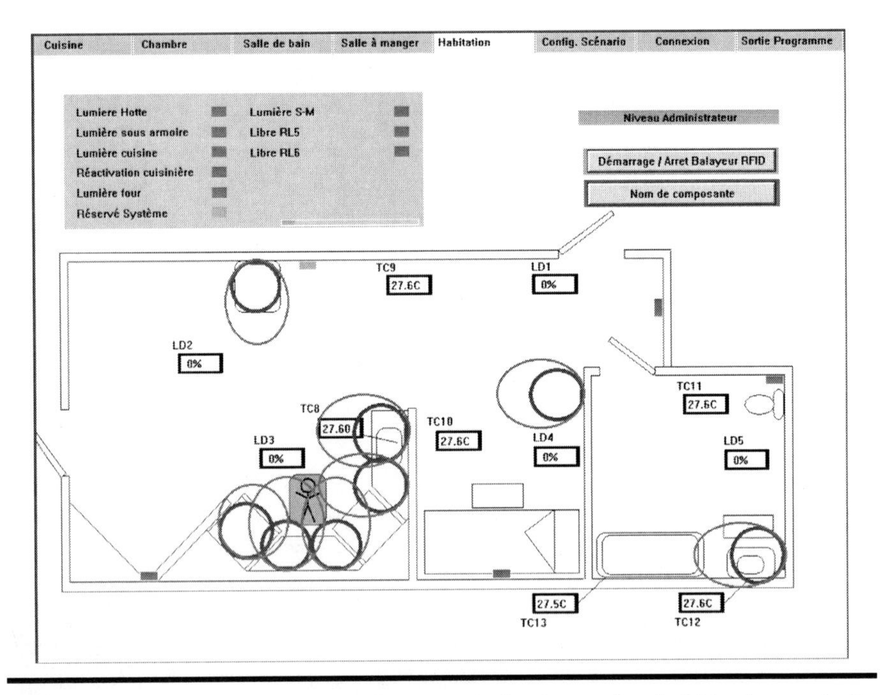

Figure 6.5 The antennas' approximate emission/reception bubbles in proximity mode (Dark gray). Gray indicates false positive areas.

6.4.1 Other Proximity Methods

Many authors have worked on the proximity method to localize entities in a smart environment. In particular, some have used technologies other than RFID to do so (e.g., infrared). In fact, proximity methods exist in a wide range of precision, technology, and coverage. Some of them extend the technique by using sophisticated filters or algorithms. One worthy of mention is the Wi-Fi-based method of Youssef et al. (2007). The idea behind their efforts was to be able to track a person in an environment equipped only with Wi-Fi access points and stations. To do so, they monitored the RSSI and analyzed the changes in the environment to correlate them with the person moving from room to room. Their concept is very promising and possesses the advantage of using cheap technology that is often already installed in working environments. Moreover, it allows following a person without the need for wearable technology. In that view, it is similar to an infrared tracking system but with a higher precision. However, many challenges remain to be addressed. Among them are the handling of noisy physical signal samples in order to obtain consistent event recognition and identification of the tracked entity. Finally, this method cannot be used to precisely track objects and thus would be better combined with another technique.

6.5 Trilateration

The method of trilateration is a well-known process to localize an object using the geometry of circles, triangles, or spheres (in 3D). Often confused with triangulation, trilateration is performed by exploiting distance measurement, in contrast to triangulation, which exploits the angle of arrival of the signal of two receivers separated by a known distance and the properties of triangles. The fundamental step to perform trilateration consists in finding the distance between the object being tracked and each antenna. The easiest way to do it with passive RFID technology is to use the RSSI. The transformation of the signal to a distance with the RSSI can be accomplished with the Friis transmission equation (Equation 6.5):

$$P_r = P_t G_t G_r \left(\frac{\lambda}{4\pi D} \right)^2 \tag{6.5}$$

where P_r, P_t, G_t, G_r, λ, and D, respectively, denote the power received by the antenna, the power of the RF emitted by the antenna, the gain of the transmitter antenna, the gain of the receiver antenna, the wavelength of the emission, and the distance from the antenna. However, in practice, this equation is often far from properly representing the wave propagation observed in smart but also busy environments. To address this situation, an alternative method could be to design a custom equation for the RFID hardware exploited. To do so, one could simply collect a data series from a tag at different positions and learn the model. One way of doing this could be to exploit mathematical regression.

The next and final step is to solve circle equations to find an intersection point. In the general case, a minimum of three circles drawn from three signals is required to find the position. However, in the special context of smart environments, antennas are often put on the walls, which make the half the surface of the circle unusable (where the second intersecting point would be). Figure 6.6 shows an example of trilateration from two antennas placed at two known positions P_1 and P_2.

Points P_1 and P_2 can be written as Cartesian coordinates, that is, $P_1 = (x_1, y_1)$ and $P_2 = (x_2, y_2)$. The aim is to find P_3 corresponding to the tracked object. The first step consists of calculating the distance between both points: $d = \sqrt{(x_2 - x_1)^2 + (y_2 - y_2)^2}$. Note that if d is bigger than the sum of both radii ($r_1 + r_2$), there is no solution, so it might be worth ensuring that the distance values are correct before performing the calculation. Then we must calculate a (or b) to find h: $a = \left(r_1^2 - r_2^2 + d^2\right)/2d$. Next, we must solve for h in order to find the center point P_4 (Equations 6.6 and 6.7), and the object position P_3 is found following Equation 6.8.

$$h = \sqrt{\left(r_1^2 - a^2\right)} \tag{6.6}$$

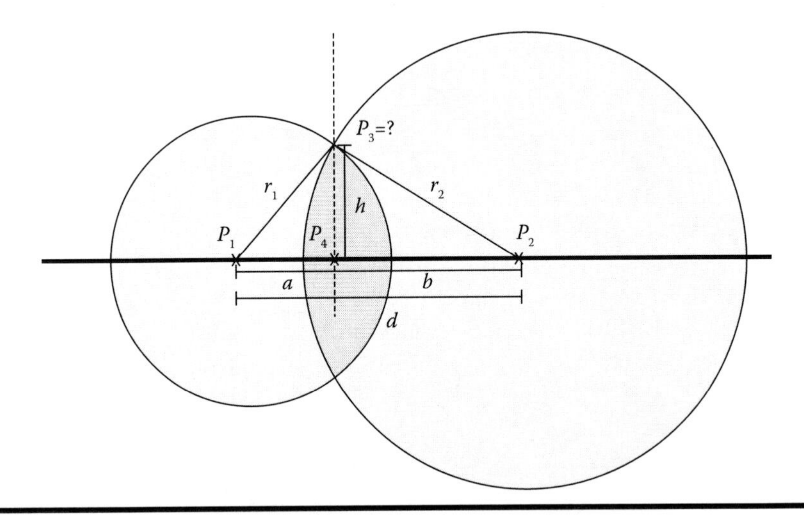

Figure 6.6 Example of trilateration with two antennas.

$$P_4 = P_1 + a(P_2 - P_1)/d \tag{6.7}$$

$$P_3 = P_4 + h(P_2 - P_1)/d \tag{6.8}$$

To conclude, when using trilateration, the system should be designed differently than with a proximity-based method. First, the antennas should be set up in such a way to ensure that at least two of them can detect the tracked objects at all times. The final configuration and arrangement of the antennas in the smart environment have a high impact upon the performance of the trilateration method and should thus be taken seriously into account. The antennas can generally be used at, or near, their full capacity (sensitivity, power). In that way, they can cover a much larger area (in our case, up to 3 meters in front of them). Additionally, trilateration is not very accurate with passive RFID. Therefore, we suggest speeding up the system as much as possible to have more data, which can be used to average the objects' trajectories (for real-time tracking). We first tested trilateration at 200 ms, but our system now supports reading cycles up to 20 ms.

6.5.1 Elliptic Trilateration

In our previous work (Bouchard et al. 2014), we proposed a method to perform an elliptical trilateration as a way to address a problem we had with circular trilateration. In fact, since our antennas are directional, the strength of the signal decreases as a function of the angle of emission. This is to be expected since the radiation pattern of such antennas looks like a sausage. In such a situation, the elliptical propagation model better approximates the real behavior and is thus more appropriate. Moreover, it is still simple to implement and possible to use in real time with a standard computer. The trilateration is based on the equation of an ellipse (Equation 6.9):

$$\frac{(x-h)^2}{A^2}+\frac{(y-m)^2}{B^2}=\mathbf{1}. \tag{6.9}$$

In this equation, A and B are the values of the major and minor axes of the ellipse and the variables h and m are the coordinates of the center of the ellipse. To compute A and B, we first have to establish the equations corresponding to the distance in function of the RSSI when the object moves away perpendicularly (major axis) and when it moves away from the side (minor axis) of the antenna.

To do this, we use the method of regression, as explained previously, which enables us to find both equations. The first one returns the value of the major axis (M_a) and the other returns the value of the minor axis (m_a), depending on the RSSI. In our case, the polynomial regression had higher correlation coefficients than the linear case (respectively, R_M^2 = 0.908 and R_m^2 = 0.909) and thus we kept the quadratic equations (Equations 6.10 and 6.11).

$$M_a(RSSI)=0.1833\times RSSI^2+8.5109\times RSSI+104.3 \quad (R_M^2=0.974) \tag{6.10}$$

$$m_a(RSSI)=0.0462\times RSSI^2+0.8155\times RSSI+104.3 \quad (R_m^2=0.937) \tag{6.11}$$

From these equations and from the respective positions of each antenna, we are now able to establish the different equations of the ellipse of the corresponding antenna simply from the RSSI received. If at least two antennas on the same wall or three on different walls detect the same tag according to the principle of trilateration, the object should hypothetically be where those ellipses intersect. In our particular context, the elliptical trilateration was implemented for the kitchen area with a set of four antennas disposed as seen in Figure 6.7. The tracked object is always seen by the four antennas, so we decided to perform trilateration for each possible pair of ellipses (a total of six). To find their intersection points, we have to solve an equation of the second or fourth degree. This depends on which pair of antennas is involved.

On one hand, when the two selected antennas are on the same wall (A1–A2, A3–A4), the equation of intersection is quadratic and is easy to solve. On the other hand, when we try to find the intersection points of two antennas located on opposite walls (A1–A3, A1–A4, A2–A3, A2–A4), we have to solve a quartic equation, and to this end we implemented the well-known Ferrari method. Therefore, we end up with five possible situations. For each pair of ellipses, we obtain between zero and four points of intersection. Obviously, the calculation of intersection points is dependent on the configuration of the antennas in the smart environment where the trilateration is being performed. Therefore, this part of the method will change depending on how many antennas there are and how they are placed.

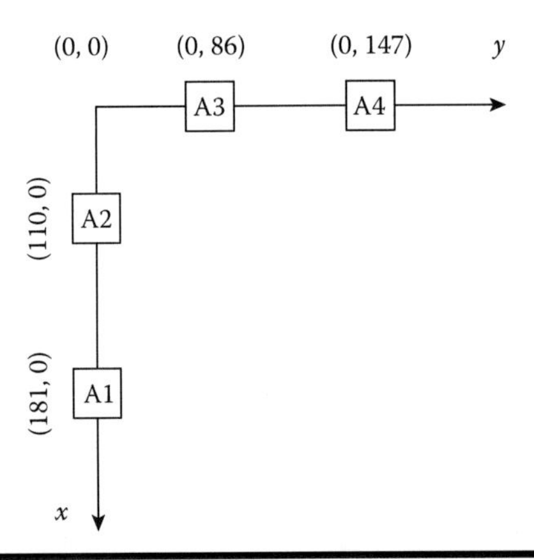

Figure 6.7 The antennas' placement and their respective Cartesian coordinates.

6.5.2 Nonintersecting Problem

One problem that might happen while performing trilateration is the lack of intersection between a pair of antennas (or more). This situation could be addressed simply by progressively increasing or decreasing the size of each ellipse or circle. The problem is that not all ellipses or circles are equally accurate. Generally, those from the antenna receiving a stronger signal are much more reliable than those with a weaker signal. Therefore, the rate of variation for each antenna should depend on the received signal strength. One thing that should be considered when progressively increasing or decreasing the ellipses or circles is that a very small increment may be time-consuming. We designed a small algorithm called the *delta filter* (Algorithm 6.1), which treats these situations and works with both standard and elliptical trilateration.

This filter treats all possible situations: the case where the major and minor axes of the two ellipses have to be increased in order to obtain a point of intersection and also the one where an ellipse covers another ellipse (or circles, which are simply a special case of ellipses). In this case, we must reduce the shape of one ellipse and increase the shape of the other one. In brief, the delta filter is used to eliminate a situation where there was no point of intersection by modifying a pair of ellipses or circles until they intersect and thereby create at least one point of intersection. The application of this filter results in one to two points of intersection for each pair of circles or one to four with ellipses. The points that correspond to a complex number or those outside the eligible area can be eliminated. If there is still more than one possible value, an arithmetic mean can be calculated to create a unique point. If the pair of antennas is not on the same wall, it might be better to keep both for the final selection or to use other criteria to select the most promising position.

Input: Two ellipses or two circles (Ellipsel and Ellipse2)
Output: One or more points of intersection

```
get delta variation value for Ellipsel(V_1) and Ellipse2(V_2)
initialize an empty table of point : P[]
repeat till P[] is empty
    Δ_1 = Δ_1 + V_1
    Δ_2 = Δ_2 + V_2
    compute intersection point for Ellipsel+Δ_1 and Ellipse2+Δ_2 and add them to P[]
    if P[] is empty then
        compute intersection point for Ellipsel-Δ_1 and
        Ellipse2+Δ_2 and add them to P[]
    end
    if P[] is empty then
        compute intersection point for Ellipsel+Δ_1 and
        Ellipse2-Δ_2 and add them to P[]
    End
End
return P[]
```

Algorithm 6.1 The delta filter.

6.5.3 Selecting the Final Position

Because of all the uncertainty from the collection of the RSSI for each antenna to the conversion either to a circular or elliptic equation, it is quite improbable that three or more ellipses or circles will converge in a unique intersection point. The simplest solution to this issue would again be the simple average. This is a method that should work fine. However, in our experiments, we observed that the intersection points obtained from antennas that have received stronger signal strength are more accurate. In order to improve the methodology, we suggest using the weighted average that we exploited in our past experiments. This average is performed through a filter called *multipoint location*, which returns a weight for each hypothetical point and is determined by the following method.

The first step is to attribute a rank to each antenna so that the antenna receiving the strongest signal is assigned the first position and so on. Next, for each pair of antennas, we set the weight depending upon the positions received for each one. Table 6.3 shows how the weights are given according to the positions assigned.

Table 6.3 The Multipoint Location Filter Matrix

Position	*1*	*2*	*3*	*4*
1		1.00	0.80	0.40
2	1.00		0.40	0.20
3	0.80	0.40		0.00
4	0.40	0.20	0.00	

For example, if a point is obtained by the intersection of two ellipses that have the strongest RSSI (Positions 1 and 2) then this one will receive a weight of 1.00.

Finally, these points and their respective weights are used to compute the actual (final) location of the tracked object. This calculation is done by using the function $f_{location}$ (Equation 6.12). In this function, $p[i]$ represents each point (x, y) and $f_{weight}(i)$ is the weight assigned to it taken from the matrix above.

$$f_{location}\left(p[\cdot]\right) = \frac{\sum_{i=0}^{n} p[i] * f_{weight}(i)}{\sum_{i=0}^{n} f_{weight}(i)} \tag{6.12}$$

6.5.4 Validation and Discussion

As we have mentioned earlier, we conducted many experiments with RFID, and we tested several forms of trilateration in our smart home. We think it would be appropriate to present some results in order to convince the reader of the relevance of trilateration. Most of our tests took place in the kitchen area and more specifically on the kitchen counter. In this part of the smart home, there are four RFID antennas that should all detect an object, wherever it is, at all times (when configured for trilateration). We chose this zone because of the nature of our other research that involved fine-grained activity recognition. We established a detailed experimental protocol to obtain meaningful results. This protocol has been divided into several steps. First, we placed the object in the center of a zone, and then we recorded all the data from the database (the RSSI returned by each tag) for at least 80 seconds (400 iterations at 200 ms). We repeated this recording process for each of the available zones of the kitchen. The data recording allowed us to compare more precisely the different algorithms by eliminating the variations in the antenna readings. Secondly, we compared the efficacy of each method with and without filters by exploiting the saved data. The results are displayed in Figure 6.8.

The first thing to note is the great enhancement enabled by the elliptical trilateration. It can be explained by the fact that our antennas are directional and

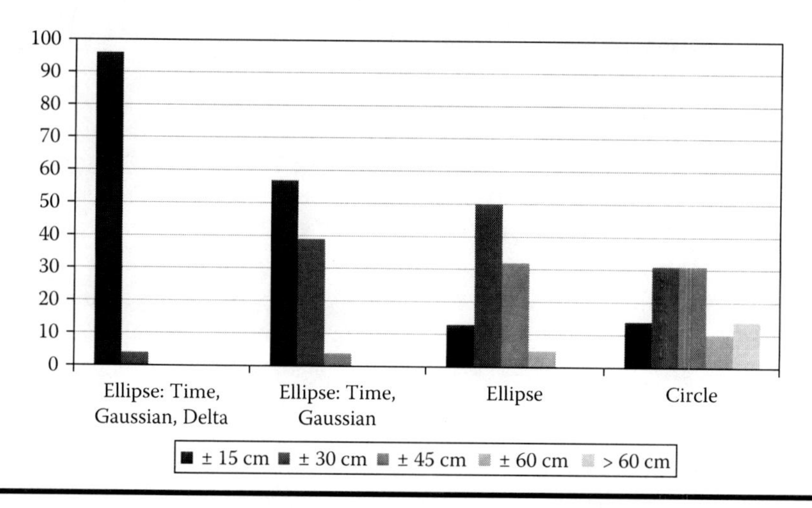

Figure 6.8 Accuracy under various configurations of the trilateration.

thus transmit their waves such that the signal loss is greater when the object moves away laterally from it. Therefore, in the lateral areas, the location success rate is really low with circles. As evidence, over more than 1,600 iterations and with an elliptical model, we obtained an average accuracy of ±14.12 cm; with the same settings but with circular trilateration, the accuracy was reduced to ±32.52 cm, which proves the effectiveness of our elliptical model. The filters added to the trilateration also greatly improve the accuracy. Often, the ellipses are very close to each other, but they do not intersect. Therefore, if we do not use the delta filter, these points would be ignored, and the location would be less accurate. On the other hand, the multipoint location filter assigns different weights to the points of intersection; it modulates their value according to their accuracy, making it possible to choose the best hypothetical position. After a full analysis of these results, we can conclude that each of these components is effective.

These experiments confirmed to us that there are many advantages to exploiting trilateration in a smart environment. First, it offers a much higher precision than proximity with only slightly inferior accuracy. Moreover, it requires about the same hardware to implement and, as a result, it generates approximately the same cost. Second, unlike proximity, trilateration enables real-time tracking of objects inside a building. Hence, it is theoretically possible to recognize ongoing actions or gestures performed with an object. Finally, a trilateration system is easy to adapt to a new environment or even new hardware. In the latest, all that will be required is to compute again the equation that converts the RSSI to a distance measure. There are also a few negative sides to trilateration. For example, the initial implementation and configuration are tedious and require an expert. It also requires much higher computation power because some mathematical equations have to be solved in real time, and the reading rate is faster.

6.6 Learning-Based Localization

In the previous sections, we presented two opposite alternatives for object localization in smart environments. Each of them had their pros and cons. There is a multitude of ways to design a localization system using proximity or trilateration, but they both require complex configuration steps only an expert can achieve. Indoor localization has been largely studied, and we are far from having covered every *family* of methods. However, there is a last method we want to discuss in this chapter: learning-based localization. That other possibility is a very different approach because it does not require elaborating a complex algorithm to transform and filter the data. The learning method takes the data without any pretreatment from the RFID system and uses the learned model to *decide* where the tag is currently. Figure 6.9 illustrates an example of how such a method can work.

Before implementing the method, the environment must be configured for this special purpose. As opposed to trilateration, the space does not need to be represented by a Cartesian space; it must be divided into logical zones. The zones should be small enough to obtain an acceptable precision, but large enough to have good accuracy. This might be the complicated part if the approximate precision is completely unknown. Furthermore, the learning phase will require more time if zones are too small. A good zone size for a commercial passive RFID system could be around 30 cm by 30 cm. Once this part is done, the learning process can begin. The first step consists of collecting supervised data on the system. To do so, a tagged object should be put in the center of a zone and the RSSI received by the antennas should be recorded. The goal is to collect labeled data samples for every defined

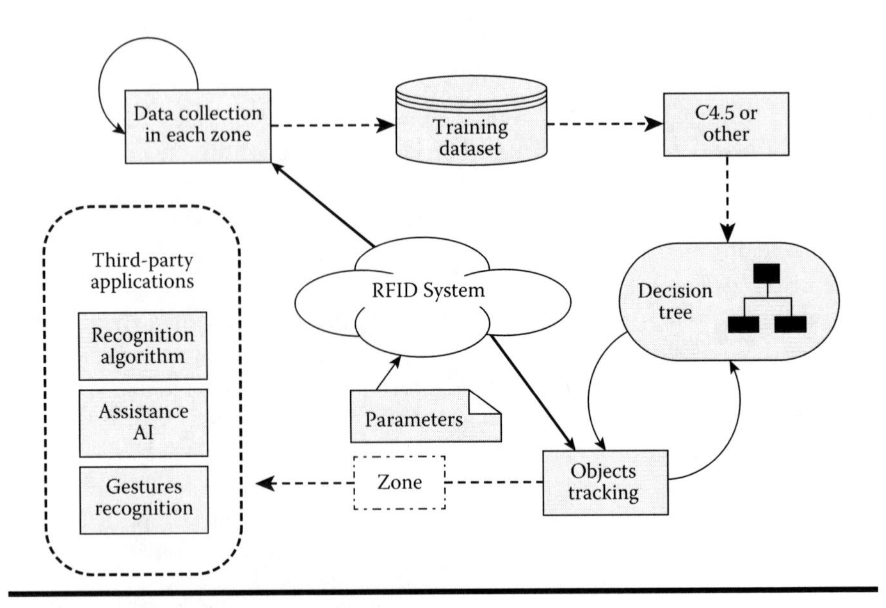

Figure 6.9 The overall learning-based localization process.

logical zone. Once this is done, the remaining tasks are fairly straightforward. It remains to choose a classifier algorithm such as C4.5 (Quinlan 1993) to automatically create a decision tree from the data. One other advantage of this method lies in the fact that it is unnecessary to implement the data-mining algorithm, since there are many versions on the Web, such as the one implemented in Weka (Hall et al. 2009). The last step in localizing an object consists of employing the extracted decision tree with the RSSI information as input. The decision tree finds the most likely area according to its built-in criteria.

The hardware setup for learning-based localization should be approximately the same as what we explained for trilateration. At least two antennas should see an object at all time, and the antennas' sensitivity and emission can be set to near the maximum level. It is less important to have a fast reading cycle, but it could nonetheless be useful to improve accuracy and enable high-level knowledge extraction. If the localization is not accurate, it might be that the logical zones are too small, so there is not enough precision in the system to build a decision tree that distinguishes one from another. There are obviously other learning methods that could be exploited, and we think that there is still a lot of work to be done on such approaches. The next subsection describes a small experiment conducted at the LIARA laboratory on an RFID localization method using C4.5 (Rocher et al. 2011).

6.6.1 Validation and Discussion

Our earliest attempt to localize objects with passive RFID technology involved a machine-learning technique (Rocher et al. 2011). This work was more generally about activity recognition, and the goal was to build a system that could be brought directly into senior citizen centers to conduct clinical trials. To do so, a support was built to place the antennas on any table in order to divide the area into a few zones. The left part of Figure 6.10 shows the physical setting of the experiment. As can be seen, four logical zones were defined where the objects could be moved during an activity. The activity to be performed by the subjects was *preparing coffee*. For every test, the objects were initially placed in the zones A0 and A1. The area A2 was where the patient was performing the activities. The area A3 was the "buffer" zone. It was made up of areas not used by other areas and aimed to reduce zoning errors. To complete an activity, the patient used the objects by taking them between the different areas.

The first phase to test this system was to exploit C4.5 on a big dataset recorded at the lab. To this end, each step of the activity (involving one of the tagged objects) was realized 10 times, and the resulting data were recorded in a database. Thereafter, 85 tests were performed for all the possible steps of the activity. In these tests, we used cheaper RFID hardware and no filter at all on the output information. Although we did not compute the localization accuracy for these experiments, the success rate of basic action recognition correlated directly with the success of localization. The success rate was excellent for actions involving one object (99%).

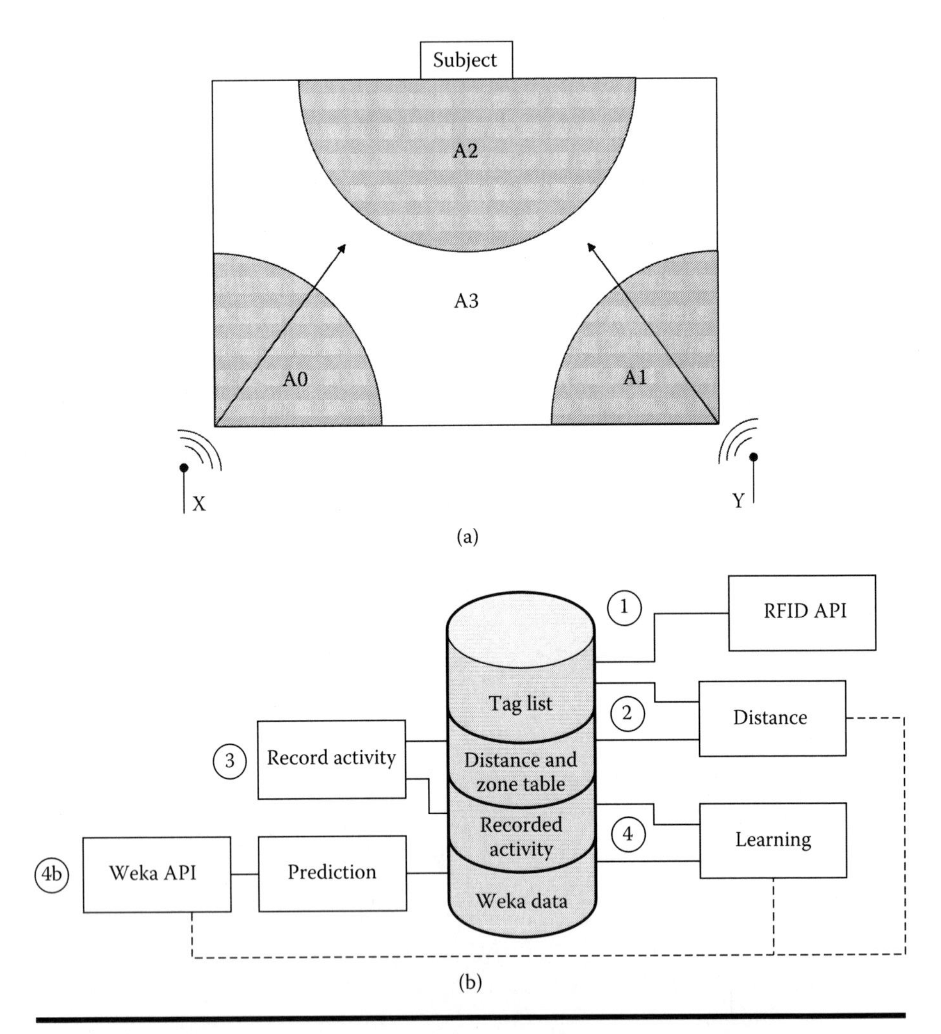

Figure 6.10 (a) The experimental setting; (b) the software architecture of the recognition platform.

However, we saw a huge decrease in performance as the number of objects increased. With two objects, it dropped to about 89% and with three objects to about 25%. This can be explained by many factors. First, the learning did not take into account the interference caused by the proximity of many objects. Secondly, the RFID system was low-end and no pretreatment was done on the data. The right part of Figure 6.10 illustrates the overall software architecture for the whole recognition platform including the learning part.

That first experiment with RFID and machine learning was interesting, mostly because it was a new avenue of solution to this important challenge in smart environments research. The main advantage of a learning-based localization model is that very few adjustments are required in order to make the localization operational. Indeed, the method is somewhat like a black box where one enters a training dataset and an intelligent agent capable of doing the work is obtained. Consequently, a nonexpert could do the installation simply by following a learning protocol. However, the learning phase can take a considerable amount of time, especially for the data collection phase. Moreover, the learning has to be conducted every time new hardware is installed and every time the localization environment changes. By contrast, one can assume that trilateration would work anywhere with the same configuration (at least if using the same hardware). A lot more work has to be done in this field to get an interesting learning-based solution. In particular, the development of an unsupervised method would provide new benefits since it would not require labeling of the learning dataset.

6.7 Modeling and Inferring Information from Sensor Output

The localization of objects or persons in a smart environment usually hides a more important goal (activity recognition, assistance, service delivery, context awareness) that cannot be reached without interpreting the meaning of the data collected. An important aspect that is generally neglected is thus the discussion surrounding the type of knowledge to be extracted from the basic information provided by each family of methods. Depending on the purpose of the localization, some of them might not be useful and other might not be expressive enough. In this section, we will provide an overview of some aspects of representation of the information. Toward that goal, Table 6.4 summarizes the types of usable information for the families of methods that are presented throughout this chapter.

6.7.1 Main Types of Information

The first information type is the position in Cartesian coordinates. To obtain it, a relative Cartesian space must be established in the smart environment. It is generally a good idea to make the units correspond to real distance values. For example, 1 meter in the environment might be 100 units in the virtual Cartesian space. That position information is very important because it can serve as the basis for calculation of many others such as the acceleration and the speed of an object. With trilateration, the Cartesian position is usually the normal output. The two other methods can simulate a real Cartesian position in a limited way. For the proximity method, it can be done by converting the RSSI to approximate the distance in front of the antenna that sees the object. In the case of the learning method, it would correspond to the center of the zone the object was allocated to. Consequently, for these methods, speed and acceleration could also be estimated, but it would not be accurate for the proximity method. These particular data can be exploited to extract high-level knowledge. For example, if the speed of an object is not null

Table 6.4 Comparison of Knowledge Extractable from Each Method

Method	*Real Position*	*Zones*	*Acceleration*	*Speed*	*Relative Position*	*Relative Distance*	*Direction*	*Gestures*
Proximity		Q						
Learning		Q	Q	Q	Q	Q	Q	
Trilateration		Q						

Notes: Dark gray = impossible, gray = yes, light gray = limited, Q = qualitative only.

(with a degree of uncertainty), it means the object is currently used (activity inference). However, if the speed is very high and the acceleration is changing a lot on one or more object, that could mean the resident is stressed. Analysis of the patterns in speed and acceleration could lead to a better understanding of the moods of humans and thus the context of activities.

Association with a zone might also be used with trilateration and has a few positive effects. First, associating an object with a zone will make its position more stable over time and will therefore make tracking patterns easier to analyze. For example, direction, relative position, and relative distance can be computed with the real position but are more meaningful and interesting when used qualitatively. The relative position describes whether the object has moved but would rarely be equal to zero in a quantitative format. Relative distance can be used to compare the topological relationships of objects. A decreasing distance between two objects could mean grouping is underway and thus that both object are implied in the current activity (Bouchard et al. 2011).

The remainder of this section discusses some initial work on RFID for gesture recognition (Asadzadeh et al. 2012). If this becomes possible, that new knowledge could lead to a wide range of applications such as human–smart environment nonverbal interaction, better activity recognition, or even analysis of performance in handling some objects. We also think that for many of these types of information fuzzy logic (Chen et al. 2010; Zadeh 2008) could bring a significant contribution. The section "Gesture Recognition" shows an applicative example implemented at the LIARA laboratory.

6.7.2 Gesture Recognition

One application of localization from passive RFID would be to try to recognize gestures performed while handling an object. This capability would have several applications in the field of human–computer interaction but also for control of and interaction with smart environments. Moreover, this would lead to a better understanding of the current activities of users (employees, residents, etc.) and thus would enable an intelligent agent to predict but also to react promptly whenever appropriate. Unfortunately, there are very few studies in the literature addressing this challenge. Moreover, those that can do it exploit different positioning methods than we have presented throughout this chapter. We still want to present what exists currently, in particular the work of Asadzadeh et al. (2012), and to discuss the future surrounding gesture recognition from passive RFID.

Asadzadeh et al. (2012) investigated a partitioning localization technique with reference tags. Approaches that use reference tags are the most popular due to their great precision and because RFID localization first started in robotics. With three antennas on a desk, they monitored an 80 cm by 80 cm area, which was divided into 64 equally sized square cells (10 cm by 10 cm). To recognize a gesture made by a user, they make a few assumptions about the sequence of traversed cells. First, the system is fast enough to never miss any cell of the sequence; that is, the tracked object cannot move farther than one cell away in between two readings.

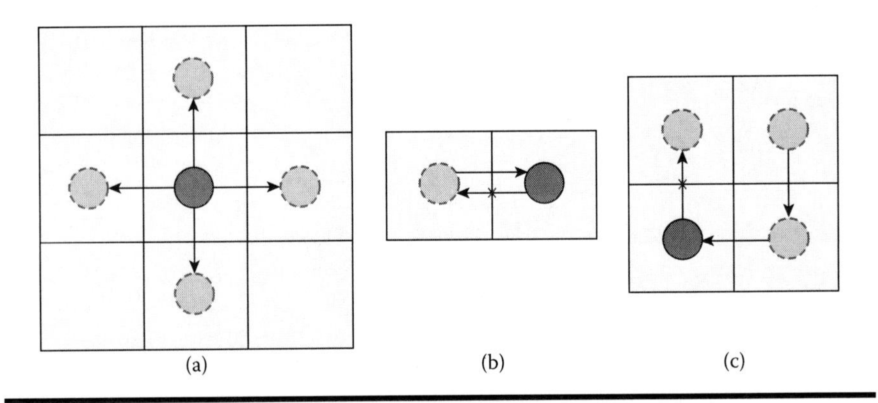

Figure 6.11 Gesture recognition: (a) legal move; (b, c) illegal moves.

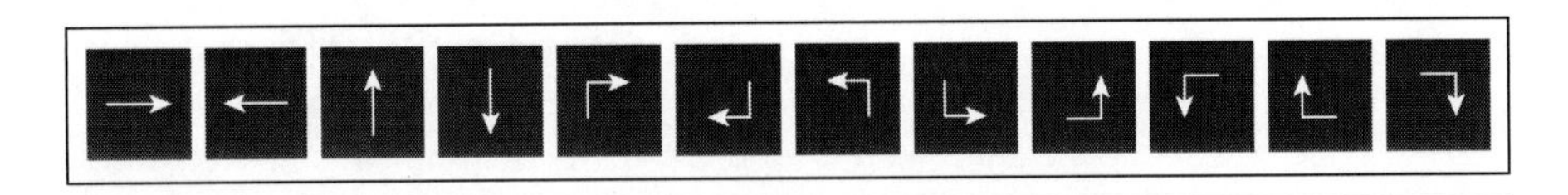

Figure 6.12 The twelve gestures recognized in their system.

Second, they assume that only forward local moves are possible. Figure 6.11a shows a legal move and Figure 6.11b and c shows illegal moves.

From the sequence of crossed cells, their algorithm generates a list of hypotheses by developing the possibilities into a tree structure. Next, a gesture matcher, GESREC, looks into a dictionary and finds the gesture that best matches the sequence. Their algorithm cannot recognize two consecutive gestures (no segmentation) but works well (93% recognition) on a dictionary of twelve gestures. Figure 6.12 shows the dictionary of gestures.

Their work showed that there is potential for gesture recognition with passive RFID, even if the gestures recognized were very simple. As the techniques and the technology improve, we predict that more research teams will further investigate this avenue.

6.7.3 Zone Association with Fuzzy Logic

We have already justified why attributing a zone to an object (instead of exploiting the real Cartesian position) could be interesting. The assignment of an object to a zone is truly intuitive; if the position is inside zone Z_a then the object is inside Z_a. However, due to the uncertainty of localization, there are many cases where the associated zone will be wrong. For example, it is not clear that an object placed at coordinates (29, 0) of zone $Z_a = [0 \leq x < 30, 0 \leq y < 30]$ is not in the next zone. In fact, if we have two propositions "$O_1 \subseteq Z_a$" and "$O_1 \subseteq Z_b$" it is not clear that they are completely true or completely false. Such situations where propositions are affected by a degree of truth are easily dealt with by fuzzy logic.

To do so, we suggest the implementation of Mamdani's fuzzy inference method. This method uses three fuzzy linguistic variables (FLVs) that enable us to reason about zone membership. The first variable represents the knowledge we are trying to infer. The *likeliness* of an object's membership in a zone ranges from 0 (not likely) to 100 (very likely). The reasoning behind the likeliness is made from the two other FLVs. The second shows how strongly an object is inside a zone. This is given by comparing the approximate Cartesian position determined from trilateration with the Cartesian coordinates of the center of the zone for which the likeliness is being calculated. For example, if the object is positioned directly in the center of zone Z_a, we can say that it is *near* at a 100% degree of truth. The last FLV uses information about the last appearance of the tracked object in a zone (new, recent, old). We propose to use that information because a tag that has been recently seen in a zone will more probably be in that zone at the next iteration than in a zone it never went to before. Figure 6.13 shows these FLVs.

To concretely use the FLVs within a fuzzy inference engine, we defined nine rules that cover all possible situations.

Rule 1: IF near AND new THEN very_likely
Rule 2: IF near AND recent THEN very_likely
Rule 3: IF near AND old THEN likely
Rule 4: IF middle AND new THEN very_likely
Rule 5: IF middle AND recent THEN likely
Rule 6: IF middle AND old THEN not_likely
Rule 7: IF far AND new THEN likely
Rule 8: IF far AND recent THEN not_likely
Rule 9: IF far AND old THEN not_likely

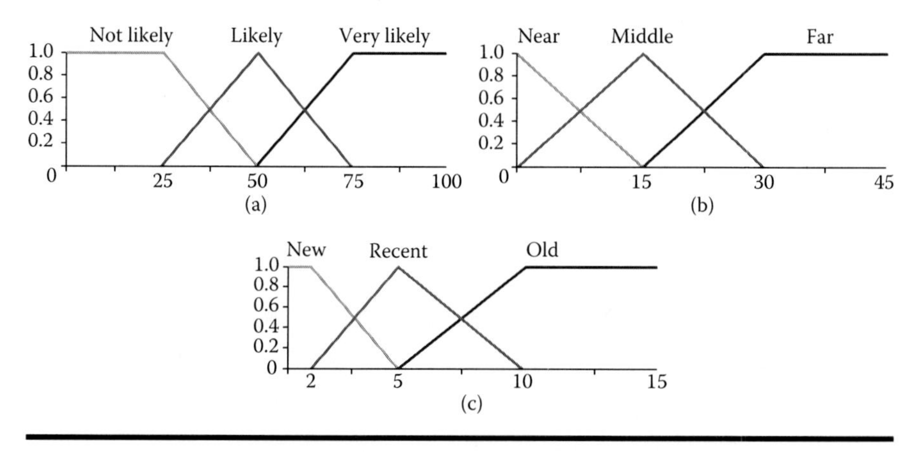

Figure 6.13 The fuzzy linguistic variables (a) likeliness, (b) distance, and (c) last detection of an object.

6.7.3.1 Evaluation of the Rules

For each logical zone, we evaluate the likeliness of the object being inside it. If performance is a concern, it is easy to design a heuristic to eliminate the very unlikely zones, considering that an object cannot travel more than a certain distance between two iterations. The steps are the same for each zone. The first step is input fuzzification. This is completed by taking the given inputs (crisp values) and evaluating the antecedent of each rule. To illustrate this process, let us suppose we are tracking O_1 and we want to know the likeliness of it being inside $Z_c = [30 \leq x < 60, 30 \leq y < 60]$. The center of the zone is positioned at (45, 45) of the origin. The result of trilateration on O_1 has given the hypothetical coordinates (45, 52.5) and its last appearance in the area dates back to four iterations. The distance from the zone's center is 7.5 cm. For the first rule, the membership value of 7.5 in the *near* set is 0.5 and the one of the *new* set with the value 4 is 0.2. If the membership equals zero, we say that the rule does not fire. Because the rules use the AND operator, we take the minimum of these values. Given the consequent of each rule and the antecedent value obtained in Step 1, we apply a fuzzy implication operator (minimum) to obtain a new fuzzy set. Figure 6.14 illustrates the complete process.

We repeat this for the other eight rules. For example, for the second rule the membership value of 7.5 in the *near* set is 0.5 and the membership of 4 in the *recent* set equals 0.8. Therefore, the rule fires with the smallest value (*Very_likely* = 0.5). For that example, the inference process would give us the associative matrix in Table 6.5.

6.7.3.2 Aggregation of Conclusions and Defuzzification

Once each of the rules has been evaluated, the conclusions obtained have to be aggregated into a single fuzzy set. To do so, a fuzzy aggregation operator is used. In our case we use the maximum operator. This will result in a set combining

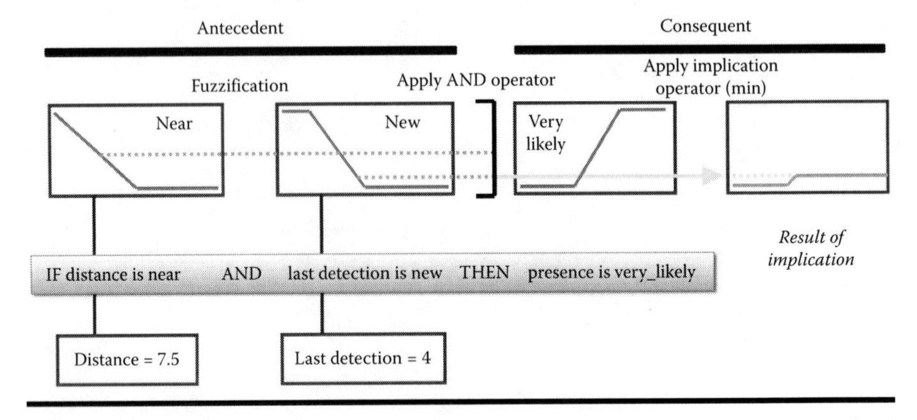

Figure 6.14 Evaluation of the first rule with the crisp value *distance* = 7.5 cm and *last detection* = 4.

Table 6.5 Fuzzy Associative Matrix

	Near	*Middle*	*Far*
New	Very Likely 0.2	Very Likely 0.2	Likely 0
Recent	Very Likely 0.5	Likely 0.5	Not Likely 0
Old	Likely 0	Not Likely 0	Not Likely 0

the values *Very_likely* = 0.5, *Likely* = 0.5, and *Not_likely* = 0. However, because we are trying to solve a decision problem, we want to obtain a crisp value (number) and not a fuzzy set. Therefore, the result of the aggregation of conclusion is transformed by proceeding to defuzzification. That can be accomplished with various methods such as the centroid, constraint decision defuzzification, and so on. For this example, we chose to use the average of maxima (MaxAv) method, which provides a good tradeoff between precision and calculation complexity. The equation for this method is shown in Equation 6.13.

$$MaxAv() = \frac{\sum rp \times \text{confidence}}{\sum \text{confidence}} \tag{6.13}$$

This method uses the average representative value (rp) of each fuzzy set of a FLV. For a triangular set, it is simply the peak. For a shoulder, it is the average value between the minimum value of the plateau and the maximum. Therefore, in our case, from the first FLV shown on Figure 6.11, $Very\_likely_{rp} = 87.5$, $Likely_{rp} = 50$, and $Not\_likely_{rp} = 12.5$. By using the confidence values of the fuzzy associative matrix (Table 6.5), the likeliness of O_1 belonging to Z_c would be given by Equation 6.14.

$$Likeliness = \frac{87.5 \times 0.5 + 50 \times 0.5}{0.5 + 0.5} = 68.75 \tag{6.14}$$

This short example shows how simply Mamdani's fuzzy inference works. For further explanation, see a reference manual such as that by Mendel (2000).

6.8 Summary

RFID technology has emerged as a winning combination for the implantation of advanced services within what we call *smart environments*. Used adequately, this technology could be used for tracking everyday life objects in real time because a wide range of passive tags, are the size of a label and can be stuck anywhere. These tags and the technology surrounding them are not expensive and offer many advantages such as robustness that make them withstand daily usage for years. As they rise in popularity, they still suffer from an underlying imprecision, which causes much uncertainty to arise while using their services. In particular, there is instability in the readings of the tags and in the returned RSSI. This chapter has explored a few of the methods that seek to partially address these challenges. It also discussed three of the most interesting methods that can be applied for localization of objects in smart environments without requiring the installation of references tags everywhere: proximity, trilateration, and learning-based localization. To increase the value, we include experiments that were conducted within a smart home and we review the positive and negative elements for each method. This chapter also discusses the various ways RFID technology and, more specifically, localization can be exploited to increase knowledge of the environmental context. Indeed, knowledge is crucial toward better service delivery and solving important challenges of the smart environment field of research.

Acknowledgments

The authors would like to thank their main financial sponsors: the Natural Sciences and Engineering Research Council of Canada, the Quebec Research Fund on Nature and Technologies, and the Canadian Foundation for Innovation. The authors would also like to acknowledge the contribution of the Centre de Santé et Services Sociaux of La Baie, the Maison Le Phare of Jonquière, and their regional Alzheimer Society for helping them recruit the participants. Finally, special thanks to our neuropsychologist partner and her graduate students, who indirectly worked on this project by supervising the clinical trials with patients.

References

Addlesee, Mike, Rupert Curwen, Steve Hodges, Joe Newman, et al. 2001. Implementing a Sentient Computing System. *IEEE Computer* 34 (8):50–56.

Asadzadeh, Parvin, Lars Kulik, and Egemen Tanin. 2012. Gesture Recognition Using RFID Technology. *Personal Ubiquitous Computing* 16 (3):225–234.

Bouchard, Kevin, Bruno Bouchard, and Abdenour Bouzouane. 2011. Qualitative Spatial Activity Recognition Using a Complete Platform Based on Passive RFID Tags: Experimentations and Results. In *9th International Conference On Smart Homes and Health Telematics*, edited by B. Abdulrazak, S. Giroux, B. Bouchard, H. Pigot, and M. Mokhtari. Montreal: Springer.

Bouchard, Kevin, Dany Fortin-Simard, Sebastien Gaboury, Bruno Bouchard, et al. 2014. Accurate Trilateration for Passive RFID Localization in Smart Homes. *International Journal of Wireless Information Networks* 21 (1):32–47.

Brusey, James, Christian Floerkemeier, Martyn Fletcher, and Mill Lane. 2003. In Workshop on Reasoning with Uncertainty in Robotics at IJCAI, pp. 23–30, 2003.

Chawla, Kirti, and Gabriel Robins. 2011. An RFID-based Object Localisation Framework. *International Journal of Radio Frequency Identification Technology and Applications* 3:2–30.

Chen, Chih-Yung, Jen-Pin Yang, Guang-Jeng Tseng, Yi-Huan Wu, et al. 2010. An Indoor Positioning Technique Based on Fuzzy Logic. *Paper read at MultiConference of Engineers and Computer Scientists,* March 17–19, Hong Kong, China.

Choi, Byoung-Suk, and Ju-Jang Lee. 2009. Mobile Robot Localization in Indoor Environment using RFID and Sonar Fusion System. In *Proceedings of the 2009 IEEE, RSJ International Conference on Intelligent Robots and Systems.* St. Louis, MO: IEEE Press.

Cook, Diane J., and M. Schmitter-Edgecombe. 2009. Assessing the quality of activities in a smart environment. *Methods of Information in Medicine* 48:480–485.

Demerism, George, Brian K. Hensel, Marjorie Skubic, and Marilyn Rantz. 2008. *Senior Residents' Perceived Need of and Preferences for "Smart Home" Sensor Technologies.* Vol. 24. Cambridge: Cambridge University Press.

Fishkin, Kenneth P., Bing Jiang, Matthai Philipose, and Sumit Roy. 2004. I Sense a Disturbance in the Force: Unobtrusive Detection of Interactions with RFID-tagged Objects. *Intellectual Property* 3205:268–282.

Friis, Harald T. 1946. A Note on a Simple Transmission Formula. *Proceedings of the IRE* 34:254–256.

Fu, Qing, and Retscher Guenther. 2009. *Active RFID Trilateration and Location Fingerprinting Based on RSSI for Pedestrian Navigation.* Vol. 62. Cambridge: Cambridge University Press.

Hahnel, Dirk, Wolfram Burgard, Dieter Fox, Ken Fishkin, and Matthai Philipose. 2004. Mapping and Localization with RFID Tags. *Paper read at Procof the IEEE International Conference on Robotics Automation ICRA,* New Orleans, LA.

Hall, Mark, Eibe Frank, Geoffrey Holmes, Bernhard Pfahringer, et al. 2009. The WEKA Data Mining Software: An Update. *SIGKDD Explorations Newsletter* 11 (1):10–18.

Hauschildt, Daniel, and Nicolaj Kirchhof. Improving indoor position estimation by combining active TDOA ultrasound and passive thermal infrared localization. In 8th Workshop on Positioning Navigation and Communication (WPNC). *IEEE* 2011:94–99.

Heesung, Chae, and Han Kyuseo. 2005. Combination of RFID and Vision for Mobile Robot Localization. Paper read at Intelligent Sensors, Sensor Networks and Information Processing Conference, 2005. *Proceedings of the 2005 International Conference on,* 5–8 December. 2005.

Hekimian-Williams, Cory, Brandon Grant, and Piyush Kumar. 2010. Accurate Localization of RFID Tags Using Phase Difference. *2010 IEEE International Conference on RFID IEEE RFID* 2010:89–96.

Hoey, Jesse. 2006. Tracking Using Flocks of Features, with Application to Assisted Handwashing. *British Machine Vision Conference BMVC* 1:367–376.

Jin, Guang-yao, Xiao-yi Lu, and Myong-Soon Park. 2006. An Indoor Localization Mechanism Using Active RFID Tag. In *Proceedings of the IEEE International Conference on Sensor Networks, Ubiquitous, and Trustworthy Computing—Vol. 1 (SUTC'06)—Volume 01: IEEE Computer Society.*

Joho, Dominik, Christian Plagemann, and Wolfram Burgard. 2009. Modeling RFID Signal Strength and Tag Detection for Localization and Mapping. In *Proceedings of the 2009 IEEE International Conference on Robotics and Automation.* Kobe, Japan: IEEE Press.

Kim, Kwangsoo, and Myungsik Kim. 2012. RFID-based Location-sensing System for Safety Management. *Personal Ubiquitous Computing* 16 (3):235–243.

Lei, Yang, Cao Jiannong, Zhu Weiping, and Tang Shaojie. 2012. A Hybrid Method for Achieving High Accuracy and Efficiency in Object Tracking using Passive RFID. Paper read at Pervasive Computing and Communications (PerCom), *2012 IEEE International Conference on*, 19–23 March 2012, Lugano, Switzerland.

Mendel, Jerry M. 2000. *Uncertain Rule-Based Fuzzy Logic Systems: Introduction and New Directions*. Prentice Hall, Upper Saddle River, NJ.

Milella, Annalisa, Donato Di Paola, Grazia Cicirelli, and Tiziana D'Orazio. 2009. RFID Tag Bearing Estimation for Mobile Robot Localization. *Paper read at International Conference on Advanced Robotics* (*ICAR*), 22–26 June 2009, Munich, Germany.

Ni, Lionel M., Yunhao Liu, Yiu Cho Lau, and Abhishek P. Patil. 2004. LANDMARC: Indoor Location Sensing Using Active RFID. *ACM Wireless Networks* 10 (6):701–710.

Parr, Andreas, Robert Miesen, Fabian Kirsch, and Martin Vossiek. 2012. A Novel Method for UHF RFID Tag Tracking based on Acceleration Data. *Paper read at IEEE International Conference on RFID* (*RFID*), 3–5 April 2012, Orlando, FL.

Patterson, Donald J., Dieter Fox, Henry Kautz, and Matthai Philipose. 2005. Fine-Grained Activity Recognition by Aggregating Abstract Object Usage. In *Proceedings of the Ninth IEEE International Symposium on Wearable Computers: IEEE Computer Society*.

Phua, Clifton, Jit Biswas, Andrei Tolstikov, Victor Siang Fook Foo, et al. 2009. Plan Recognition based on Sensor Produced Micro-Context for Eldercare. *Paper read at Proceedings of The First International Workshop on Context—Awareness in Smart Environments: Background, Achievements and Challenges* (*CASEbac 2009*), Toyama, Japan.

Quinlan, J. Ross. 1993. *C4.5: Programs for machine learning*. Morgan Kaufmann, San Mateo, CA.

Ramos, Carlos, Juan Carlos Augusto, and Daniel Shapiro. 2008. Ambient Intelligence: The Next Step for Artificial Intelligence. *IEEE Intelligent Systems* 23 (2):15–18.

Ricquebourg, Vincent, David Durand, Logé Delahoche, David Menga, et al. 2006. The Smart Home Concept: Our Immediate Future. *Paper read at 1ST IEEE International Conference on E-Learning in Industrial Electronics*.

Robles, Rosslin John, and Tai-Hoon Kim. 2010. Applications, Systems and Methods in Smart Home Technology: A Review. *International Journal of Advanced Science and Technology* 15:37–48.

Rocher, Pierre-Olivier, Bruno Bouchard, and Abdenour Bouzouane. 2011. A New Platform to Easily Experiment Activity Recognition Systems Based on Passive RFID Tags: Experimentation with Data Mining Algorithms. *International Journal of Smart Homes* 1–18.

Sample, Alanson P., Craig Macomber, Jiang Liang-Ting, and Joshua R. Smith. 2012. Optical Localization of Passive UHF RFID Tags with Integrated LEDs. *Paper read at RFID (RFID), 2012 IEEE International Conference on*, 3–5 April 2012.

Song, Jongchul, Carl T. Haas, and Carlos H. Caldas. 2007. A Proximity-based Method for Locating RFID Tagged Objects. *Advanced Engineering Informatics* 21 (4):367–376.

Vorst, Philipp, Sebastian Schneegans, Yang Bin, and Andreas Zell. 2008. Self-Localization with RFID Snapshots in Densely Tagged Environments. *Paper read at IEEE/RSJ International Conference on Intelligent Robots and Systems* (*IROS*), 22–26 September 2008, Nice, France.

Weiser, Mark. 1991. The Computer for the 21st Century. *Scientific American* 265 (3):66–75.

Youssef, Moustafa, Matthew Mah, and Ashok Agrawala. 2007. Challenges: Device-free Passive Localization for Wireless Environments. In *Proceedings of the 13th Annual ACM International Conference on Mobile Computing and Networking*. Montréal, Québec: ACM.

Zadeh, Lotfi A. 2008. Is There a Need for Fuzzy Logic?*Information Science* 178 (13):2751–2779.

Zhang, Yimin, Moeness G. Amin, and Shashank Kaushik. 2007. Localization and Tracking of Passive RFID Tags Based on Direction Estimation. *International Journal of Antennas and Propagation* 2007:17426.

Zhou, Junyi, and Jing Shi. 2009. *RFID Localization Algorithms and Applications—A Review*. Vol. 20. Heidelberg: Springer.

Chapter 7

Smart Homes: Practical Guidelines

Kevin Bouchard, Bruno Bouchard, and Abdenour Bouzouane

Contents

7.1 Introduction

Smart homes have become a very active topic of research in ambient intelligence in the last decades primarily due to advances in engineering and technology (Ramos et al. 2008). Scientists and private corporations around the world are working on this paradigm, either to make life easier for habitants or to provide services to a particular target audience. Many researchers (Boger et al. 2006; Nugent et al. 2007; Roy et al. 2009), including our team at the Laboratoire d'Intelligence Ambiante pour la Reconnaissance d'Activités Laboratoire d'Intelligence Ambiante pour la Reconnaissance d'Activités (LIARA) (Bouchard et al. 2012b), believe that smart homes could be exploited to provide support services to vulnerable persons and those stricken by cognitive impairment. However, smart homes that are implemented and used in real contexts are actually not that "smart" (Augusto and Nugent 2006). That is, artificial intelligence (AI) is one of the key components missing to turn that technology into useful services (Robles et al. 2010). For example, it is difficult to interpret raw data perceived from environmental sensors in order to target the assistance needs of a resident and to be able to select an appropriate technology adapted to assist him or her considering the individual's profile (Van Tassel et al. 2011). It is also difficult to monitor the smart home itself to keep its network running smoothly at all times. To accomplish the goal of smarter smart homes, researchers need to conduct larger experiments on prototypes inside real smart homes, but they also have to improve their collaboration with teams around the world. The design and implementation phase of a smart home is quite complex. These very expensive projects are generally completely built from scratch and are often the first attempt of laboratories (lacking experience) resulting in repetition of design errors.

We believe that a basic guide discussing the main issues would be useful to allow researchers to promptly begin the design of their smart home and to give them a new perspective on certain issues. Our team, composed primarily of AI specialists, has implemented projects through the years on four major smart environment infrastructures worldwide (Bouchard et al. 2008; Phua et al. 2009; Roy et al. 2010), including our own recently built full-size smart home. From that experience, we propose, in this chapter, a set of guidelines for designing and implementing an efficient smart home architecture, both hardware and software, specifically oriented for cognitive assistance, our main applicative background. The objective of this chapter is to facilitate research by the following:

1. Reviewing hardware/software for rapid prototyping
2. Giving a smart home use case construction and design choices
3. Promoting collaboration and exchange for incremental advances

The objective of these guidelines is to help research teams getting started with the design phase but also to prevent the repetition of design errors that might have profound consequences. A predicted second effect will be to shorten the conception phase, which may prove to be a tedious and expensive operation. The chapter studies and details the case of our newly built intelligent home infrastructure. It also describes the software developed to operate it and the most important experiments conducted at the laboratory from the contextual goal of this chapter. The architectural choices are described along with our own technological options and software techniques, which were selected to enable rapid prototyping of AI and networking algorithms. The remainder of the chapter is structured as follows. The next section ("Smart Homes") introduces important notions tied to smart home research and the goals of these guidelines. The section "Smart Home Hardware" describes the challenges to face in the design of the hardware portion of smart homes. In particular, the criteria to take into account when choosing sensors are discussed, accompanied by a description of the advantages and disadvantages of each type. It also gives information about communication technologies and about the importance of the effectors. The section "Smart Home Software" introduces the crucial software functionalities that should be implemented toward the goal of rapid prototyping focusing particularly on AI. Some important criteria for software development for smart homes are also discussed. The section "The LIARA Smart Home Case Study" fully describes our smart home from both the hardware and software sides. Finally, the section "Promoting Collaboration between Researchers" discusses the importance of promoting international exchange and collaboration between researchers and describes our open tools project, which serves that goal (Bouchard et al. 2012).

7.2 Smart Homes

The development of guidelines for designing smart homes is not a simple task. A large number of factors must be taken into account. Why is the smart home being built? How much money is available? What type of services need to be provided (automation, assistance, etc.)? Nevertheless, before even trying to answer all these questions, a definition of the term *smart home* must be provided. Originally, a smart home was simply a house with automated environmental systems such as lightning and heating control features. The word *smart* was widely used for any technological feature in a house that could automate simple tasks. Nowadays, almost any electrical house components can be included in the system (Robles et al. 2010), and a wide range of sensors are now within reach of public buildings and residential houses.

Smart homes are used for several purposes. They can improve comfort at home, reduce energy consumption, and enable automation of household chores. They can provide better quality entertainment by adapting to the preferences of residents. However, many scientists, like us, believe that the field of smart homes will reach its full potential by providing health assistance to impaired or frail persons. This very interesting application could help residents to remain autonomous in their homes for an extended time. It would reduce the workload of caregivers and ease the anxiety of families not able to monitor activities. Such technology would not only help the resident directly but also produce reports for physician or allow instant monitoring.

Many teams, such as ours at the LIARA, try to build assistive smart homes more precisely for people with cognitive impairments such as Alzheimer's disease (AD). This is a particularly challenging application for smart homes because a resident may act incoherently with respect to his or her goals and may need to be assisted in the activities of daily living (ADLs). Moreover, it cannot be assumed that the technologies will be used correctly and therefore foolproof systems and networks are needed. Although the precise context of our research might be considered one of the most challenging instances of assistive smart homes, the researchers are linked by the same issues that prevent the practical implementation of these projects in the real world. Smart homes need to be more than an aggregation of sensing technologies. They need to become smarter (Augusto and Nugent 2006) in a broad sense. It can be synthesized by the following questions: How can we create algorithms able to recognize a resident's goals and ongoing activities of daily living (Bouchard et al. 2007; Tim van Kasteren et al. 2010)? How can the appropriate moment be identified for providing help and adapted guidance (Van Tassel et al. 2011)? How can the system automatically adapt to the resident's habits (Lapointe et al. 2012)? How can the network and the technology be built to support failure and to be robust?

In this chapter, we adhere to the definition of a *smart home* as an enhanced house designed to assist and help a resident in his or her everyday activities. However, we focus on the assistance of cognitively impaired persons because we are pursuing the goal of building a home with helping and teaching capabilities that will enable longer autonomous living. The guidelines are still kept general and applicable to other smart home contexts. It is based upon the experience our researchers accumulated over the past decade with four major smart home infrastructures worldwide (Bouchard et al. 2008; Phua et al. 2009; Roy et al. 2010): in Sherbrooke (Canada), in Toronto (Canada), in Singapore, and our newly built smart home in Chicoutimi (Canada) (Bouchard et al. 2012).

7.3 Smart Home Hardware

To build a smart home, it is necessary to first choose the kind of technology and hardware to integrate into it. Of course, there are many ongoing intelligent house projects around the world such as the MavHome (Cook et al. 2003), the eHome (Kaila et al. 2008), and the House_n project (Intille 2002) from which knowledge can be gained. The problem is that none clearly provides information about their creation process and hardware choices. Furthermore, it is unrealistic to assume, as many teams do, that smart homes will only be deployed via the construction of new houses. Older houses may present different challenges and are harder to deal with. In fact, to spread this technology to the residential market, it must be possible to implement it in existing houses. In this section, the guidelines focus on the material aspects while keeping that criterion in mind.

7.3.1 Sensor Selection

In order to design better smart homes, a very important phase is the selection of sensors and technology that need to be exploited. To do so, the criteria to judge the selection by must first be identified. It should be evaluated from both the user's point of view and from the system's perspective. On the user side, it is certainly preferable to implement a smart home at a reasonable price, so low-priced technology (*cost*) should be prioritized. On the other hand, the resident obviously does not want an unreliable system (*robustness*). Using bottom-of-the-range residential home automation equipment is therefore not an option. Rugged sensors that can withstand daily use are a better conception choice, even more so in a prototyping smart home that is going to be used by many students and subjects under unpredictable conditions.

On the system side, it is best to have easy-to-install sensors that can be put into any housing without considerable difficulty (*installation*). This is important for flexibility since real smart homes will often be installed in old buildings. Finally, the precision of the sensors and the complexity of the information they transmit must be taken into account. It is evident why the first is important, but the justifications for the second are somewhat more obscure. Data complexity is important for two reasons. First, the objective of a sensor is to get useful information to use in our AI algorithms. If data are complex to interpret, rapid prototyping will be impossible to achieve and might even be partly unusable. Second, it is important for a smart home to act fast when the user needs its assistance. If the system is too slow, it will be more harmful than helpful. If the data are too complex, it is very likely that the algorithms processing them will be calculation hungry. A rule of thumb for complexity is that the data of the sensors should be easily usable instantaneously on a human timescale—in other words, in less than a second.

The following subsections delve into two important elements that could easily be ignored at the final step of sensor selection for integration into a smart home and that deserve to be discussed in more than a line or two.

7.3.1.1 Energy Efficiency

Something that many researchers fail to recognize is the importance of energy management. There are many reasons to choose sensors and devices that minimize energy consumption. First, it matters for the resident. Of course, if researchers aim to spread smart home adoption in the consumer market, they will need to be proven as an economically viable technology. Residents care a great deal about the cost of their electrical bills and furthermore may want to reduce their environmental footprint. As a consequence, technologies that optimize electricity consumption should be prioritized and those that use disposable batteries should certainly be avoided whenever possible (no user likes to buy and change batteries). Moreover, the latter is a big issue for assistive smart homes, because it is expected that the smart homes will remain completely independent and autonomous. For example, if a smart home needs to exploit radio frequency identification (RFID) technology, passive tags are preferred to their active counterparts.

7.3.1.2 Resident's Perception of Sensors

Another point that is often minimized is the resident's perception of the sensors and habitat. Various research has shown over time that if residents feel observed and invaded in their private life, they have a lower quality of life (Weiser 1991). In addition, if a resident suffers from a cognitive affliction, his state might worsen significantly as a consequence. That is directly in contradiction to the goal we try to achieve by assisting Alzheimer's subjects with smart home technologies. Therefore, it is important to carefully choose the sensors and the effectors of a smart home in order to minimize the negative impacts of invasiveness. Sensors should also be installed with special care to hide them from the view of the house's resident.

7.3.2 Sensor Type Description

We compiled information on the most common types of sensors that are usually deployed in smart homes. Table 7.1 summarizes the main characteristics to allow a

Table 7.1 Comparison of the Most Common Sensors

	Video cameras	*Microphones*	*Smart power analyzer*	*RFID*	*Electromagnetic contacts*	*IR motion sensors*	*Pressure mats*	*Light sensors*	*Flow switch*	*Temperature sensors*	*Ultrasonic sensors*	*Loadcells*	*Accelerometers*
Cost	C-E	B-D	B-C	B-C	A-B	A	C	B	B	A	B-D	B	A
Robustness	B-C	A	A	B	A	A	B	B	B	A	B	A	A
Precision	B	B	B-C	B-D	A	C	A	B	B	B	B-D	B	B
Invasiveness	E	C	A	B	A	B	B-C	A	A	A	A	A	A-B
System installation	B-D	B-C	A	C-E	B	A	A	C	B-C	C	C-E	B-C	B
Data complexity	E	D	C	B-C	A	A	A	A	B	A	C-D	B-C	B-C

quick comparison between them. However, a few of them merit further consideration to properly evaluate their characteristics. The following subsections describe each of them and highlight both their advantages and disadvantages. Figure 7.1 shows an aggregation of images from many types of sensors.

7.3.2.1 Video Cameras and Microphones

Smart homes are often equipped with cameras in the scientific literature (Nguyen et al. 2005; Hoey et al. 2010). These offer the advantage of being able to play the role of a large number of different sensors. This is indeed the type of sensor that offers the greatest expressivity. Cameras are available at a considerable variety of prices and most models are sufficiently robust to withstand continuous employment in a smart home. However, they are highly invasive and the processing of their data is complex. For instance, recognizing simple shapes under a wide range of lighting conditions, orientations, and colors requires fairly elaborate AI algorithms (Patterson et al. 2007). One consequence is the difficulty

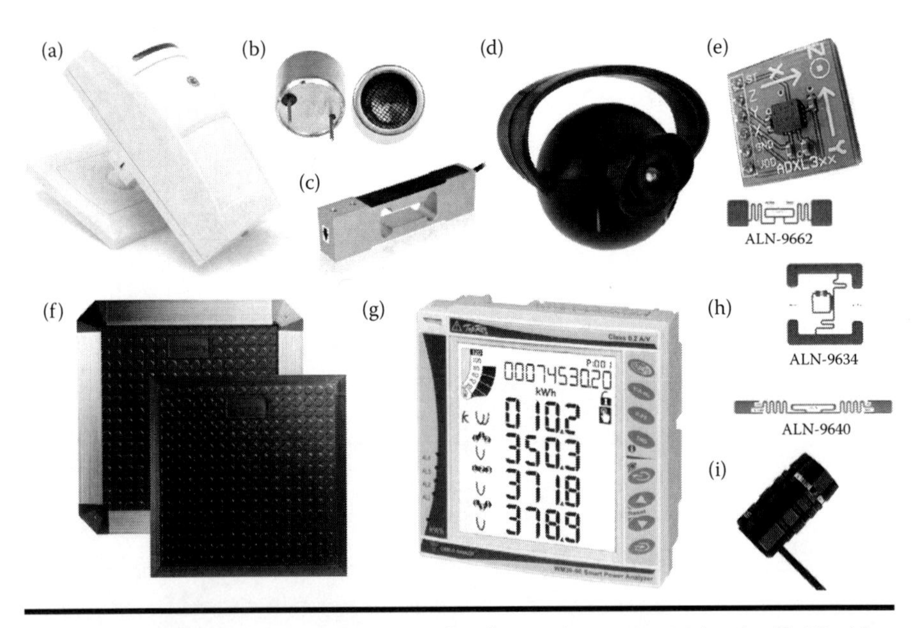

Figure 7.1 (a) Infrared motion sensor; (b) ultrasonic sensor; (c) load cell; (d) video camera; (e) accelerometer; (f) pressure mat; (g) smart power analyzer; (h) RFID tags; (i) microphone.

in building a generalized smart home solution that can be straightforwardly installed in any house.

Microphones share similar characteristics with cameras. Recognizing ADLs from the sound is possible and very interesting (Giannakopoulos et al. 2008; Ntalampiras et al. 2009), but it is not stable enough to be used alone because a high decibel background sound will prevent it from working (e.g., dishwasher). Microphones may also be rejected by the resident and family.

7.3.2.2 Smart Power Analyzer

Our team recently explored the use of a smart power analyzer that enables the reading of electrical outlets throughout the house (Belley et al. 2013). With a low-end model, in our experiments, we developed an algorithm was able to recognize all the electrical devices of the smart home and many simple ADLs. These devices are available as many industrial models, but the cost is generally not under a thousand dollars. However, only one is required to cover the electrical box of an entire house. Thus, it is an inexpensive price to pay for the quantity of information it gives. It is very robust, and the installation is simple. Its major drawback comes from the fact that each electrical device must be labeled manually (since they have a unique signature).

7.3.2.3 Radio Frequency Identification

Another interesting technology is RFID. The base cost of an RFID system is generally significant (mainly due to the software and firmware on the collection module), but the supplementary tags and antennas are cheap. There are two main families: passive RFID and active RFID. Active RFID can be useful for tracking a resident around the house but requires batteries and is more costly and invasive than its counterpart. It is preferable not to rely on battery-powered devices because they require punctual maintenance. Passive tags are much cheaper, ranging from one or two dollars to only a few pennies. They are less precise but small enough to be hidden on or in objects. The technology is very robust, although the initial installation might be difficult depending on the algorithm implemented. The simplest way to use them is through proximity-based localization (see the section "Object Tracking" for more detail).

7.3.2.4 Ultrasonic Sensors

Ultrasonic sensors are often used to partially replace video cameras or RFID. These sensors work very well and can give a clear image of an environment in 3D. However, they are usually slow and so their information is unreliable (not up to date). Moreover, they suffer from a line-of-sight problem. They are not very invasive due to their small size and can be hidden easily. Literature on their use in smart homes is scarce, but in experiments they have proven themselves very useful. Our assessment is that they could be combined with other technologies to improve the base services of smart homes (e.g., localization, movement, etc.). More research is required on this technology in the future.

7.3.2.5 Other Sensors and Comparison

There is still a wide variety of types of sensors on the market that can perform more or less specific tasks that have not been covered here but are described in Table 7.1. For instance, infrared motion sensors are a cheap solution for tracking a resident in the house. However, they are very imprecise. Light sensors and others can be combined to improve precision. They are also very useful for checking whether a light has been left on and can be exploited to optimize energy consumption. Finally, to be able to adequately weight the information presented in Table 7.1, it is necessary to provide precision on some criteria.

Data complexity: Previously, the importance of this aspect was mentioned. However, one cannot only choose sensors with very low data complexity because it is directly linked with the expressivity of the information. Therefore, researchers must try to strike a balance between having data that are too complex and a lack of information.

Cost: The cost varies as a function of two things: the product retail price change and the quantity required. For example, a video camera may be cheaper than a smart power analyzer but many more than one are required to cover a smart home.

Robustness: It does not take into account that some sensors such as cameras may be out of reach of the resident because behaviors are often unpredictable with an AD subject.

System Installation: It covers the initial installation aspects (e.g., calibration). For example, adding RFID tags is very easy and simple, but setting the module and the antennas correctly might be complex.

7.3.3 Communication Technologies

Another key element in the design of smart homes is the choice of communication technologies to implement (de Vicente et al. 2006). There is currently still no consensus among manufacturers on the choice of an universal standard. Consequently, smart home builders often need to deal with compatibility problems. The most extended technologies in smart home networks are wired technology where X10 dominates as a power line carrier (PLC) standard. A PLC uses the home's existing electric wiring to operate. The disadvantage is that not all components are compatible with the same PLC technology. Moreover, installing these technologies in older buildings is often complex and thus wireless technology should be prioritized whenever possible. Notably, many are familiar with Wi-Fi, Bluetooth, and RFID. There is also UWB, the wireless version of USB (Ricquebourg et al. 2006), which allows communication between current USB objects at short distance (approximately 10 meters). Finally, one of the most promising wireless standards is certainly ZigBee (Poole 2004), an open platform based on IEEE specifications for personal networking. It takes its name from messages transmitted like bees (in a zigzag), looking for the best path to the receiver. Note that the increasing reliance on radio frequency (RF) in new technologies has raised concerns regarding the impact on health for continued exposure. Researchers need to keep an eye on the health aspect of their technologies, because it could have consequences for ambient intelligence. Many researchers have studied the effects of long-term exposure to various types of radio waves. An expert workgroup created by the World Health Organization concluded that

> Despite unavoidable uncertainty, current scientific data are consistent with the conclusion that public exposures to permissible RF levels from mobile telephone and base stations are not likely to adversely affect human health. (Valberg et al. 2007)

Moreover, they draw attention to the fact that populations have been exposed to RF from radio and television broadcast waves for more than 50 years with little evidence of deleterious health consequences.

7.3.3.1 Centralized or Decentralized Processing

When designing a new smart home, there is a choice between using classical centralized communication through a server or trying to decentralize communication as in the vision of ubiquitous computing. In a centralized system, components are dumb; they transmit their input directly to a server. By contrast, in a decentralized system, components communicate with each other to try to make decisions and collaborate on services.

There are many researchers working toward the development of a decentralized auto-deployment system (Coyle et al. 2007; Evensen and Meling 2009). These systems would be able to adapt their services to the appearance or disappearance of a new component. Their major drawback is their design complexity. Therefore, we propose to put our efforts first toward the creation of a working centralized solution and to create a decentralized version when the research is advanced enough. It is common sense to work iteratively by first building the simplest solution and to gradually improve it once we have a working model. Furthermore, nothing prevents someone from conceiving a decentralized solution on the software side (based on a multi-agent paradigm) on a centralized hardware architecture.

7.3.4 Choosing the Right Effectors

Collecting information from sensors about resident activities in a smart home is very important, but it would be for naught without methods to react or to provide assistance. As shown by Lancioni et al. (2009), the improvement in performance achieved by participants prompted adequately by assistive technology seems to counter their growing failure to complete tasks, their frustration, and withdrawal. Moreover, in the case of an Alzheimer's afflicted resident, good assistance can slow the progression of the disease.

The smart home literature predominantly uses verbal prompts with little knowledge about their effectiveness (Mihailidis et al. 2004). Deeper research revealed that it was generalized in research for assistive technology to persons with Alzheimer's disease (Lancioni et al. 2009). To be effective, it is important to use prompts that are optimized with the profile of the resident and the characteristics of the tasks. For instance, a verbal prompt would have little effect on a person with Wernicke's aphasia, a language comprehension disorder. That is why part of our team is investigating the effectors' efficiency. Experience has shown that each type of prompt has a contextual specialty. Therefore, we recommend including enough effectors in smart homes to be able to send all types of prompts everywhere it is relevant.

Additionally, in an effort to contribute to solving this important issue, practical guidelines that could be used by smart home researchers toward assistance of cognitively impaired persons were proposed by Van Tassel et al. (2011). Through the next subsections, the context of this work is described along with its important aspects, the resulting guidelines, and the experiments conducted to validate its accuracy.

We strictly concentrate on the part that interest us: the conception of a smart home for cognitive assistance, but more details are available in Van Tassel et al. (2011).

7.3.4.1 Categories of Prompts

There are three main categories of prompts: auditory, visual, and video. Auditory prompts can be verbal (instructions, feedback, questions, etc.), sound (alerts, reminders, etc.), or musical (with or without lyrics). They can be exploited with various equipment such as speakers, portable audio devices, handheld systems, and so on. Visual prompts may be photographic (e.g., colors, shapes, images, pictures), textual (e.g., keywords, sentences, textual descriptions), or light (variations of intensity, color, blinking, and direction). They are mostly used via screens but also with the use of projectors, lightbulbs, LEDs, or laser pens. Video prompts are pictorial (i.e., an auditory prompt and a visual prompt together) or modeling (i.e., video of a person performing a task with an auditory prompt) and require the use of the same equipment as the auditory and visual prompts. They, of course, require the use of screens or projecting devices. Sound and music (without lyrics) prompts are not explicit enough to be used as guidance. On the contrary, light prompts, which are also not explicit, offer a number of minor but essential benefits (e.g., increasing attention by directing where to focus one's energy) not seen with other prompts (Giroux et al. 2009). These advantages are made possible by varying the intensity or color of the lighting and by using the light for pointing out objects or for flashing.

7.3.4.2 Guidelines for Choosing Prompts

In order to partially solve the important issue of choosing the right effectors and the right type of prompts to assist a resident afflicted by AD, comprehensive guidelines were built by our team and structured as a decision tree in Figure 7.2 (Van Tassel et al. 2011). This decision tree was designed by combining the knowledge and experimental results gathered from many disciplines (e.g., psychology, computer science, medicine, education). The advantage of this simple structure is the ease of implementation in an AI algorithm (which again strengthens our goal of rapid AI prototyping).

The guidelines base the selection of prompts accordingly on the problem implied in the current context. AD is well known for the memory impairment it provokes. The consequence is that the person may not remember something from general knowledge (e.g., the correct steps to make a cake) or from events. Aphasia and agnosia are, however, much less known. Aphasia is the difficulty for a person to produce or understand spoken language or written language even if he or she hears and sees correctly through intact organs. Agnosia is the general difficulty to recognize stimuli while still being able to perceive them. AD also often leads to ideational apraxia, which is characterized by a lack of knowledge about the sequence of actions (e.g., substitution of objects, omissions, mistakes in the sequence of actions) needed to accomplish a certain task. Finally, persons afflicted by AD also often

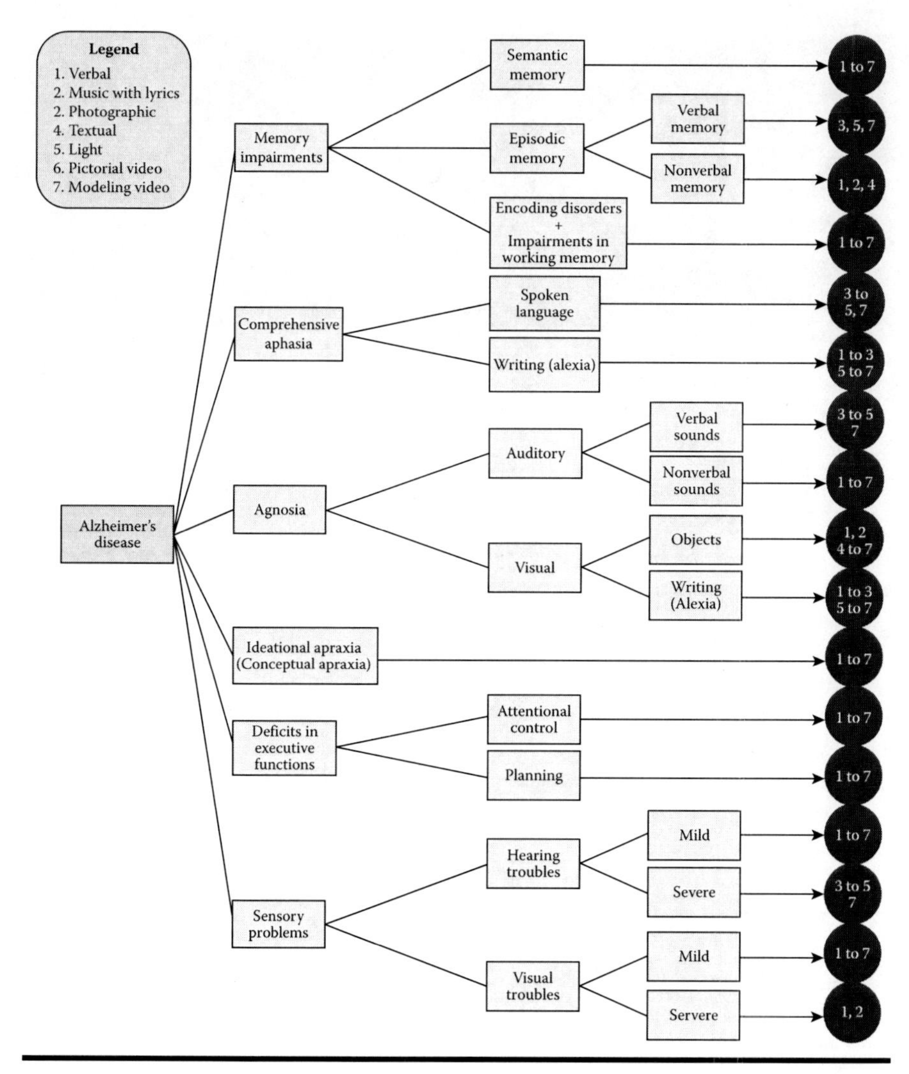

Figure 7.2 Decision tree for choosing which prompts are appropriate for an impairment.

suffer from deficits in executive functions and sensory problems that can be the simple consequence of aging.

7.3.4.3 Validation of the Guidelines

To validate the guidelines, the team recently conducted clinical trials that aimed directly to evaluate the effects of the different type of prompts. It was decided to exploit a well-known test called the *Naturalistic Action Test* (NAT) (Schwartz et al. 2002), which evaluates the performance of individuals with neurological afflictions. This test,

Figure 7.3 An Alzheimer's disease subject doing the NAT.

consisting of three activities of daily living, was held at the laboratory. All participants were required to travel to the scene. We needed both normal subjects and patients with mild to moderate AD. Thus, to implement the whole process, forms needed to be filled out and a hearing was held with the ethics committee, which is responsible for ensuring that experiments comply with human rights and pose no danger to the well-being of individuals.

Every activity was recorded on video camera for future analysis and a prompting software specially developed for that purpose was used (see the section "Prompting Tool"). While the experiments are still ongoing, the first phase had enough participants to outline a trend. The verbal prompt had an efficacy rate of 51% on average. The modeling video prompt, meanwhile, had an average efficiency of 58%. In addition, results indicate that the modeling video without sound is less effective for the participants we encountered. For more details on the results, see Lapointe et al. (2012). Figure 7.3 shows a patient performing the task *preparing coffee with toast.*

7.4 Smart Home Software

Building a smart home for rapid AI prototyping is more than choosing which technologies it should implement. In between the sensors, the architecture, and the effectors, another crucial step remains. Software applications should be included to provide an abstraction layer to work with the infrastructure. This step improves usability by removing the need to redo the communication with all the heterogeneous components. This layer can also provide students and researchers with useful services to enhance the control flexibility over the smart home. Some research teams might be tempted to skip this step because it is time-consuming (i.e., it necessitates a lot of programming without any scientific achievement). However, we advise against this because it could later become a burden for the experiments.

In this section, we provide an overview of the main dilemmas and choices faced when building the software part of a smart home. General tips are provided for both software developed for research and that built for the final user. The first part discusses middleware and the possible alternatives to it. The second part argues why reducing the calculation workload is important. The third subsection talks about the importance of the resident in the design process, and the fourth part broaches the subject of energy efficiency.

7.4.1 Addressing Heterogeneity

The software side of a smart home has the important role of creating uniformity in the various heterogenic technologies of the house. Traditionally, this problem is addressed by the development of middleware, which processes input from various sources and changes it into uniform output (Coyle et al. 2006). However, middleware is not the best choice for every situation. Its development is a tedious task. For that reason and others, our infrastructure is not based on such middleware.

In our case, the database plays the role of providing a uniform access point to data. It is true that a database requires slow writing and reading to hard drives. Yet it is not significant because the volume of data is generally not very high in comparison with the amount of calculations required to compute it and for AI execution. Furthermore, the reading and writing of information from all sensors can be synchronized when exploiting a database. The choice of whether to use middleware become even more important when thinking about future software development (e.g., services, AI algorithms, assistive programs, etc.). Despite the uniformity of data provided by middleware, the lack of concrete files to access data might limit the choice of languages to use in development of prototypes. By contrast, a database makes it possible to use multiple programming languages and development platforms.

7.4.2 Calculation Complexity

The hardware section already covered the topic of data complexity. However, from the software point of view, it is rather more important to pay attention to it. In particular, the AI of a smart home must be responsive and very fast to be respected by the residents and to be regarded as intelligent. A slow system results in delayed interactions with the resident. Let us imagine a case where we have an Alzheimer's patient and the system always prompts him long after he has already committed a mistake due to his impairment. It will certainly not be very helpful, and it might confuse him even more.

It is to avoid such situations that the computational load in the design needs to be evaluated. The usage of non-calculation-hungry techniques must be maximized to reduce it. In fact, a smart home AI should be able to process all the information almost instantaneously on a human timescale. It is even more important in the context of assisting technology. Of course, one could argue that program performance is not a significant concern since computer power is relatively inexpensive, but if calculation complexity is greater than quadratic, adding more processing power might not be enough. Moreover, it is desirable to minimize the space required for computer systems at home because it might be limited in some existing buildings.

7.4.3 Designing for the Resident

One of the most important things in designing good smart home software is the effect on its resident and the perception it gives. Cognitively impaired persons need to be challenged in order to stimulate their brain activity and slow down the degradation of their state. A smart home should encourage its resident to perform tasks by himself or herself rather than automating them (that would be often easier to accomplish). For instance, if a window needs to be closed because it has begun raining outside, the house should prompt the resident to go to the window and close it. Moreover, tasks can be facilitated for elders without automating them (e.g., easy-to-push buttons to close windows, voice control for the execution of tasks, etc.).

Another point that should be considered when conceiving smart home applications is the notion of control. It is important that the resident feel empowered by the smart home in his or her activities but that the utmost control remains in his or her hands (Rodin 1986). It must not feel like decisions are made by the smart home and all that remains for the resident is to execute them. Moreover, in the long-term, an AI program that makes suggestions can degrade gracefully but an algorithm that makes decisions will most probably end up being overlooked as dumb (Intille 2002).

We must also consider the characteristics of residents targeted by our habitat in the design process. As pinpointed in the section "Choosing the Right Effectors," a person's profile may require different types of prompting or may need an adaptation of the effectors (e.g., a higher audio volume). For the elderly, control interfaces should always be intuitive and simple. The graphical user interface (GUI) should be conceived with big buttons, a legible typeface, and high-contrast colors. A software GUI built for the elderly and persons with cognitive degeneration should be carefully evaluated (Shirogane et al. 2008) because it might greatly influence the service's efficiency.

7.4.4 Energy Management

Throughout the previous section, the importance of choosing energy-efficient hardware was emphasized. For the resident, it might be a crucial issue and an important argument to convince him or her to approve the installation of a smart home infrastructure. Therefore, we should go a step further by incorporating advanced concepts of energy management. A smart home should also possess an AI that tries to save energy (Intille 2002) in everyday activities. For example, the AI of a smart home should predict the weather and optimize the A/C or heater activation. If the resident is sleeping, the AI could let cold air enter to optimize the temperature as much as possible during the day. Similarly, the AI could automatically open the blinds to naturally warm the house with sunlight while the resident is away. Again, the point here is that it might be important to prepare the ground for rapid AI prototyping by designing software that allows abstraction of the complexity of controlling different devices.

7.5 The LIARA Smart Home Case Study

Our team at the LIARA laboratory recently conceived and implemented a new cutting-edge smart home infrastructure that is about 100 square meters and possesses around 100 different sensors and effectors. Among the sensors, there are infrared sensors, pressure mats, electromagnetic contacts, various temperature sensors, light sensors, a smart power analyzer, and eight RFID antennas. We also have many effectors, including an Apple iPad, many IP speakers around the apartment, a flat-screen HD television, a home theater, and many lights and LEDs hidden in strategic positions. Figure 7.4 shows a cluster of images from different parts and orientations of our smart home. The main image is the kitchen. At the bottom from left to right, the following can be seen: a tagged cup (RFID tag), the dining room, an RFID antenna, and the HD television. From top right to bottom, the following can be seen: the server, the bathroom, and the library. The server is a Dell industrial, Round Rock, TX blade computer, and it is the one in command of processing the information. The smart home is also equipped with an AMX system, Dallas, TX to control multimedia hardware such as the DVD player, the television, and the IP speaker.

As shown in Figure 7.4, the iPad is embedded in the refrigerator. It controls the habitat for the experiments and can be used to test the equipment or to assist the resident with the help of videos when he or she is located near the kitchen. The television can be remotely controlled from a computer (or AMX) for that same purpose. Our offices and a meeting room are built around the intelligent habitat. In addition, the inside of the apartment can be seen from outside through mirrored windows specially designed for experiments with subjects.

7.5.1 Hardware Design Choices

The LIARA smart home hardware architecture follows the proposed guidelines. It has been conceived to be sturdy enough to support intensive daily use. For that purpose, industrial-grade material was installed while trying to keep the cost as

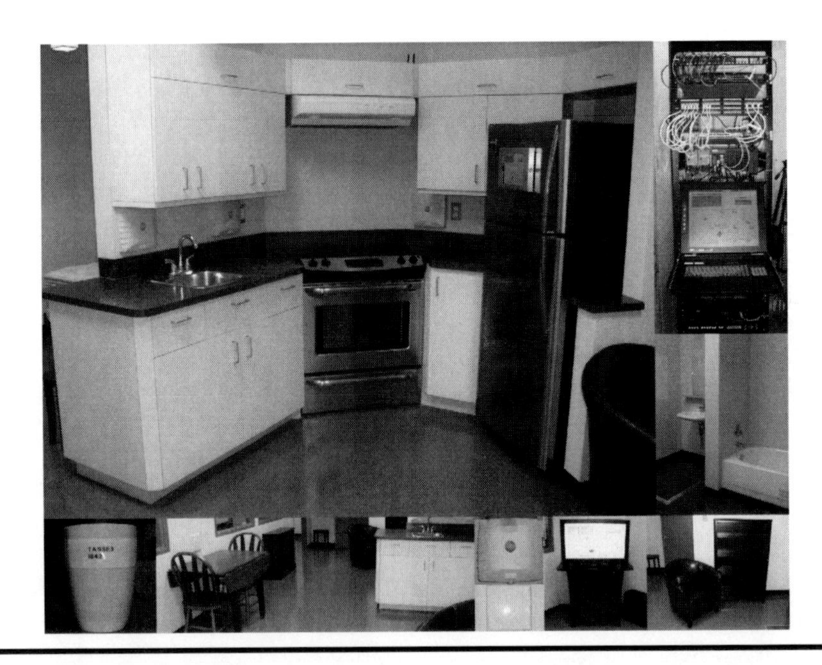

Figure 7.4 The LIARA's smart home.

low as possible. Hazardous situations need to be avoided as much as possible. For example, cheaper automation systems for controlling the house lighting can provoke undesired results when a problem occurs. Imagine a situation where the resident cannot turn on the light due to a system fail. That is undoubtedly to be avoided.

In our architecture, the various sensors and RFID antennas are connected to four independent fault-tolerant islands. If a block falls, only the sensors of that zone are affected. An APAX-5570 automata collects the information in real time and sends it on to the central computer to an SQL Server database. Thereby, this transfer hides the heterogeneity of the information coming from sensors and resolves potential communication incompatibilities between various standards exploited by the manufacturers of the sensors.

On the server of the habitat, an application reads this data every 100 ms. The application reads RFID antennas every 500 ms because this is the optimal refresh time for them. RFID generates a lot of information that could slow down the system if not configured carefully. In our case, all the objects in the environment are tagged. Multiply this by the eight antennas and that sums up to a lot of information. It could still be read generally under 200 ms, but we prefer to trade off reading speed for stability in that case. By default, the central database offers no persistence in our system. The automata simply overwrites the existing information each time. Again, this was chosen to limit the potential stability problem and to get greater stability for our smart home. Figure 7.5 shows the hardware architecture of the laboratory.

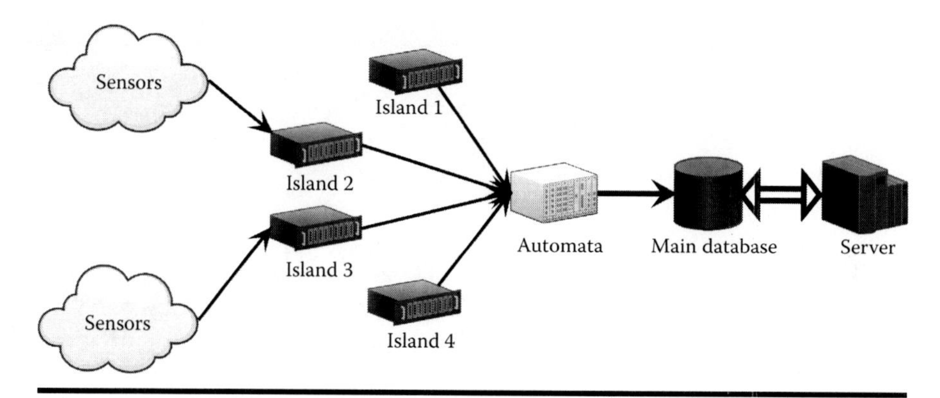

Figure 7.5 Hardware architecture of the LIARA's smart home.

7.5.1.1 LIARA's Sensors and Effectors

In the design stage of our new smart home, the various criteria detailed in the previous sections were taken into account to choose the technology to integrate in our infrastructure. First, we chose to exploit classical infrared motion sensors that provide simple-to-process binary information about the presence (or absence) of activity in a zone. These sensors are not only minimally invasive (due to their physical appearance) but also cheap to buy.

We also decided to use electromagnetic contacts that give binary information about the state of the two parts of the sensor (touched or not). Although they are wired, they can be completely hidden from the user's view due to their very small size. They are used mostly on doors and panels. Moreover, they are wired into an island that is hidden nearby. Although the electromagnetic contacts must be wired, the island can be wireless.

At a few strategic locations, we decided to integrate pressure mats that also give binary information (pressed or not). The cost of this type of sensor is significantly higher, so we tried not to rely on them too much. In addition, the installation requires the alteration of the environment (the floor), which may be unacceptable to some residents. Some were integrated into our prototype smart home, but we would rather eliminate them in a future real-world deployment.

Among other types of sensors, we used light detection, temperature, and RFID technology but no video camera or look-alike technology. That is because cameras are too invasive and almost always rejected by people (Demerism et al. 2008). That is true even when the residents are told that only the system will ever access the image. Moreover, computer vision is far from having the capacity to obtain all the information from a complex video camera output in a reasonable computational time (Patterson et al. 2003). Some researchers have drawn attention to exploiting Microsoft Kinect to replace cameras (Stone and Skubic 2011), but it is still not easy to process the output data. Moreover, to reduce invasiveness it would have to be limited to usage in the living room (on the television), which is not a location of interest for assistive technology. Kinect should be used for specific services, and the main smart home AI should not rely on that more than on video cameras.

In our new smart home, we have taken control over the complete light system by installing simple three-way switches. The lighting system (including the LED) can be controlled thanks to the installed wires. For upgrade compatibility purposes, we decided that all wired effectors (light system, speaker, television, etc.) and sensors would implement the Ethernet protocol, with no further explanation. Using that protocol ensures compatibility with a wide range of networking products. Moreover, any Ethernet devices can easily be converted to wireless at any time. Thus, it is realistic to consider that our system could operate in old buildings.

7.5.1.2 RFID Technology

Because we work in assistive technology, the ADLs of the resident need to be recognized. To do that, passive RFID technology is used to detect object location and movement. It is not an easy task to choose the right set of RFID and configure them for best performance. We chose to use passive RFID tags because we needed to put them on everyday life objects. These are small tags that are cheap to buy (often less than a dollar) and require no other power than a radio pulse from a nearby antenna. Of course, when compared to their active counterparts, which use their own power to emit signals and are always awake, passive tags have a reduced range and precision. Nevertheless, with proper adjustment, they give good results, and they are robust (sometimes even washable).

The quantity of visible tags must also be limited as much as possible. This can be accomplished by shielding the storage areas (such as cabinets) with simple aluminum foil. Finally, a good smart home should minimize pollution from radio waves by turning off radio emission when not required (also to reduce energy consumption). For instance, when the resident is sleeping, obviously tracking the objects in the kitchen is not required.

7.5.2 LIARA's Software Architecture

Following discussion about the LIARA's smart home hardware, it is only natural to discuss the software that was implemented. In the previous section, the description was stopped at the database level where the software takes over control. An application was designed to control the smart home to enhance flexibility and robustness. This software reads the database in real time and copies the data on a second identical database for communication with AI (AIDB). This is important because it protects the real data from being modified by third-party users (malfunctioning program, students doing experimentation, etc.). Nevertheless, the reason we implemented this was mostly to allow easy rerouting of the data source. In consequence, the source could be changed without the third-party applications ever noticing.

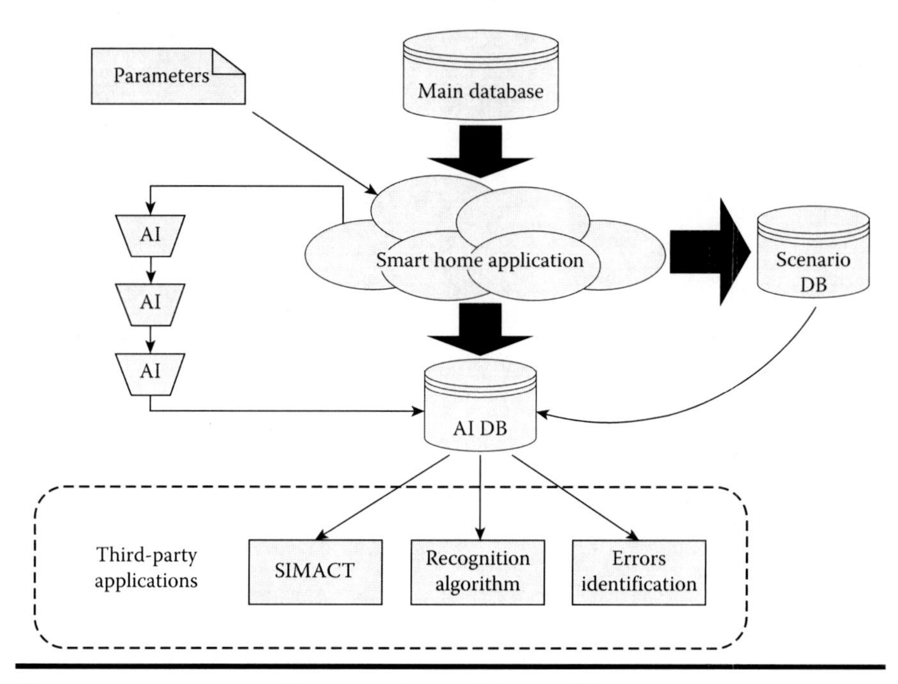

Figure 7.6 Software architecture. AI, artificial intelligence.

It would also work to the contrary: route the main data to another place. Moreover, that multilayered architecture allows us to add an AI module that can provide services to transform raw data into high-level information. The software architecture can be seen in Figure 7.6. More details will be given in the following sections.

7.5.2.1 Smart Home Visualization Tool

In order to facilitate testing, we developed smart home visualization software. A screenshot of this software showing the overall smart home can be seen in Figure 7.7. The graphic interface of this software allows us to see different parts of the smart home or the overall picture. In each of these interfaces, we can see the state of many sensors such as infrared sensors, light sensors, and so on. We also can see an approximate location of the objects in the smart home (rounded rectangle; only appears if RFID antennas are activated) and the current position of the resident (in front of the kitchen counter on the right part of Figure 7.7). These functionalities are very useful when conducting experiments because they allow us to analyze what went wrong by reproducing the sensor activation and double-checking whether the material works properly. In addition, it allows manual testing of the effectors of the smart home, including the television, the oven, and the audio system.

On a more practical side, the application also allows us to remotely control the smart home from a tablet PC. It is particularly useful when we organize

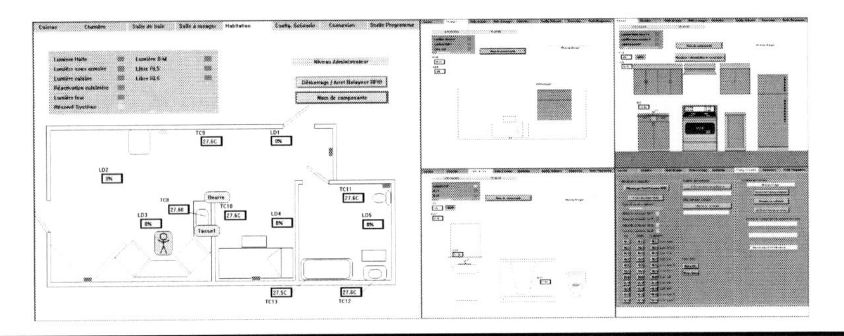

Figure 7.7 The smart home visualization software.

demonstrations for other scientists, health-care professionals, and others. We can easily walk around the smart home and show them the effectors and the sensors. Finally, such an application might seem superficial, but it greatly improves our productivity and allows the students and researchers to concentrate on the most important part of their work.

7.5.2.2 Scenario Recording

An imperative application of our system is unequivocally the scenario recording functionality enabled by the multilayered architecture of our database. In fact, in the visualization software, real-life scenarios can be created from a single button (see the bottom right corner of Figure 7.7). To do so, one has only to enable recording and ask a participant to perform the desired ADL normally in the apartment. When recording is activated, the smart home central application copies the data gathered to a third database layer that is identical to the main one. It does not stop the redirection to the AI database; it is really only a copy (with data persistence). As a result, third-party software (the AI, the visualization tool, etc.) can continue their proper execution. Once recorded, scenarios can be replayed from the visualization tool. To do so, the user needs only to select his or her scenario's name, and the central application will retrieve the record and redirect the flux of data from the scenario DB to the AIDB (see Figure 7.6). In other words, from the third-party perspective, there is no difference between being in playback mode and normal activity. This functionality is especially useful to compare different algorithms because it makes it possible to test them on exactly the same execution sequence.

It is also worthy to note that for a larger data recording, the third party can scan through the scenario database. Still, we generally prefer to do it offline by making a local copy of all the data (in case of a problem, the scenarios will not be corrupted). The main reason for all these precautions is that we prefer to not have to monitor which student does what and when with the data of the smart home. It is very important that the smart home always work perfectly and that the students not be restrained in their activities to achieve rapid AI prototyping.

7.5.2.3 Clinical Trials at the Laboratory

Perhaps the value of the recording/playing functionalities for the scenarios is not so obvious. In the section "Validation of the Guidelines," we talked about experiments conducted with several persons in order to evaluate their cognitive impairment, to identify common mistakes they make in their everyday life, and to evaluate the effects of different types of prompts. This test, consisting of three ADLs, was held in the laboratory and all participants were required to travel to the scene. We needed both normal subjects and patients with mild to moderate stage AD. Thus, to implement the whole process, we had to fill out forms and be screened by the ethics committee, which is responsible for ensuring that experiments comply with human rights and pose no danger to the well-being of individuals. Moreover, we had to find partners (public/private center for long-term care) to help us recruit interested persons who corresponded to the characteristics of our tests. Preparing and conducting such experiments is tedious and requires considerable time, as you can imagine.

We were able to reuse the results of these experiments in other testing of AI prototypes (Bouchard et al. 2011) only because we had recorded them on video. However, it was impossible to test our latest algorithms directly on the actions of real subjects; it was always via an extrapolation of the records that we were able to proceed. Of course, it is unrealistic to hope to bring patients to the laboratory to validate each of our prototypes. Now that our new smart environment incorporates features for recording, the action sequences of the subjects who came to the LIARA could be repeated infinitely in order to benchmark the various prototypes. Moreover, two algorithms could be compared on exactly the same data sequences. That is why we say that this feature is inestimable and that any research team should consider its integration in the construction process of an intelligent home.

7.5.3 Supplementary AI

As was discussed earlier, it is possible to add AI modules to the server to process the transformation of raw data into useful high-level knowledge (see Figure 7.6). That functionality is due to the choice of implementing a multilayered database. On the current system, two additional modules provide general services to third-party applications. The first deals with the raw information from RFID tags in order to alter them into useful data. For example, it associates the unique identifier of the tag with a meaningful name (cup X) and gives the rough position of the associated physical object in the home. For its part, the second module is a small inference engine that intercepts information from motion sensors to infer the approximate position of the resident in the home. The advantage of these services is twofold. On the one hand, the prototypes can either be developed and tested quickly using the services or they can be implemented using the raw data from the sensing devices. In addition, it allows more flexibility because the algorithms on the server can be easily modified without affecting the service provided and therefore without affecting third-party applications that use it.

7.5.3.1 Tracking the Resident

One of the most active problems of smart homes is the localization/tracking in real time of the resident in the house. To begin with, many use wearable devices for this purpose (Patterson et al. 2003). It is unrealistic to expect the resident to always wear them, especially if he or she is afflicted by a type of cognitive impairment. To create a tracking system, our needs in the matter of precision must first be defined. For most smart home applications, approximate position is enough (at the scale of large part of room). This is why our system is divided into logical zones for this service. Therefore, an application can directly interrogate it to know whether there is a presence in a zone.

Our system is primarily based on motion sensors, which are moderately slow and cannot cover every part of a zone (there are blind spots). The consequence is that we cannot always locate the resident with 100% certainty. However, it is not necessary. We improve the certainty when there is detected activity in the smart home by considering every sensor's activation in the house. If the system loses track of the resident, it assumes that he or she has not moved from the last known location.

There is also the issue of multiple presences. It is very hard to track multiple persons who are not carrying a wearable device (and without video cameras). It is also an issue for other challenges regarding cognitive assistance such as activity recognition (Gu et al. 2009). Although we strongly believe that future work should focus on systems that support multiple residents, the subsequent problems are far from solved. Therefore, we propose a simple interim solution. If the resident receives a visitor, the system should just decrease its intervention level. It can be supposed in the context of assistance of an AD resident that the visitor would be able to help him or her and so the smart home would be less necessary. It is easy to note the presence of one more person. The smart home does not need to differentiate between the two persons; it might assist either one that needs it.

7.5.4 Object Tracking

As mentioned before, our research relies heavily on RFID technology. One of the best purposes is to track various objects around the apartment. For general prototyping of AI software, we use a proximity-based localization algorithm. With this technique, the object is simply considered in a zone if it is detected by an antenna. Consequently, to obtain good precision, the antenna signal power (range) must be adjusted carefully. The smaller the range, the better the precision. Of course, that also means that to enable localization throughout the environment dense deployment of antennas is required. However, objects do not need to be precisely located everywhere in the smart home so it is possible to limit the number of antennas to the critical zones (we use eight antennas).

Alternatively, we are working on localization algorithms based upon the trilateration method (Fortin-Simard et al. 2012). The advantages are the higher precision obtained and the possibility to monitor the movement of the objects. The main principle behind this technique is to use the received signal strength indication to estimate the distances from at least two antennas on the same wall or three on different walls and use these distances as a radius to draw a circle centered on each antenna. The estimated position of the object is then the intersection point of the circles. Of course, to achieve reasonable performance a variety of ameliorations must be done. In our work, we exploit a multifilter approach that tries to address each of the problems arising from RFID trilateration independently. The method is only used punctually in our smart home for now, but good experimental results that we obtained recently have proven that it could be permanently applied in a smart home. A hybrid version between proximity localization and trilateration should be considered when building a new smart home.

7.6 Promoting Collaboration between Researchers

At the beginning of this chapter, we stated that one of our objectives was to promote collaboration and exchange for incremental advances in smart homes researches. Many research teams are working in agreement with that goal. For example, Cook and Schmitter-Edgecombe (2009) from Washington State University shared the data set from their smart home and Giroux et al. (2009) shared their real case scenarios. On our side, to achieve this goal, we are pursuing a project of collaborative tools in synergy with our research that we want to distribute freely to help researchers in their work. Moreover, these tools would not only increase collaboration between researchers but also help them to speed the prototyping process and the validation of theories.

7.6.1 SIMACT

Recently, we created a smart home simulator in the Java programming language that was named *SIMACT* (Bouchard et al. 2012a). This simulator was based on a recent 3D engine and was designed to easily conceptualize a smart home in 3D. To do so, the user only needs to upload a 3D drawing to the software via an editor. To simplify the life of the user, SIMACT comes with a smart home kitchen that was built from the Sketchup free object library of modeling software from Google. SIMACT can be seen in Figure 7.8.

SIMACT was primarily made to successfully experiment without requiring a real smart home infrastructure. We had this idea because many researchers work at small universities or simply do not have the financial support to build a full-sized home. However, it was built with the goal of providing high flexibility, and it could certainly be used with little modification for smart home sensor visualization.

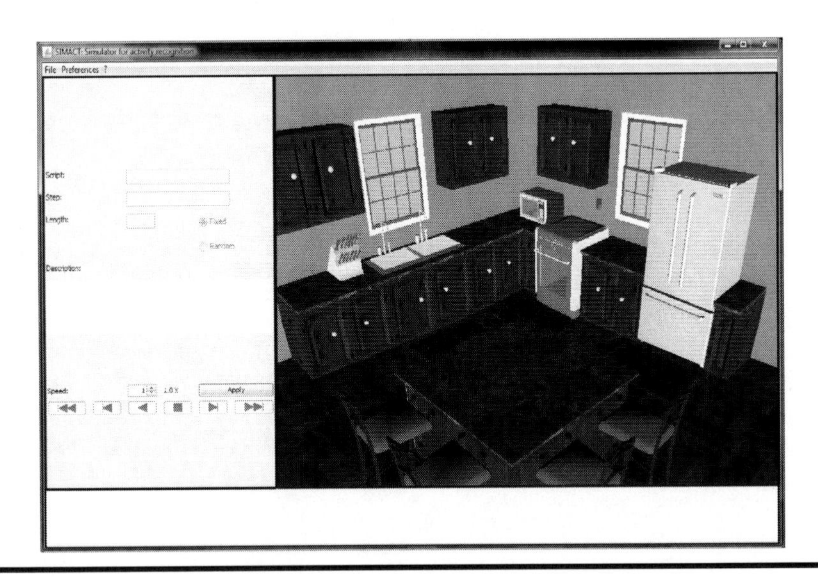

Figure 7.8 SIMACT in action.

Because we are aware that we cannot build the software to answer the precise needs of every team in the world, it was decided to give the code to the community. That is, researchers can do whatever they want with the code except commercialize it.

7.6.1.1 SIMACT Architecture

SIMACT was designed in such a way as to separate the code from every dynamic aspect of the software. Basically, everything is configurable in the software. It works similarly to a multimedia player but use activities as input instead of audio/video files. When the 3D environment is loaded (done automatically at startup), it allows the user to load scenarios that will then be played by the simulator. These scenarios are normal ADLs or ADLs with mistakes that will simulate the actions exactly as it would do in a real smart home; by firing sensors. The difference here is that the sensors are virtual conceptions. However, the information is recorded in the same way our material smart home does it. SIMACT will save the sensors firing into a database accessible from third-party applications. The consequence is that communication with a real smart home is simple to realize. SIMACT could put information from a scenario into the smart home database or read information from it and show what happened in the 3D frame.

The scenarios are special because they are built from simple-to-use XML language. The software works as an interpreter and recognizes the different tags of a scenario. If a scenario is grammatically correct, many actions can be performed automatically in the 3D environment such as object movement or rotation.

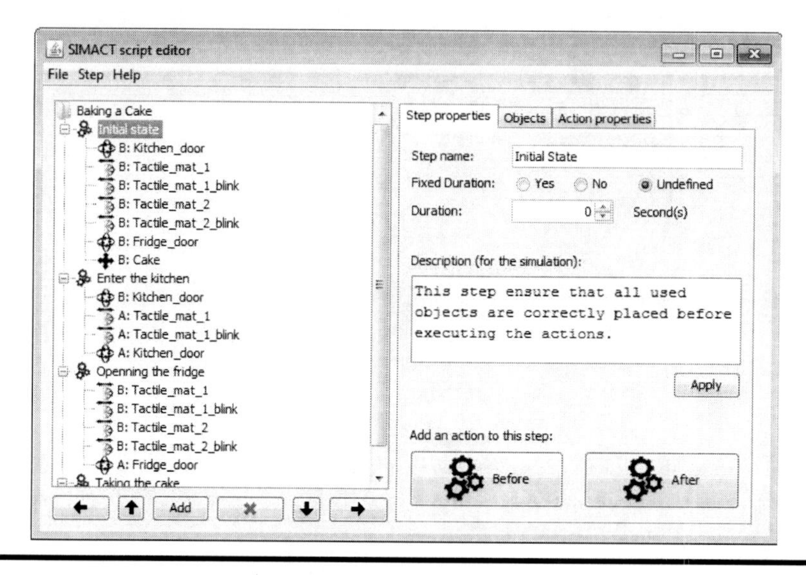

Figure 7.9 SIMACT's scenario editor.

In addition, like a multimedia player, the scenario execution can be played, paused, stopped, and even rewound. Figure 7.9 shows a scenario editor we are working on for future versions of SIMACT. Additionally, we plan on adding materials to the virtual environment to multiply the possibilities.

7.6.2 Open Database

With SIMACT, we had the idea of building an open database for ADL scenarios. We already built few scenarios that are distributed with SIMACT, and we are working to create a few more. We want to provide real case scenarios from our AD experiments in order to have a fundamental basis for experiments and result comparison of activity recognition algorithms and assisting algorithm. The experiments conducted at the laboratory were conducted with this second goal in mind. Every bit of information was recorded to build realistic scenarios and share them with the community. Similarly, we hope that teams will also contribute to this open knowledge by sharing their own experiments. A second part of this project will be the sharing of our experimental results and analysis. Of course, this is already done via scientific publication, but due to limited space it is often reduced and lacking detail.

Furthermore, the smart home data from activity recording is becoming a major stake due to the rapid development of activity recognition approaches based on data mining techniques. Many researchers have started to share their data, but it is often unstructured and partial information. For example, some report sharing months of data but only few hundred lines of data are given on the Internet. We plan on doing larger-scale experiments in the next year and sharing the entire data set.

We encourage research teams to pursue efforts to share their data sets because it will allow more persons to work on data mining solutions for activity recognition. In addition, it will establish a common foundation for comparing different algorithms to each other.

7.6.3 Prompting Tool

To experiment with real Alzheimer's subjects, a prompting tool named *ART* (Activity Report Tool for the cognitively impaired) was created. This tool is simple software that enhances the evaluation methodology of a well-known cognitive test such as the NAT (Schwartz et al. 2002). It allows an assistant to send a prompt from a distant computer whenever required (e.g., if the participant uses the wrong utensil). With a simple click of a button, the chosen form of prompt (i.e., verbal, modeling video without sound, or modeling video with sound) can be sent for a specific step of the task through a computer screen and speakers placed in front of the participant.

Moreover, the software allows the evaluator to comment on the results, the erroneous steps, and the type of problem (e.g., omission, inaction, substitution) through a simple dynamic text box. It is also possible to get a percentage of the task completed. Moreover, it allows us to save each session separately. The type of prompt sent, the completion time of the task, the notes, and other information that is relevant is recorded for further consultation. From these sessions, ART is able to generate a text report. The ART software is already available for teams that are interested in it, and we created a short practical guide to enable a neophyte to begin rapidly. Figure 7.10 shows the GUI of the software.

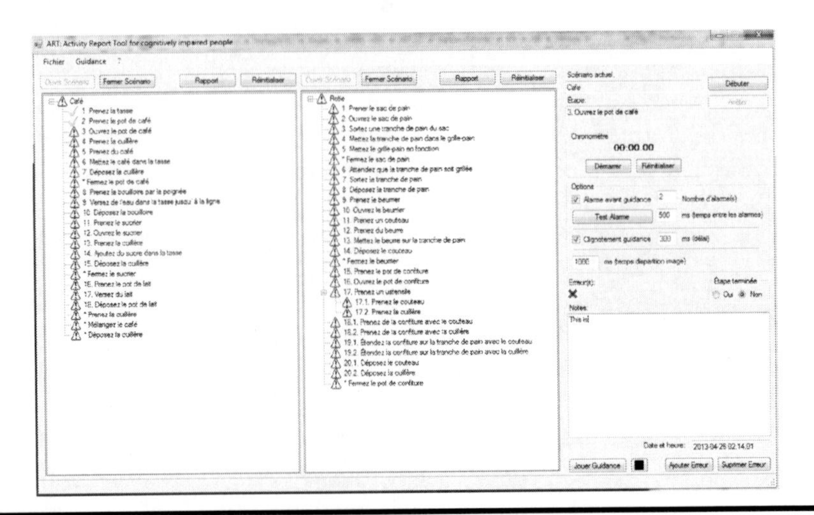

Figure 7.10 Activity Report Tool for the cognitively impaired.

7.7 Summary

Recent advances in sensor networks, engineering, and ambient intelligence have brought us toward the dream of creating smart homes that can help humans in their everyday lives. Many agree that current technologies are sufficient to achieve this vision but that major breakthroughs are needed on the software side (activity recognition, security, context awareness, opportunistic networks, etc.). In fact, it is the key component for enabling sensors and effectors to provide valuable services. Consequently, it is important for researchers to prototype and experiment on new algorithms and models in a realistic setting (e.g., smart home). In order to face this challenge, many research teams have to build new experimental infrastructures without any background experience, guidance, or even a real idea of their research needs and issues.

In this chapter, we proposed a set of guidelines for the conception of smart homes for cognitive assistance from both the hardware and software perspectives. The main idea is to provide basic information about how to efficiently design and implement a smart home infrastructure for faster prototyping (focusing on AI). Our own architectural and design choices that were focused toward that goal were presented. Of course, the building of smart homes is not an exact science, and thus our choices are not universal. However, as was shown with our own smart home prototype, our various applications, and our experiments, the proposed guidelines are practical for designing and building smart homes. The chapter is aimed directly at researchers who plan to conceive new smart home infrastructures or even for enthusiasts who would like to enhance their houses. In that way, we want to help the community to make well-informed choices in their design. We hope that it will lay a foundation for the establishment of an incremental process toward smart home implementation.

A secondary goal of this chapter was to promote collaboration and exchange between research teams. We hope to convince the reader of the importance of sharing knowledge to advance farther in the field with the Open Tools project of the LIARA and the given examples. By achieving those two points, faster and easier experimentation and collaboration between laboratories, smart home assistance will evolve as a real applicable solution in the consumer market. In future work, we invite other researchers to improve these guidelines with their own experience such as construction of smart homes, use of a specific technology, and so on. Each detailed description of the conception of a new smart home is a valuable piece of information to the field.

Acknowledgments

The authors would like to thank their main financial sponsors: the Natural Sciences and Engineering Research Council of Canada, the Quebec Research Fund on Nature and Technologies, and the Canadian Foundation for Innovation.

The authors would also like to acknowledge the contribution of the Centre de Santé et Services Sociaux of La Baie, the Maison Le Phare of Jonquière, and their regional Alzheimer Society for helping them recruit the participants. Finally, special thanks to our neuropsychologist partner and her graduate students, who indirectly worked on this project by supervising the clinical trials with patients.

References

Augusto, Juan C., and Chris D. Nugent. 2006. Smart homes can be smarter. In *Designing Smart Homes: Role of Artificial Intelligence*. Berlin: Springer-Verlag.

Belley Corinne, Sebastien Gaboury, Bruno Bouchard, and Abdenour Bouzouane. 2013. Efficient and inexpensive method for activity recognition within a smart home based on load signatures of appliances. *Journal Pervasive and Mobile Computing* 12:58–78.

Boger, Jennifer, Jesse Hoey, Pascal Poupart, Craig Boutilier, Geoff Fernie, and Alex Mihailidis. 2006. *A planning system based on Markov decision processes to guide people with dementia through activities of daily living.* Washington, DC: IEEE Transactions on Information Technology in Biomedicine 10 (2):323–333.

Bruno Bouchard, Sylvain Giroux, and Abdenour Bouzouane. 2007. A keyhole plan recognition model for Alzheimer's patients: First results. *Journal of Applied Artificial Intelligence* 21 (7):623–658.

Bouchard, Bruno, Patrice Roy, Abdenour Bouzouane, Sylvain Giroux, and Alex Mihailidis. 2008. An activity recognition model for Alzheimer's patients: Extension of the COACH Task Guidance System. In *ECAI*, edited by Malik Ghallab, Constantine D. Spyropoulos, Nikos Fakotakis, and Nikos Avouris, IOS Press.

Kevin Bouchard, Amir ajroud, Bruno Bouchard and Abdenour Bouzouane. 2012a. SIMACT: A 3D open source smart home simulator for activity recognition with open database and visual editor. *International Journal of Hybrid Information Technology* 5(3):13–32.

Bouchard, Kevin, Bruno Bouchard, and Abdenour Bouzouane. 2011. Qualitative spatial activity recognition using a complete platform based on passive RFID tags: Experimentations and results. In *9th International Conference On Smart homes and health Telematics*, edited by Bessam Abdulrazak, Sylvain Giroux, Bruno Bouchard, Hélène Pigot, and Mounir Mokhtari. Montreal: Springer.

Kevin Bouchard, Bruno Bouchard, and Abdenour Bouzouane. 2012b. Guideline to efficient smart home design for rapid AI prototyping: A case study. Paper read at International Conference on PErvasive Technologies Related to Assistive Environments, Crete Island, Greece.

Cook, Diane J., and M. Schmitter-Edgecombe. 2009. *Assessing the quality of activities in a smart environment.* Vol. 48. Methods of Information in Medicine. Schattauer Publishers, Stuttgart, Germany.

Cook, Diane J., Michael Youngblood, III, Edwin O. Heierman, et al. 2003. MavHome: An agent-based smart home. In *Proceedings of the First IEEE International Conference on Pervasive Computing and Communications*, IEEE Computer Society.

Coyle, Lorcan, Steve Neely, Gatan Rey, et al. 2006. Sensor fusion-based middleware for assisted living. In *Proceedings of 1st International Conference On Smart homes \& heath Telematics (ICOST'2006) "Smart Homes and Beyond"*, IOS Press.

Coyle, Lorcan, Steve Neely, Graeme Stevenson, Mark Sullivan, Simon Dobson, and Paddy Nixon. 2007. Sensor fusion-based middleware for smart homes. *International Journal of Assistive Robotics and Mechatronics* (*IJARM*) 8 (2):53–60.

Demerism, George, Brian K. Hensel, Marjorie Skubic, and Marilyn Rantz. 2008. *Senior residents' perceived need of and preferences for "smart home" sensor technologies*. Vol. 24. Cambridge: Cambridge University Press.

de Vicente, Antonio J., Juan R. Velasco, Ivan Marsa-Maestre, and Alvaro Paricio. 2006. A proposal for a hardware architecture for ubiquitous computing in smart home environments. Paper read at Proceedings of the I. International Conference on Ubiquitous Computing: Applications, Technology and Social Issues, Alcala de Henares, Madrid, Spain, June 7–9, 2006.

Evensen, Pal, and Hein Meling. 2009. Sensor virtualization with self-configuration and flexible interactions. In *Proceedings of the 3rd ACM International Workshop on Context-Awareness for Self-Managing Systems*. Nara, Japan: ACM.

Fortin-Simard, D., K. Bouchard, S. Gaboury, B. Bouchard, and A. Bouzouane. 2012. Accurate passive RFID localization system for smart homes. In *3rd IEEE International Conference on Networked Embedded Systems for Every Application*. Liverpool, UK: IEEE.

Giannakopoulos, Theodoros, Aggelos Pikrakis, and Sergios Theodoridis. 2008. A novel efficient approach for audio segmentation. In *International Conference on Pattern Recognition*. Tampa, FL: IEEE.

Giroux, Sylvain, Tatjana Leblanc, Abdenour Bouzouane, Bruno Bouchard, Hélène Pigot, and Jérémy Bauchet. 2009. The praxis of cognitive assistance in smart homes. In *Behaviour Monitoring and Interpretation*, edited by B. Gottfried, and H. K. Aghajan. BMI Book: IOS Press, pp. 183–211, Amsterdam, The Netherlands.

Gu, Tao, Zhanqing Wu, Liang Wang, Xianping Tao, and Jian Lu. 2009. Mining emerging patterns for recognizing activities of multiple users in pervasive computing. Paper read at 6th Annual International Conference on Mobile and Ubiquitous Systems: Computing, Networking and Services, Toronto, ON, Canada.

Hoey, Jesse, Pascal Poupart, Axel von Bertoldi, Tammy Craig, Craig Boutilier, and Alex Mihailidis. 2010. Automated handwashing assistance for persons with dementia using video and a partially observable Markov decision process. *Computer Vision and Image Understanding* 114 (5):503–519.

Intille, Stephen S. 2002. Designing a home of the future. *IEEE Pervasive Computing* 1 (2):76–82.

Kaila, Lasse, Jussi Mikkonen, Antti-matti Vainio, and Jukka Vanhala. 2008. The eHome—A practical smart home implementation. Paper read at Pervasive 2008, Sydney, Australia.

Lancioni, Giulio E., Maria L. La Martire, Nirbhay N. Singh, et al. 2009. Persons with mild or moderate Alzheimer's disease managing daily activities via verbal instruction technology. *American Journal of Alzheimer's Disease and Other Dementias*, 23 (6):552–562. Thousand Oaks, CA: SAGE.

Jessica Lapointe, Bruno Bouchard, Julie Bouchard, and Abdenour Bouzouane. 2012. Smart homes for people with Alzheimer's disease: Adapting prompting strategies to the patient's cognitive profile. In *5th ACM International Conference on Pervasive Technologies Related to Assistive Environments*. Crete Island, Greece: ACM Publisher.

Mihailidis, Alex, Joseph C. Barbenel, and Geoff Fernie. 2004. *The efficacy of an intelligent cognitive orthosis to facilitate handwashing by persons with moderate to severe dementia*. Hove: Psychology Press.

Nguyen, Nam T., Dinh Q. Phung, Svetha Venkatesh, and Hung Bui. 2005. Learning and detecting activities from movement trajectories using the hierarchical hidden Markov models. In *Proceedings of the 2005 IEEE Computer Society Conference on Computer Vision and Pattern Recognition (CVPR'05)—Volume 2—Volume 02*, IEEE Computer Society.

Ntalampiras, Stavros, Ilyas Potamitis, and Nikos Fakotakis. 2009. An adaptive framework for acoustic monitoring of potential hazards. *EURASIP Journal on Audio, Speech, and Music Processing* 2009:1–15.

Nugent, Chris, Maurice Mulvenna, Ferial Moelaert, et al. 2007. Home based assistive technologies for people with mild dementia. In *Proceedings of the 5th International Conference on Smart Homes and Health Telematics*. Nara, Japan: Springer-Verlag.

Patterson, D., H. Kautz, and D. Fox. 2007. Pervasive computing in the home and community. In *Pervasive Computing in Healthcare*. Boca Raton, FL: CRC Press.

Patterson, Donald J., Lin Liao, Dieter Fox, and Henry Kautz. 2003. Inferring high-level behavior from low-level sensors. In *Ubicomp 2003: Ubiquitous computing*, edited by A. K. Dey, A. Schmidt, and J. F. McCarthy. pp. 73–89, Berlin: Springer-Verlag.

Phua Clifton, Biswas Jit, Tolstikov Andrei, et al. 2009. Plan recognition based on sensor produced micro-context for eldercare. Paper read at Proceedings of The First International Workshop on Context—Awareness in Smart Environments: Background, Achievements and Challenges (CASEbac 2009), Toyama, Japan.

Ian Poole. 2004. What exactly is ZigBee? *Communications Engineer* 2 (4):44–45.

Ramos, Carlos, Juan Carlos Augusto, and Daniel Shapiro. 2008. Ambient intelligence: The next step for artificial intelligence. *IEEE Intelligent Systems* 23 (2):15–18.

Ricquebourg, Vincent, David Menga, David Durand, Bruno Marhic, Laurent Delahoche, and Christophe Loge. 2006. The smart home concept: Our immediate future. Paper read at 1ST IEEE International Conference on E-Learning in Industrial Electronics.

Robles, Rosslin John, and Tai-hoon Kim. 2010. Applications, systems and methods in smart home technology: A review. *International Journal of Advanced Science and Technology* 15:37–48.

Rodin, Judith. 1986. *Aging and health: Effects of the sense of control*. Vol. 233. Washington, DC: American Association for the Advancement of Science.

Roy, Patrice, Bruno Bouchard, Abdenour Bouzouane, and Sylvain Giroux. 2009. A hybrid plan recognition model for Alzheimer's patients: Interleaved-erroneous dilemma. *Web Intelligence and Agent Systems* 7 (4):375–397.

Roy, Patrice, Bruno Bouchard, Abdenour Bouzouane, and Sylvain Giroux. 2010. Challenging issues of ambient activity recognition for cognitive assistance. *Handbook of Research on Ambient Intelligence and Smart Environments* 320–345.

Schwartz, Myrna F., Mary Segal, Tracy Veramonti, Mary Ferraro, and Laurel J. Buxbaum. 2002. *The naturalistic action test: A standardised assessment for everyday action impairment*. Vol. 12. Hove: Psychology Press.

Shirogane, Junko, Takashi Mori, Hajime Iwata, and Yoshiaki Fukazawa. 2008. Accessibility Evaluation for GUI Software Using Source Programs. In *Proceedings of the 2008 Conference on Knowledge-Based Software Engineering: Proceedings of the Eighth Joint Conference on Knowledge-Based Software Engineering*, IOS Press.

Stone, Erik E., and Marjorie Skubic. 2011. Evaluation of an inexpensive depth camera for passive in-home fall risk assessment. Paper read at Pervasive Computing Technologies for Healthcare (PervasiveHealth), 2011 5th International Conference on, 23–26 May 2011.

Tim van Kasteren, Gwenn Englebienne, and Ben Krose. 2010. Activity recognition using semi-Markov models on real world smart home datasets. *Journal of Ambient Intelligence and Smart Environments* 2 (3):311–325.

Valberg, Peter A., T. Emilie van Deventer, and Michael H. Repacholi. 2007. Workgroup report: Base stations and wireless networks-radiofrequency (RF) exposures and health consequences. *Environmental health* perspectives. Vol. 115. pp. 416–424.

Van Tassel, Mike, Julie Bouchard, Bruno Bouchard, and Abdenour Bouzouane. 2011. *Guidelines for increasing prompt efficiency in smart homes according to the resident's profile and task characteristics*. Edited by Bessam Abdulrazak, Sylvain Giroux, Bruno Bouchard, Hélène Pigot, and Mounir Mokhtari. Vol. 6719, ICOST, Berlin: Springer.

Weiser, Mark. 1991. The computer for the 21st century. *Scientific American* 265 (3):66–75.

Chapter 8

Wireless Sensor Network–Based Smart Agriculture

Sarang Karim and Faisal Karim Shaikh

Contents

8.1 Introduction

Agriculture is considered one of the most ancient professions of human beings, and innovations and technologies are usually adopted to find solutions for agricultural problems and issues. This could be done for improving production, minimizing the use of fertilizers and pesticides, reducing human effort, and most importantly optimizing the budget needed to produce crops. The main source of livelihood for mankind worldwide is agriculture. Cotton, wheat, sugarcane, and vegetables are the major cash crops grown around the world (Shinghal et al. 2010; Abdullah and Barnawi 2012).

Farmers around the world have not yet discovered the full power of technology to avail the necessary information for increasing the quality and productivity of their crops. For example, farmers lack information on managing the required water for crops for achieving high productivity. They use an inefficient amount of water, which might affect the land's fertility and have negative effects on product quality (Jiber et al. 2011). As the population increases, food accessibility is becoming difficult for the majority of the population. Thus, we must find new techniques to maximize crop production. Inevitably, this will involve using modern technology in agriculture. New systems will be created to assist farmers by giving them access to all required and up-to-date information about crop status to enhance productivity.

Hence, in this way, farmers can also minimize their losses and the cost of crops. Information and communications technology (ICT) can be used in the agriculture sector for improving production efficiency, production quality, the post-harvest process, and also to monitor and control environmental parameters that affect crop yield. The term *smart agriculture* can be justified for automation in agriculture using ICT (Shinghal et al. 2010) for measuring, monitoring, and detecting agricultural parameters. The main objectives of smart agriculture (Jiber et al. 2011) are the use of new techniques and technologies, providing up-to-date information about the crops, management perceptions, and decision-making. In order to achieve these objectives, a profound technology that supports many applications, including smart agriculture, is wireless sensor networks (WSNs) (Baggio 2005; Lea-Cox et al. 2007; Yoo et al. 2007; Ahonen et al. 2008; Balendonck et al. 2008; D. Anurag et al. 2008; Panchard et al. 2008; Das et al. 2009; Lima et al. 2010; Merrett and Tan 2010; Shinghal et al. 2010; Angelopoulos et al. 2011; Jiber et al. 2011; Alberts et al. 2013).

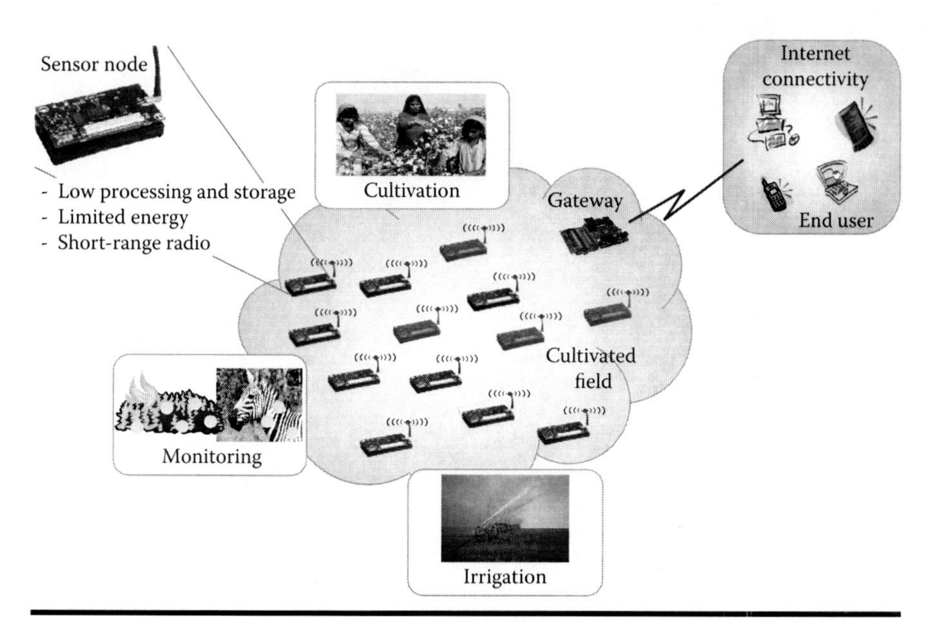

Figure 8.1 Deployment of wireless sensor network–based smart agriculture.

A WSN is composed of a number of tiny devices called *sensor nodes* that sense the physical environmental data and either store or forward it to a central node called a *base station* (BS) for further processing (Akyildiz et al. 2002a, 2002b; Reichenbach et al. 2006; Garca-Hernndez et al. 2007; Xiao and Guo 2010; Kapoor et al. 2011). A sensor node is equipped with different sensors; limited processing, memory, and power capabilities; and a transceiver (Akyildiz et al. 2002a; Karl and Willig 2003; Mampentzidou et al. 2012). Sensors used for agricultural WSNs capture temperature, humidity, soil pH, soil moisture, electrical conductivity, and so on. The quantity of sensor nodes may vary according to the application scenario from a few to hundreds or thousands (Reichenbach et al. 2006; Garca-Hernndez et al. 2007). From the BS, the data are sent to the end user, where appropriate decisions can be made. The complete scenario of smart agriculture using WSNs is depicted in Figure 8.1.

In agriculture, there are various key parameters for which WSNs can be deployed, for example, environmental (temperature, humidity, CO_2) (Yoo et al. 2007; Ahonen et al. 2008; Corke et al. 2010), terrestrial (insect detection, leaf chlorophyll) (Buratti et al. 2009; Ruiz-Garcia et al. 2009b), underground (soil temperature, soil humidity, soil moisture, etc.) (Akyildiz and Stuntebeck 2006; Garcia-Sanchez et al. 2011; Yu et al. 2012), and irrigation (water flow and level) (Balendonck et al. 2008; Lima et al. 2010; Shinghal et al. 2010; Angelopoulos et al. 2011; Jiber et al. 2011; Singh and Bansal 2011). Figure 8.2 shows the key parameters monitored and controlled by WSNs impacted by various underground

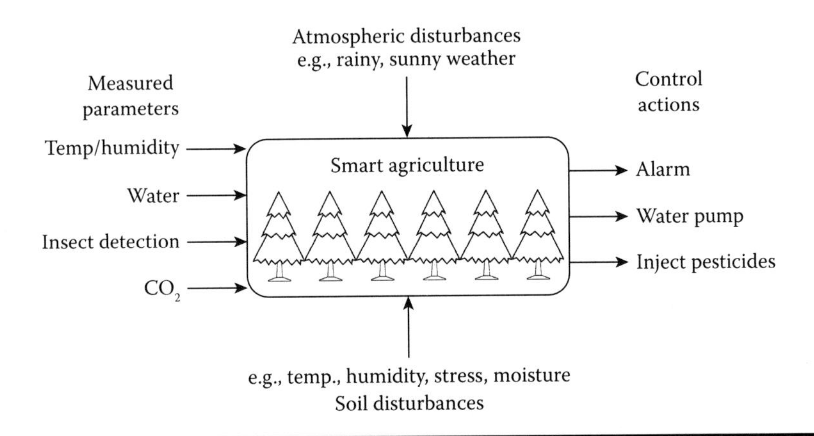

Figure 8.2 Control of environmental parameters by smart agriculture.

and aboveground environmental changes. By monitoring and controlling these parameters, we will be able to improve production efficiency, product quality, and post-harvest operations.

The rest of the chapter is organized as follows. The section "WSN Applications in Smart Agriculture" presents the different application domains of smart agriculture where WSN can be deployed to get more benefits. In the section "WSN-SA: A Generic WSN-Based Smart Agriculture Architecture," we propose a simple yet generic WSN-based smart agriculture architecture that captures the whole cycle to help the end user make managed and timely decisions. A case study of WSN deployment for cotton crops is detailed in the section "WSN Deployment for Smart Agriculture: A Case Study," which makes use of the proposed architecture. The section "The Road Ahead" highlights the prospective future of smart agriculture, followed by the chapter conclusion.

8.2 WSN Applications in Smart Agriculture

There are number of areas where WSNs can be deployed for smart agriculture. Some of the major smart agriculture applications where WSNs play an important role are precision agriculture, irrigation systems, greenhouses, weather stations, and underground monitoring (Figure 8.3).

8.2.1 Precision Agriculture

Precision agriculture is needed to minimize the labor of farmers and increase their crop profitability and productivity (Ruiz-Garcia et al. 2009b; Jiber et al. 2011). The process of using different techniques, new technologies, and management practices in agriculture can be referred to as *precision agriculture* (or *precision farming*).

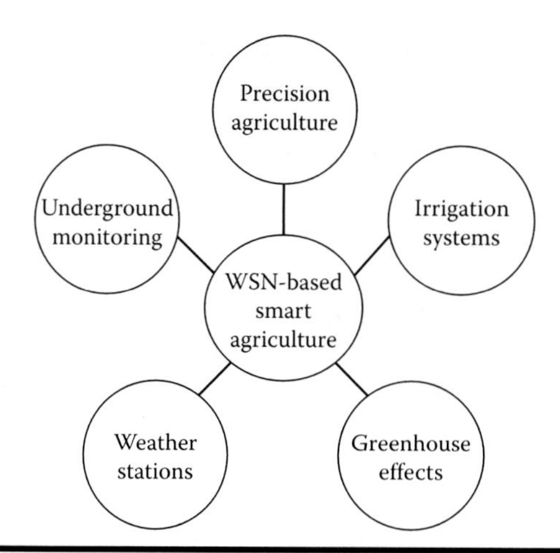

Figure 8.3 Major areas for deployment of wireless sensor networks in agriculture.

Precision agriculture is an environment for observing crops, and measuring and controlling the pre- and post-duties of farmers of field-wide scenarios. Precision agriculture can be illustrated as a methodology of applying inputs (water, insecticides, fertilizers, etc.) at the best time and location for improving the production quality and quantity of the crops (Shinghal et al. 2010). Precision agriculture aspects can be obtained easily by using WSNs (Garcia-Sanchez et al. 2011). The sensors can monitor temperature, humidity, soil properties, plant growth status, and the day-to-day conditions of the field, and necessary actions can be taken even remotely.

Garcia-Sanchez et al. (2011) proposed distributed crop monitoring and video surveillance for precision agriculture. Sensor nodes are used for data collection (soil moisture, pH, salinity, and temperature), and cameras and motion detectors are utilized for surveillance. Video surveillance provides facility to the farmers for protecting their fields and farm machinery. Jiber et al. (2011) developed the iFram framework system for agricultural monitoring of processes such as irrigation systems, pest detection, and environmental monitoring. This system was designed for increasing crop productivity and to give solutions for parameters affecting crops. The iFram system provides solutions for precision agriculture, microclimates, smart irrigation management, conservation, and environmental factors. This is a WSN-based system consisting of soil moisture, temperature, humidity, and water content sensors. The iFram system is easy to deploy and understand, and it is flexible for agricultural monitoring.

8.2.2 Irrigation Systems

Generally, water plays an important role in the agriculture sector. Due to this fact, researchers have focused on adequate ways to organize the proper usage of water (de Lima et al. 2010; Shinghal et al. 2010; Angelopoulos et al. 2011).

The proper amount of water should be applied to crops in order to avoid over- or underirrigation. An irrigation system is used for controlling and managing the irrigation schedule. When crops are being irrigated, two characteristics of the soil must be observed (Shinghal et al. 2010): (1) the water absorption rate of the soil and (2) the amount of water stored by the soil.

Many sensors are available on the market to capture data for irrigation (Technical Guide 2013). The sensors needed for irrigation include liquid flow and level sensors, soil moisture sensors, soil depth sensors, and humidity sensors. In addition, actuators and a valve controller are needed to control the flow and level of water.

Balendonck et al. (2008) have worked on irrigation management and introduced a system called *FLOW-AID* (Farm-Level Optimal Water Management Assistant for Irrigation under Deficit). The main objective of FLOW-AID is to organize and optimize the use of water under deficit conditions, when water resources are limited. This system has focused on irrigation issues and contributed to managing those issues. In this system, various irrigation controllers are used and distributed over the agricultural fields for the management of irrigation scheduling. These controllers are programmed to run autonomously. WSN has been employed for monitoring and controlling the whole system scenario, for which sensor nodes and repeaters are used.

In another related effort, Angelopoulos et al. (2011) developed a smart watering system for irrigation of plants using WSNs. The architecture of the system includes sensor nodes (including soil humidity sensors), and electro-valves are used for controlling water flow. The collected data from the field are stored in a MySQL database at the base station.

8.2.3 Greenhouses

There are various important measuring parameters to monitor and control in greenhouses, such as light, temperature, humidity, level of CO_2, and adequate irrigation (Lea-Cox et al. 2007; Ahonen et al. 2008). Continuous measuring of these parameters and proper adjustment will help with understanding the requirements of greenhouse plants for average growth and with management for optimum production and profit. In addition, controlling and monitoring parameters such as dew and moisture is also important. A WSN-based greenhouse is more effective for fine-grained monitoring of parameters like temperature, relative humidity, light, CO_2, and so on. For proper monitoring of greenhouses, the position, orientation, and gap between the sensor nodes have significant impacts on the results. Therefore, careful attention is needed to the deployment of WSNs for greenhouses.

Seong-eun Yoo et al. (2007) performed work on a real deployment of an automated agriculture system (A^2S) in a greenhouse. This system is composed of a WSN (25 sensor nodes, 3 base stations, and 1 actuator), gateways (TCP/IP), and a subsystem for management (database, application server, and remote access a web server).

This system was utilized for controlling the environmental parameters and monitoring the growth process of melons and cabbage in a greenhouse.

Ahonen et al. (2008) proposed a WSN-based system for measuring and controlling the environmental parameters of greenhouses. The system comprises sensors measuring luminosity, temperature, humidity, and CO_2 because these factors play an important role in the growth of the plant, as well as its quality and productivity. Four sensor nodes were deployed at the corners of the greenhouse for continuous monitoring.

8.2.4 Weather Stations

Among other applications for smart agriculture, monitoring weather factors such as wind speed, temperature, humidity, atmospheric pressure, rainfall, dew point, and so on (Akyildiz et al. 2002a; Corke et al. 2010; Kapoor et al. 2011; Yu et al. 2012) are also essential. For example, rainfall information is important for irrigation systems, whereas temperature and humidity are necessary to measure for the application of pesticides. When the humidity is high, more pesticide needs to be applied, whereas at low humidity levels a moderate amount can be applied. In this area, again deployment of WSN will give fine-grained information from different areas of the field and different decisions can be made accordingly.

8.2.5 Underground Monitoring

Currently, major research in smart agriculture is based on terrestrial information systems. Surface deployment is more simple and convenient than underground deployment. For collecting underground information, sensors and systems are completely buried in the ground at a certain depth (Akyildiz and Stuntebeck 2006; Heidemann et al. 2006; Yu et al. 2012; Technical Guide 2013). Underground information systems are used to find out soil moisture, soil tension, water depth, soil water content, and so on. The purpose of an underground system is to collect the data and then transmit that data to the surface for further processing. Thus, for better productivity the soil type and its properties must be measured and known (Akyildiz and Stuntebeck 2006).

8.3 WSN-SA: A Generic WSN-Based Smart Agriculture Architecture

In order to cover the entire spectrum of smart agriculture applications using WSNs, it is necessary to have a generic architecture that covers almost all aspects. Accordingly, the general architecture of WSN-based smart agriculture (WSN-SA) consists of three basic generic components: system input, system output, and a system communication component (as shown in Figure 8.4). The different components

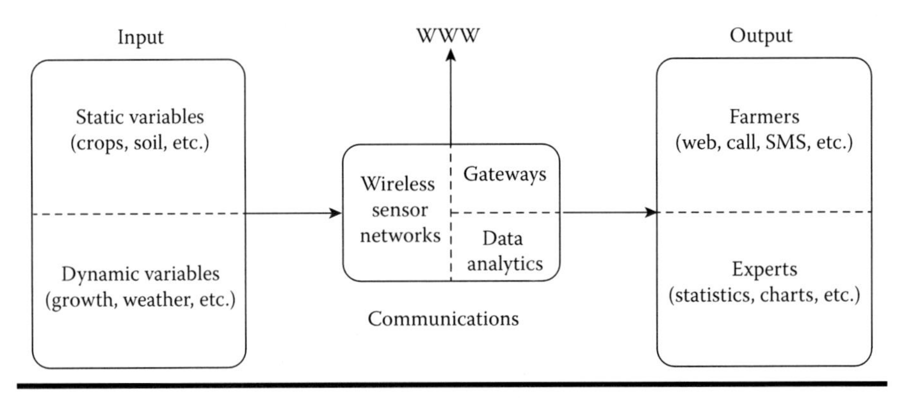

Figure 8.4 Generic smart agriculture architecture.

are flexible enough to adapt according to the different application requirements. Following is a description of the three components of WSN-SA.

8.3.1 System Input Module

This module contributes to sensing and collecting data from the field and comprises a variety of different sensors. There are two types of input variables: static input (e.g., soil properties) and dynamic input (e.g., plant growth, temperature, humidity, moisture, etc.). The system input module comprises different sensors that are used to collect the data. The system input module a highly complex arrangement of multiple integrated digital and analog systems and technologies. Figure 8.5 shows an overview of the main submodules of the input module: the sensing unit, processing unit, transceiver unit, and power unit.

8.3.1.1 Sensing Unit

The smart agriculture sensing unit consists of one or more sensors and analog-to-digital converters (ADC). The sensor senses a physical phenomenon and produces an analog signal, then the ADC converts it into digital format for feeding into the processing unit. There are as many as thousands of different types and categories of sensors available on the market with varying classifications. According to White (1987), sensors can be classified on the basis of (1) measurement, (2) technology, (3) detection of phenomena, (4) conversion, (5) sensor packaging, and (6) application. Based on this classification, Table 8.1 summarizes sensors based on class, type, and application area.

8.3.1.2 Processing Unit

This section is composed of two other subunits: the processor (CPU) and memory (volatile and nonvolatile). Here, the processor manages the sensor collaboration

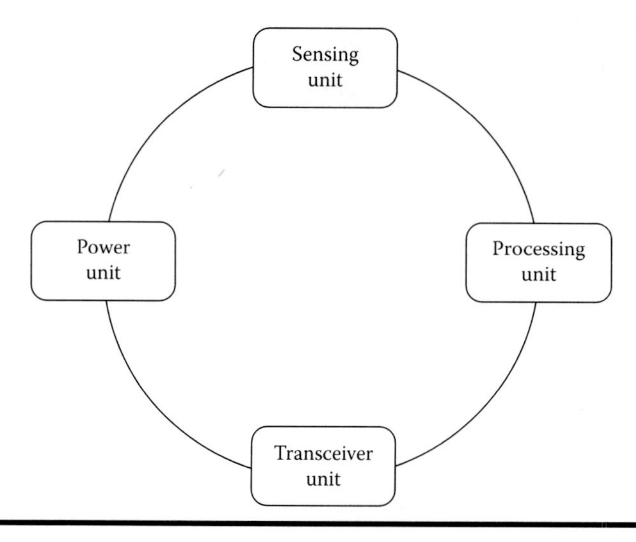

Figure 8.5 Overview of WSN-SA system input module.

Table 8.1 Different Sensor Classes, Types, and Application Areas

Class	*Type*	*Application*
Physical sensors	Temperature sensor	Plant, soil, and environment
	Humidity sensor	Plant, soil, and environment
	Watermark sensor	Soil humidity and electronic adaption
	Leaf wetness sensor	Plants, crops, and trees
	Terrestrial sensor	Surface of the field (monitoring and detection)
	Underwater sensor	Underground
	Underground sensor	Soil moisture, soil compaction, and salinity
	Water level sensor	Irrigation
	Luminosity (LDR)	Sunlight
Mechanical sensors	Insect detector	Pest/bug detection (greenhouse and agriculture)
	Pressure sensor	Atmospheric
	Wind sensor	Speed and direction of air
	Water level sensor	Ground and underground
Chemical sensors	Biosensor	Glucose, ascorbic acid, and lactic acid
	Gas sensor	Air, CO_2, and O_2

Table 8.2 ZigBee, Bluetooth, and Wi-Fi Protocols

Protocol	*Standard*	*Frequency*	*Communication Range*	*Data Rate*	*Power Consumption*
ZigBee	802.15.4	2.4 GHz	<100 m	250 Kbit/s	Low
Bluetooth	802.15.1	2.4 GHz	>8 m	1 Mbit/s	Medium
Wi-Fi	802.11	2.4 GHz	>50 m	11 Mbit/s	High

process to accomplish the assigned tasks. The memory is required for code and data storage purposes.

8.3.1.3 Transceiver Unit

There are three standard wireless technologies widely used in smart agriculture systems: ZigBee, Bluetooth, and Wi-Fi (Buratti et al. 2009; Ruiz-Garcia et al. 2009b; Technical Guide 2013). The standard, frequency, range, data rate, and power consumption of these wireless technologies are mentioned in Table 8.2.

8.3.1.4 Power Unit

The power unit can be supported by either a DC source or rechargeable batteries. Currently, there are many efforts to harvest energy from the ambient environment (e.g., solar energy). Generally, the sensing and processing units consume less power compared with the transceiver unit.

8.4 Typical WSN Platforms to Enable Smart Agriculture

Various WSN platforms have been designed for smart agriculture. Libelium (Technical Guide 2013) introduced a smart agriculture board (Figure 8.6a) and the Waspmote Plug and Sense module (Figure 8.6b) that allows monitoring of various environmental parameters. On the board, multiple sensors have been provided such as soil temperature and humidity, air temperature and humidity, soil moisture, solar radiation, rainfall, wind speed, wind direction, atmospheric pressure, leaf wetness, and dendrometers. Autonomous Systems Lab (CSIRO ICT Centre) has designed the Fleck platform for WSN (Sitka et al. 2007; Corke et al. 2010) (Figure 8.7a). For outdoor environmental monitoring, the Fleck family is a very suitable candidate. CSIRO Centre has also designed a versatile WSN device of small size and low cost, called Fleck Nano Wireless In-rumen (Kusy 2012) (Figure 8.7b). This tiny device is equipped with many sensors, such as temperature, pH, and CO_2, that can be easily deployed for environmental monitoring in smart agriculture applications.

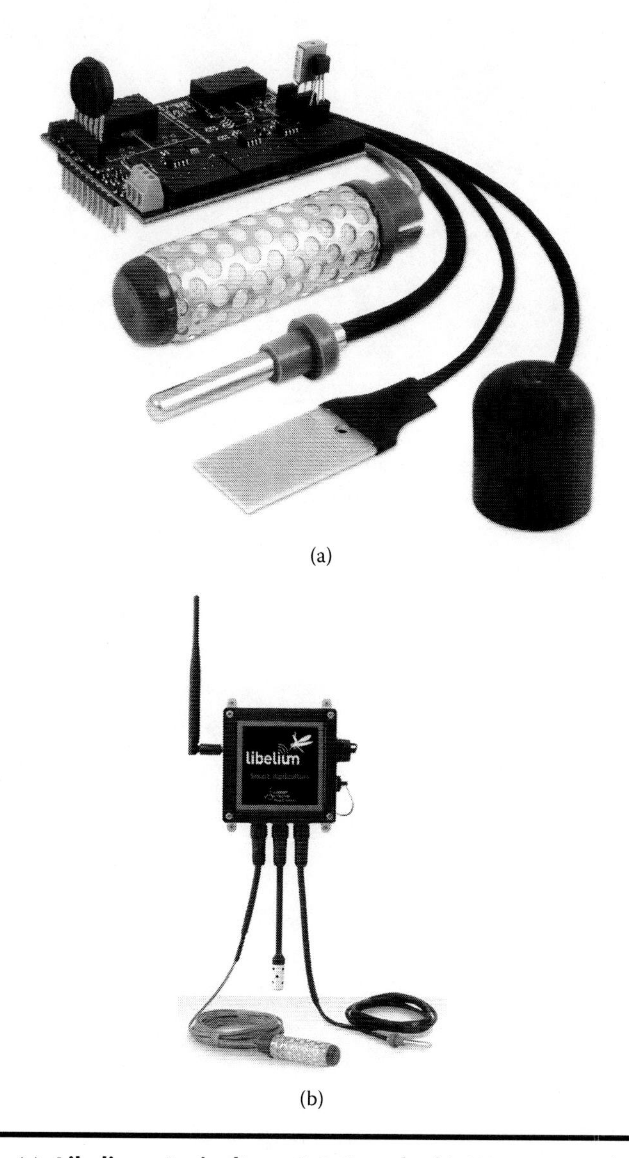

Figure 8.6 (a) Libelium Agriculture 2.0 Board; (b) Waspmote Plug & Sense module. (From Technical Guide, http://www.libelium.com/uploads/2013/02/agriculture-sensor-board_2.0_eng.pdf, 2013.)

The IRIS mote (Figure 8.8a) is a widely used WSN mote platform developed by Crossbow. The MDA300 Data Acquisition Board (Figure 8.8b) is commonly used with the IRIS mote in order to derive sensor output. This platform can be deployed in the agriculture sector for monitoring a variety of parameters. The smart watering system (Angelopoulos et al. 2011) is a real-time application for IRIS mote deployment.

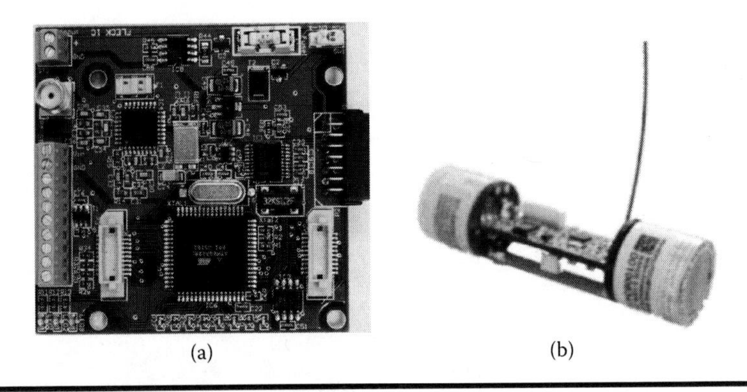

(a) (b)

Figure 8.7 (a) The Fleck node (Sitka et al. 2007); (b) Fleck Nano In-rumen sensor. (From Kusy, B., *Opal sensor node: Computation and communication in WSN platforms*. Autonomous Systems Lab, CSIRO ICT Centre, 2012.)

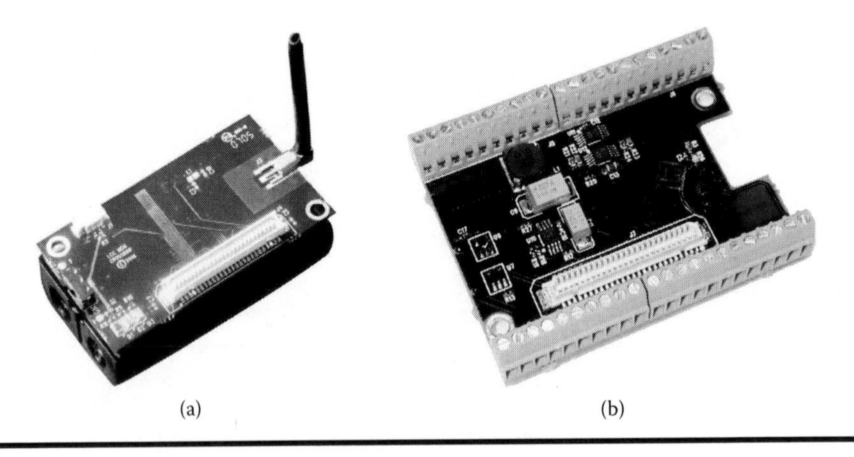

(a) (b)

Figure 8.8 (a) MDA300 data acquisition board; (b) IRIS mote.

8.4.1 System Communication Module

This module is responsible for transmitting the data from a WSN to the base station and providing the end user with access to the data via the Internet. The data can also be stored in the database via the base station on a database server. The data can be transmitted using different techniques between the WSN and base station (ZigBee, Bluetooth), between the base station and gateway (Bluetooth, Wi-Fi), and between the gateway and database (Wi-Fi, wired). Beyond the base station, high-end Internet connectivity is generally available for communication, and the approaches are fairly well defined (e.g., TCP/IP protocol stack). To enable WSN-based smart agriculture, the sensor nodes are deployed in such a way that they

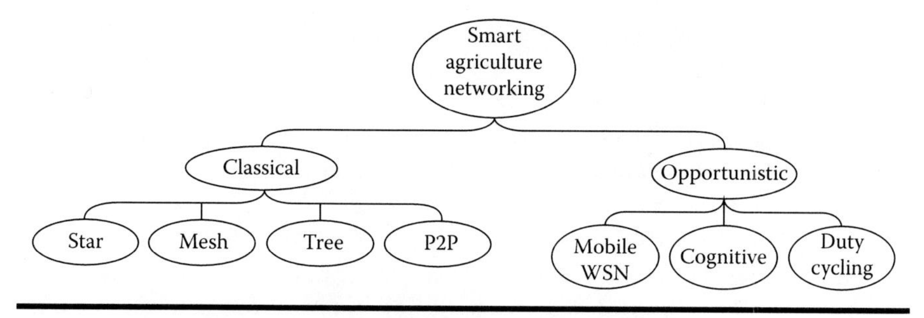

Figure 8.9 Hierarchy of networking in smart agriculture.

will cover the field and meet all the information needs (Konstantinos et al. 2007). Accordingly, the topology of the network (established using sensor nodes) plays an important role for nodes to properly send the data to the base station and ultimately to the end users.

Sensor node deployment in the field can be either fixed (predefined and calculated positions) or random (e.g., nodes dropped from the air into the field). After deployment the sensor nodes start to communicate with each other, forming a wireless network. Data transfer among the sensor nodes can utilize classical network architecture or can exploit opportunistic networking. Based on this, the networking of wireless nodes in smart agriculture is divided into two main classes: classical and opportunistic (Figure 8.9). It should be noted that wireless links between the sensor nodes are generally dynamic in nature.

8.4.1.1 Classical Networking

The classical networking class includes well-defined networking architectures from the wired and wireless domains. The classical networking class is further divided into various subclasses: star, mesh, tree, and peer-to-peer (P2P) topologies.

In star topology, each node is connected or sends the data directly to the base station (Buratti et al. 2009) and the nodes have no direct communication with each other. The most important advantage of a star topology is that the failure of any sensor node will not disturb the communication of other sensor nodes with the base station, whereas if the base station fails the whole network comes down. Furthermore, the star network is suitable only for small-sized fields, because the sensor nodes are equipped with radio, which can communicate only over short distances in order to save the energy costs associated with wireless communication. Some practical examples of the star network in the agriculture sector can be found in the literature (Ruiz-Garcia et al. 2009b; Merrett and Tan 2010; Garcia-Sanchez et al. 2011).

The mesh topology is a network architecture widely used in the agriculture sector (Anurag et al. 2008; Balendonck et al. 2008; Siuli Roy and Bandyopadhyay 2008; Ruiz-Garcia et al. 2009b; Merrett and Tan 2010; Jiber et al. 2011). In the mesh network, multiple communication paths exist between the nodes

(Siuli Roy and Bandyopadhyay 2008; Ruiz-Garcia et al. 2009a). All the sensor nodes can communicate directly with each other. This network is suitable for high traffic volume between the sensor nodes. The main advantage of a mesh is that of fault tolerance—that is, the data can be routed over alternate paths if some nodes in the network start to malfunction or die out. The main disadvantages of the mesh network are its complexity and the increased processing required by the tiny sensor nodes.

In the P2P layout, the sensor nodes can share information with each other without sending it to the base station (Sahar et al. 2011; Sahar and Shaikh 2012; Sahar et al. 2013). The main advantage of P2P WSN is that the information is available and stored within the network as well as the base station. Therefore, if an intermediate sensor node requires some information it does not have to ask the base station but can extract it from the nearby sensor nodes. Garcia-Sanchez et al. (2011) and Heidemann et al. (2006) used P2P overlays for information transfer in smart agriculture.

The tree topology can be formed by combining two or more star networks. In the tree topology, different sections are routed to the main root node—the base station. None of the sections in the network can communicate directly with any other section. Similarly, the nodes also follow a path to communicate with each other. The tree topology is also called a *hybrid* or *hierarchical topology* and can be considered for deployment in large areas by transferring data using a multihop approach.

8.4.1.2 Opportunistic Networking

In an opportunistic network configuration, sensor node communication can be dynamic due to mobility in the network, switching the radios to different frequencies, and sleeping of sensor nodes in the network. Accordingly, the opportunistic network class is divided into three subclasses: mobile WSNs, cognitive radio WSNs, and duty cycle–enabled WSNs.

In order to save energy (since batteries are a precious resource) during communication, mobility inside the network can be exploited (Khelil et al. 2009). In a general scenario, sensor nodes send data towards the base station in a multihop fashion.

As the number of hops increases, the energy consumption also increases due to an increase in the number of transmissions needed to reach the base station. Alternatively, the source sensor nodes can store data and when they encounter a mobile node, they transfer the data to it in a single hop. Once the mobile node reaches the base station, it offloads the data to it, saving lot of energy by having a lower number of transmissions. In some scenarios, the mobility can be controlled, whereas in other cases it may not be possible to do so. The mobile nodes can be farmers moving in the field or specialized robots moving around in the environment (Jea et al. 2005; Jun et al. 2005; Jain et al. 2006). Furthermore, the mobile nodes can interact with the existing smart agriculture network. Real-time application of mobile WSN in smart agriculture can be found in Corke et al. (2010); Singh and Bansal (2011); Ipema et al. (2008); Valente et al. (2011); and Tinka et al. (2009).

A cognitive radio network is a wireless communication network in which the sensor nodes are aware of their ambient radio environment. The sensor nodes adapt according to the current radio environment and perform spectrum-efficient wireless networking to provide sufficient bandwidth for communication. By doing this, the sensor nodes also ensure a minimum of interference with other sensor nodes in the vicinity (Bhattacharya et al. 2011). The unique characteristics of cognitive radio are reliability, adaptivity, and dynamic coordination for data transfer (Haykin 2005; Fette and Fette 2006). Wang and Wei (2009); Dong et al. (2012); Akyildiz et al. (2009); and NICTOR (2013) provided a description of how cognitive radio can be utilized for smart agriculture.

To save energy in WSNs, duty cycling is exploited by keeping the sensor nodes in sleep mode when there is no data to send (Shaikh et al. 2013). Sensor nodes can perform duty cycling according to a schedule or on demand when user data needs to be delivered. The idea behind on-demand approaches is that the sensor node will awaken before the data arrive from the neighboring node. Generally, two different channels are used: a data channel (for normal data transfer) and a wakeup channel.

Some scheduled duty cycling strategies require that all neighboring nodes wake up at the same time (Anastasi et al. 2006; Keshavarzian et al. 2006), whereas other strategies avoid the tight synchronization and allow each node to wake up independently (Zheng et al. 2003; Paruchuri et al. 2004). When nodes are awakened synchronously, the data can traverse easily between the source and destination. In contrast, when sensor nodes are awakened independently, the forwarding path may suffer high latency because the other sensor nodes along the path may not be available during the same time period. To save energy in WSNs in smart agriculture, duty cycling has been applied in potato farming, agriculture land monitoring, and to investigate the impact of permafrost in the Swiss Alps (Beckwith et al. 2004; Langendoen et al. 2006; Camilli et al. 2007; Talzi et al. 2007).

8.4.2 System Output Module

The system output module is responsible for showing results in a simplified way that framers can easily understand, for example, by displaying alerts in a web browser, by e-mail, SMS, voice call, and so on. It should also provide detailed analysis and pictorial/technical results for experts. The main idea behind this module is to provide tools that can interact with the base station or sensors directly and present the statistics in a comprehensive manner. One of the major initiatives is MOTE-VIEW, which is an interface (client layer) between an end user and a deployed WSN. It provides the tools to simplify deployment and monitoring. It also makes it easy to connect to a database, to analyze data, and to plot sensor readings. Al Nuaimi et al. (2012) presented a good survey of Web-based tools for WSN monitoring and analysis. Despite of all these efforts, there is a need for a generic output module that can be optimized for smart agriculture.

8.5 WSN Deployment for Smart Agriculture: A Case Study

In order to deploy the WSN for smart agriculture, some basic guidelines should be followed (Mampentzidou et al. 2012). These guidelines will help in proper management of the deployed WSN and to understand the different phenomena occurring inside the network.

Predeployment: Before deployment of any technology (e.g., WSNs), it is important to know about budget constraints. The deployment location must be chosen—indoor (e.g., greenhouse) or outdoor (e.g., crop field). In addition, the requirements of the crop to be monitored must be learned. The most important thing is to fix the parameters that need to be monitored such as appropriate sensor requirements should be known.

Sensing: Before choosing the sensors, their sensitivity, precision, resolution, range, life, and power consumption must be kept in mind. There are many sensors (e.g., temperature, humidity, moisture, flow level) available on the market. Depending on the nature of the crop, the parameters and appropriate sensors need to be selected. In addition, the monitoring interval should be decided in order to monitor the crop in an efficient manner.

Hardware and Software: Select the hardware and compatible software in order to integrate the sensors properly. Various commercial platforms are available, such as from Crossbow (IRIS and Mica families), CSIRO (Fleck family), Moteiv (Tmote Sky), Shockfish (TinyNode), and many others. Among these, Crossbow is the leading supplier, with many deployments across the globe (Mampentzidou et al. 2012). In order to increase the number of sensors, it is important that the chosen hardware have sufficient expansion pins to accommodate several sensors. Generally, for software two main operating systems (OS) are competing in the market: Contiki and TinyOS. The chosen hardware will influence the choice of OS.

Communication: Depending on the area covered, the number of sensor nodes and their topology will be created and maintained. Other factors such as weather conditions, electromagnetic fields, and the water conditions around the deployment area need to be considered, since they will impact communication between the nodes.

Power: The average period of any crop is about 3–6 months; thus, ideally the WSN should work throughout this period. Accordingly, a good battery or a rechargeable option should exist in the deployment. This could be achieved by using renewable energy resources. Solar energy is a very good option. Before selecting a battery, the chemistry of the battery must be known (e.g., lead–acid or lithium–ion) and its behavior under the required conditions must be considered. The main issue is the implementation of power saving and management techniques by using different protocols and algorithms.

Safety and Maintenance: For long-term deployment, the maintenance of the system requires great consideration. The system must be protected from environmental impacts like rain, moisture, mud, and so on. Thus, it is necessary to put the system into protected casings. Furthermore, birds and insects may also damage the sensor nodes or hinder the resolution of the monitored phenomena. Thus appropriate measures should be taken before deployment. Table 8.3 provides an overview of WSN deployment for a cotton crop covering a 1,000 square meter area based on the guidelines discussed above. The WSN in Table 8.3 is an example of a WSN that can be deployed for other crops as well.

8.6 The Road Ahead

Time will decide which technologies become dominant and most suitable for specific and embedded applications, specifically smart agriculture. However, it seems that the machinery used in today's deployment scenarios is not a single technology but a combination of different competitive technologies. Focus on WSNs has been growing, as have its potential and accessibility into the future. There has been extensive use of many technologies for smart agriculture over the last few decades. Satellite remote sensing has been widely used, but now confidence is moving towards another newly emerged technology, WSNs, as discussed earlier. Confidence in the technology can be understood to mean that either (1) the technology is mature enough for larger scale adoption by all stakeholders (i.e., researchers, farmers, and industry) or (2) the technology is ready to step into an advanced phase of R&D to meet the demands of end users. Alternatively, commercialization activities will characterize the future direction for smart agriculture. Furthermore, the technology must mature for integration of large-scale deployments and must be robust with nonrecurring engineering costs. Large-scale deployments with hundreds or thousands of nodes that run for months or years will require some form of hierarchy in the network. Another major challenge for scaling up is detecting node malfunctions, which has become more probable, and recovering from these faults seamlessly.

The recent advancements in WSNs may shift the data flow, storage, and communication patterns in smart agriculture. Until recently, the WSN was predominantly viewed as a data gathering tool where the physical environment was periodically sampled, in-network processing was carried, and the data were sent to an end user. Technological improvements in sensor nodes will enable them to make decisions on the fly by fusing multiple information streams to process, compress, store, actuate, and send sensor data back to end users for recording purposes. There has also been a shift from the use of simple scalar sensors to more complex multimedia sensors, which poses new challenges such as designing appropriate triggering mechanisms and distributed coordination among multiple sensor nodes in a heterogeneous sensor network. Currently, in smart agriculture ZigBee, Bluetooth, Wi-Fi, and so on are widely used. However, in future, we foresee that cognitive radios may replace other wireless technologies. Similarly, software-defined radio,

Table 8.3 Overview of WSN Deployment for a Cotton Crop Covering a 1,000 Square Meter Area

#	*Guideline*	*WSN-SA[a] Module*	*Parameters*	*Description*
1	Predeployment	Input	Budget = good Outdoor Crop = cotton Period = 180 days Area = 1000 m^2 Grid deployment # sensor nodes = 400	Ample amount is available Field One crop season Supposing one sensor node covers 5 m
2	Sensing	Input	Temperature Humidity Water flow Soil moisture	
3	Hardware and Software	Input	Sensor node = IRIS MDA 300 POW110D3B 10HS OS[b] = TinyOS	IRIS is deployed in many other projects too Onboard sensors for IRIS including temperature and humidity Flow sensor Soil moisture sensor Widely used OS

(Continued)

Table 8.3 *(Continued)* **Overview of WSN Deployment for a Cotton Crop Covering a 1,000 Square Meter Area**

#	*Guideline*	*WSN-SA[a] Module*	*Parameters*	*Description*
4	Communication	Communication	Topology = tree, clustered Data rate = 1 msg/5 min Node-node = RF ZigBee Node-BS = RF ZigBee BS-Server = 3G/4G Server-User = Internet	Due to large area of deployment Application specific
5	Power	Input	Solar harvested rechargeable batteries	Due to more sunlight available in the deployed area
6	Safety and maintenance	Output	Node cases = plastic Cameras to avoid theft	Few are used across the field

[a] Wireless sensor network–based smart agriculture.
[b] Operating system.

supporting frequency, antenna, modulation, and data rate diversity, may also flourish and find its way into the smart agriculture domain.

Adaptive power management strategies that efficiently manage the different activities in the smart agriculture field without compromising performance quality also remain an open direction for continued investigation.

The shift towards IP support for WSN enables smart agriculture to be on the frontier of applications for the Internet of Things (IoT). The data can be accessed and the actuators controlled via the Web using IoT architecture.

Another need for the future is smart agriculture standardization. As with other maturing technologies, smart agriculture has reached a stage where standard protocols are needed to control different processes in the field.

One of the major hurdles in smart agriculture technology adaption is the need to make the whole process easy for nonspecialists. Generally, software and hardware deployment requires technological experts who are not very familiar with the physical environment. There is a great need for the software and hardware configuration to be simple enough that a common person can understand and use it without an expert in the loop. This will help wider adaptation of smart agriculture by farmers and landowners. In summary, the next wave in smart agriculture will combine commercialization of the current technology, that is, more and larger scale deployments and continued development of more advanced functionality. This leads to sensor nodes with multiple sensing capabilities and diverse radio configurations, as well as the continuous redefinition of the design space to identify and address the challenges of smart agriculture.

8.7 Conclusion

This chapter broadly discusses the architecture, available platforms, deployment guidelines, and a case study to show the usefulness of WSNs in smart agriculture. In addition, we have mentioned different approaches and techniques for data transfer in smart agriculture.

Smart agriculture has been adopted in various agricultural applications for monitoring and detecting of various agricultural parameters, such as inputs, temperature, humidity, water level and flow, and so on. As smart agriculture has already been deployed on a large scale, so there should be effective mechanisms to increase the production quality and efficiency.

Acknowledgments

This work was carried out under the financial support of the Mehran University of Engineering and Technology, Jamshoro, Pakistan, and STU, Umm Al-Qura University, Makkah, KSA.

References

10hs soil moisture sensor, Decagon Devices. (Accessed on, March 1, 2014). http://www.decagon.com/en/soils/volumetric-water-content-sensors/10hs-large-volume-vwc/.

Abdelmajid Khelil., Faisal Karim Shaikh, Azad Ali, and Neeraj Suri. 2009. gMAP: An efficent construction of global maps for mobility-assisted wireless sensor networks. In *Proceedings of the Conference on Wireless on Demand Network Systems and Services* (*WONS*), 189–196.

Abdullah, Ahsan, and Ahmed Barnawi. 2012. Identification of the type of agriculture suited for application of wireless sensor networks. *Russian Journal of Agricultural and Socio-Economic Sciences* 12(12):19–36.

AH (Bert) Ipema, D. Goense, PH (Pieter) Hogewerf, HWJ (Wim) Houwers, and H. van Roest. 2008. Pilot study to monitor body temperature of dairy cows with a rumen bolus. *Computers and Electronics in Agriculture* 64(1):49–52.

Ahonen, Teemu, Reino Virrankoski, and Mohammed Elmusrati. 2008. Greenhouse monitoring with wireless sensor network. *IEEE/ASME International Conference on Mechtronic and Embedded Systems and Application, MESA 2008*, 403–408.

Akyildiz, Ian F., and Erich P. Stuntebeck. 2006. Wireless underground sensor networks: Research challenges. *Ad Hoc Networks* 4(6):669–686.

Akyildiz, Ian F., Weilian Su, Yogesh Sankarasubramaniam, and Erdal Cayirci. 2002a. Wire-less sensor networks: A survey. *Computer Networks* 38(4):393–422.

Akyildiz, Ian F., Weilian Su, Yogesh Sankarasubramanian, and Erdal Cayirci. 2002b. A survey on sensor networks. *IEEE Communications Magazine* 42(5):102–114.

Akyildiz, Ian F., Won-Yeol Lee, and Kaushik R. Chowdhury. 2009. Crahns: Cognitive radio ad hoc networks. *Ad Hoc Networks* 7(5):810–836.

Alberts, Maris, Ugis Grinbergs, Dzidra Kreismane, Andris Kalejs, Andris Dzerve, Vilnis Jekabsons, Normunds Veselis, Viktors Zotovs, Liga Brikmane, and Baiba Tikuma. 2013. New wireless sensor network technology for precision agriculture. *International Conference on Applied Information and Communication Technologies* (*AICT2013*), 153–162.

Al Nuaimi, Klaithem, Mariam Al Nuaimi, Nader Mohamed, Imad Jawhar, and Khaled Shuaib. 2012. Web-based wireless sensor networks: A survey of architectures and applications. In *Proceedings of the 6th International Conference on Ubiquitous Information Management and Communication*, vol. 113, ACM.

Anastasi, Giuseppe, Marco Conti, Mario Di Francesco, and Andrea Passarella. 2006. An adaptive and low-latency power management protocol for wireless sensor net-works. In *Proceedings of the 4th ACM International Workshop on Mobility Management and Wireless Access*, 67–74, ACM.

Angelopoulos, Constantinos Marios, Sotiris Nikoletseas, and Georgios Constantinos Theofanopoulos. 2011. A smart system for garden watering using wireless sensor networks. In *Proceedings of the 9th ACM International Symposium on Mobility Management and Wireless Access*, 167–170. MobiWac '11, New York: ACM.

Anurag Dugar, Siuli Roy, and Somprakash Bandyopadhyay. 2008. Agro-sense: Precision agriculture using sensor-based wireless mesh networks. In *Proceeding of: Innovations in NGN: Future Network and Services, 2008. K-INGN 2008. First ITU-T Kaleidoscope Academic Conference.*

Baggio, Aline. 2005. Wireless sensor networks in precision agriculture. In *ACM Workshop on Real-World Wireless Sensor Networks* (*realwsn 2005*), Stockholm, Sweden.

Beckwith, Richard, Dan Teibel, and Pat Bowen. 2004. Report from the field: Results from an agricultural wireless sensor network. In *29th Annual IEEE International Conference on Local Computer Networks*, 471–478, IEEE.

Bhattacharya, Partha P., Ronak Khandelwal, Rishita Gera, and Anjali Agarwal. 2011. Smart radio spectrum management for cognitive radio. International Journal of Distributed and Parallel Systems (IJDPS), *CoRR* abs/1109.0257, 2 (4):12–24.

Bruce A. Fette and Bruce Fette. 2006. *Cognitive Radio Technology (Communications Engineering)*. ISBN:0750679522, Newnes.

Buratti, Chiara, Andrea Conti, Davide Dardari, and Roberto Verdone. 2009a. An overview on wireless sensor networks technology and evolution. *Sensors* 9(9):6869–6896.

Camilli, Alberto, Carlos E. Cugnasca, Antonio M. Saraiva, André R. Hirakawa, and Pedro L.P. Corrêa. 2007. From wireless sensors to field mapping: Anatomy of an application for precision agriculture. *Computers and Electronics in Agriculture* 58(1):25–36.

Corke, Peter, Tim Wark, Raja Jurdak, Wen Hu, Philip Valencia, and Darren Moore. 2010. Environmental wireless sensor networks. *Proceedings of the IEEE* 98(11):1903–1917.

Datasheet, IRIS Mote. (Accessed on August 5, 2013)a. *Wireless measurement system.* Document Part Number: 6020-0124-01 Rev A. http://www.xbow.com.

Datasheet, MDA300. (Accessed on August 5, 2013)b. *Data acquisition board.* Document Part Number: 6020-0052-04 Rev A, http://www.memsic.com.

de Lima, Gracon H.E.L., Lenardo C. e Silva, and Pedro F.R. Neto. 2010. WSN as a tool for supporting agriculture in the precision irrigation. *Sixth International Conference on Networking and Services, IEEE Computer Society,* 137–142.

Dong, Xin, Mehmet C. Vuran, and Suat Irmak. 2012. Autonomous precision agriculture through integration of wireless underground sensor networks with center pivot irrigation systems. *Ad Hoc Networks* 11(7):1975–1987.

Garca-Hernndez, Carlos F., Pablo H. Ibargengoytia-Gonzlez, Joaqun Garca-Hernndez, and Jess A. Prez-Daz. 2007. Wireless sensor networks and applications: A survey. *International Journal of Computer Science and Network Security* (*IJCSNS*) 7(2):264–273.

Garcia-Sanchez, Antonio-Javier, Felipe Garcia-Sanchez, and Joan Garcia-Haro. 2011. Wireless sensor network deployment for integrating video-surveillance and data-monitoring in precision agriculture over distributed crops. *Computers and Electronics in Agriculture* 75(2):288–303.

Haykin, Simon. 2005. Cognitive radio: Brain-empowered wireless communications. *IEEE Journal on Selected Areas in Communications* 23(2):201–220.

Heidemann, John, Wei Ye, Jack Wills, Affan Syed, and Yuan Li. 2006. Research challenges and applications for underwater sensor networking. In *Proceedings of the IEEE Wireless Communications and Networking Conference*, 228–235.

Ipsita Das, Kadayinti Naveen, Shailendra Yadav, Abhishek Kodilkar, Narendra Shah, Shabbir Merchant, and Uday Desai. 2009. WSN monitoring of weather and crop parameters for possible disease risk evaluation for grape farms—Sula vineyards, a case study. *Presented in the Geomatri09 Indian Conference, Oral Session Five.*

Jain, Sushant, Rahul C. Shah, Waylon Brunette, Gaetano Borriello, and Sumit Roy. 2006. Exploiting mobility for energy efficient data collection in wireless sensor networks. *Mobile Networks and Applications* 11(3):327–339.

Jea, David, Arun A. Somasundara, and Mani B. Srivastava. 2005. Multiple controlled mobile elements (data mules) for data collection in sensor networks. In *Proceedings of the IEEE/ACM International Conference on Distributed Computing in Sensor Systems* (*DCOSS*), 244–257.

Jos Balendonck, Jochen Hemming, Eldert J. van Henten, Luca Incrocci, Alberto Pardossi, and Paolo Marzialetti. 2008. Sensors and wireless sensor networks for irrigation management under deficit conditions (flow-aid). *Coordinated by Wageningen University and Research Centre in the Netherlands.*

Jiber, Yassine, Hamid Harroud, and Ahmed Karmouch. 2011. Precision agriculture monitoring framework based on WSN. In *IWCMC*, 2015–2020, IEEE.

Jun, Hyewon, Wenrui Zhao, Mostafa H. Ammar, Ellen W. Zegura, and Chungki Lee. 2005. Trading latency for energy in wireless ad hoc networks using message ferrying. In *Proceedings of the Third IEEE International Conference on Pervasive Computing and Communications Workshops*, 220–225.

Kapoor, Namarta, Nitin Bhatia, Sangeet Kumar, and Simranjeet Kaur. 2011. Wireless sensor networks: A profound technology. *International Journal of Computer Science and Technology* (*IJCST*) 2(2):211–215.

Karl, Holger, and Andreas Willig. 2003. *A short survey of wireless sensor networks.* TKN Technical Reports Series, TKN-03-018, Editor: Adam Wolisz, Telecommunication Network Group, Technical University, Berlin, October 2013.

Keshavarzian, Abtin, Huang Lee, and Lakshmi Venkatraman. 2006. Wakeup scheduling in wireless sensor networks. In *Proceedings of the 7th ACM International Symposium on Mobile Ad Hoc Networking and Computing*, 322–333, ACM.

Konstantinos, Katsalis, Xenakis Apostolos, Kikiras Panagiotis, and Stamoulis George. 2007. Topology optimization in wireless sensor networks for precision agriculture applications. In *Proceedings of the 2007 International Conference on Sensor Technologies and Applications*, SENSORCOMM '07, 526–530.

Kusy, Brano. 2012. (Accessed on April 7, 2014). *Opal sensor node: Computation and communication in WSN platforms.* Autonomous Systems Lab, CSIRO ICT Centre. http://www.csiro.au.

Langendoen, Koen, Aline Baggio, and Otto Visser. 2006. Murphy loves potatoes: Experiences from a pilot sensor network deployment in precision agriculture. In *20th International Parallel and Distributed Processing Symposium, IPDPS 2006*, 8, IEEE.

Lea-Cox, John D., George Kantor, Joshua Anhalt, Andrew Ristvey, and David S. Ross. 2007. A wireless sensor network for the nursery and greenhouse industry. *SNA Research Conference, Engineering, Structures and Innovations Section* 52:454–458.

Mampentzidou, Ioanna, Eirini Karapistoli, and Anastasios A. Economides. 2012. Basic guidelines for deploying wireless sensor networks in agriculture. In *2012 4th International Congress on Ultra Modern Telecommunications and Control Systems and Workshops* (*ICUMT*), 864–869, IEEE.

Moteview works. Memsic Inc. (Accessed on March 31, 2014). http://www.memsic.com/wireless-sensor-networks/WSNSoftwareDwnld.cfm.

Merrett, Geoff V. and Tan, Yen Kheng (eds.) 2010. Wireless sensor networks: application-centric design, Rijeka, Croatia: InTech, pp. 504. http://www.intechopen.com/. (Accessed on December 12, 2013).

NICTOR Sensor Network Platform. (Accessed on October 1, 2013). http://www.cse.unsw.edu.au/~sensar/hardware/pictures/Nictor.pdf

Panchard, Jacques, Seshagiri Rao, Madavalam S. Sheshshayee, Panagiotis Papadimi-tratos, Sumanth Kumar, and Jean-Pierre Hubaux. 2008. Wireless sensor networking for rain-fed farming decision support. In *Proceedings of the Second ACM SIGCOMM Workshop on Networked Systems for Developing Regions*, 31–36.

Paruchuri, Vamsi, Shivakumar Basavaraju, Arjan Durresi, Rajgopal Kannan, and S. Sitharama Iyengar. 2004. Random asynchronous wakeup protocol for sensor networks. In *Broadband Networks, 2004. Broadnets 2004. Proceedings. First International Conference on*, 710–717, IEEE.

POW110D3B, G1/2 water flow sensor. (Accessed on March 1, 2014). https://www.seeedstudio.com/G1%26amp%3B2%26quot%3B-Water-Flow-Sensor-p-635.html#

Reichenbach, Frank, Andreas Bobek, Philipp Hagen, and Dirk Timmermann. 2006. Increasing lifetime of wireless sensor networks with energy-aware role-changing. In *Proceedings of the Second IEEE International Conference on Self-Managed Networks, Systems, and Services*, 157–170. SelfMan'06, Berlin: Springer-Verlag.

Ruiz-Garcia, Luis, Loredana Lunadei, Pilar Barreiro, and Ignacio Robla. 2009. A review of wireless sensor technologies and applications in agriculture and food industry: State of the art and current trends. *Sensors* 9(6):4728–4750.

Syeda Nida Sahar, and Faisal Karim Shaikh. 2012. Data management in mobile wireless sensor networks. In *International Multi-Topic Conference*, 409–420. Springer, Berlin.

Syeda Nida Sahar, Faisal Karim Shaikh, Sana Hoor Jokhio. 2011. P2P based data management in WSNs: Experiences and lessons learnt from a real-world deployment. *Mehran University Research Journal of Engineering and Technology* 30(4):689–698.

Sahar, Syeda Nida, Faisal Karim Shaikh, and Imran Jokhio. 2013. P2P data management in mobile wireless sensor network. *Mehran University Research Journal of Engineering and Technology* 32:339–352.

Shaikh, Faisal Karim, Sherali Zeadally, and Farhan Siddiqui. 2013. Energy efficient routing in wireless sensor networks. In *Next-Generation Wireless Technologies: 4G and Beyond*, edited by Chilamkurti, Naveen, Zeadally, Sherali, and Chaouchi, Hakima. 131–157, London: Springer.

Shinghal, Kshitij, Arti Noor, Neelam Srivastava, and Raghuvir Singh. 2010. Wireless sensor networks in agriculture: For potato farming. *International Journal of Engineering Science and Technology* 2(8):3955–3963.

Singh, Iqbal, and Meenakshi Bansal. 2011. Monitoring water level in agriculture using sensor networks. *International Journal of Soft Computing and Engineering (IJSCE)* 1(5):202–204.

Sikka, Pavan., Peter Corke, Leslie Overs, Philip Valencia, and Tim Wark. 2007. Fleck-a platform for real-world outdoor sensor networks. In *3rd International Conference on Intelligent Sensors, Sensor Networks and Information, 2007. ISSNIP 2007.* 709–714.

Siuli Roy, A.D., and S. Bandyopadhyay. 2008. Agro-sense: Precision agriculture using sensor-based wireless mesh networks. In *Innovations in NGN: Future network and services, 2008. K-INGN 2008. First ITU-T Kaleidoscope Academic Conference*, 383–388, IEEE.

Talzi, Igor, Andreas Hasler, Stephan Gruber, and Christian Tschudin. 2007. PermaSense: Investigating permafrost with a WSN in the Swiss Alps. In *Proceedings of the 4th Workshop on Embedded Networked Sensors*, 8–12, ACM.

Technical Guide. 2013. (Accessed on March 14, 2014). *Agriculture 2.0 document version: v4.0–02/2013*. Libelium Comunicaciones Distribuidas S.L. http://www.libelium.com.

Tinka, Andrew, Issam Strub, Qingfang Wu, and Alexandre M. Bayen. 2009. Quadratic programming based data assimilation with passive drifting sensors for shallow water flows. In *Proceedings of the 48th IEEE Conference on Decision and Control held jointly with the 2009 28th Chinese Control Conference. CDC/CCC 2009*, 7614–7620, IEEE.

Valente, Joo, David Sanz, Antonio Barrientos, Jaime del Cerro, Ngela Ribeiro, and Claudio Rossi. 2011. An air-ground wireless sensor network for crop monitoring. *Sensors* 11(6):6088–6108.

Wang, Li-feng, and Sheng-qun Wei. 2009. Cognitive engine technology. *ZTE Comunications* 2:003.

White, Richard M. 1987. A sensor classification scheme. *IEEE Transactions on Ultrasonics, Ferroelectrics and Frequency Control* 34(2):124–126.

Xiao, Lei, and Lejiang Guo. 2010. The realization of precision agriculture monitoring system based on wireless sensor network. *International Conference on Computer and Communication Technologies in Agriculture Engineering*, 89–92, IEEE.

Yoo, Seong-eun, Jae eon Kim, Taehong Kim, Sungjin Ahn, and Jongwoo Sun-gand Daeyoung Kim. 2007. A2s: Automated agriculture system based on WSN. In *Proceedings of IEEE International Symposium on Consumer Electronics (ISCE) 2007*, Irving, TX, 20–23 June, 1–5.

Yu, Xiaoqing, Pute Wu, Ning Wang, Wenting Han, and Zenglin Zhang. 2012. Survey on wireless sensor networks agricultural environment information monitoring. *Journal of Computational Information Systems* 8(19):7919–7926.

Zheng, Rong, Jennifer C. Hou, and Lui Sha. 2003. Asynchronous wakeup for ad hoc networks. In *Proceedings of the 4th ACM International Symposium on Mobile Ad Hoc Networking & Computing*, 35–45, ACM.

Chapter 9

Cognitive Radio Networks: Concepts and Applications

S. M. Kamruzzaman, Abdullah Alghamdi, and M. Anwar Hossain

Contents

9.1 Introduction

Wireless communication is considered one of the fastest growing segments of the communications industry. Over the last two decades, there has been an exponential increase in spectrum demands due to new emerging wireless services, which has caused a shortage of allocable wireless spectrum resources [1–3]. According to the current static spectrum allocation policy, each new wireless service/protocol should be assigned a spectrum band that has never been allocated, and therefore most parts of the spectrum under 3 GHz are now allocated to specific use [4]. A number of reports have shown that there is a significantly unbalanced usage of wireless spectrum [5–7], with a small portion of spectrum (e.g., cellular band, unlicensed band) increasingly crowded while most of the remaining spectrum is underutilized. This potential resource scarcity is actually solvable, because the shortage derives from inefficient utilization of spectrum by static spectrum allocation [8].

Cognitive radio (CR) has emerged as a promising technology for maximizing the utilization of the limited radio spectrum while accommodating the increasing amount of services and applications in wireless networks [9]. Spectrum utilization can be improved significantly by introducing an opportunistic spectrum access policy to utilize the temporarily vacant part of the spectrum already licensed to primary users (PUs) by unlicensed opportunistic users or secondary users (SUs), also called *CR users* [10,11]. A CR transceiver is able to adapt to the dynamic radio environment and the network parameters to maximize utilization of the limited radio resources while providing flexibility in wireless access [12]. CR technology has been proposed as a potential solution to share scarce spectrum resources in an opportunistic way while avoiding disruptions to the legacy users of wireless networks. The CR user is allowed to use only locally unused spectrum so that it does not cause any interferences or collisions for the incumbent or PUs. Spectrum measurement reports show that a fixed spectrum policy is becoming unsuitable for today's wireless communications [5–7]. As the frequency spectrum becomes exhausted, CR is becoming a hot research topic in the wireless communications arena [13,14].

The term *cognitive radio* can be defined as follows [15]: A cognitive radio is a radio that can change its transmitter parameters based on interaction with the environment in which it operates. *Cognitive radio networks* (CRNs) refers to networks where nodes are equipped with a spectrum agile radio that has the capabilities of sensing the available spectrum band, reconfiguring radio frequency, switching to the selected frequency band, and using it efficiently without interference to PUs [16]. Cognitive radio ad hoc networks (CRANs) are emerging—infrastructureless multihop CRNs. In CRANs, CR users (nodes) can communicate with each other through ad hoc connections [17].

This new area of research foresees the development of CRNs to further improve spectrum efficiency. The basic idea of CRNs is that the CR users need to vacate the frequency band once the PU is detected. CRNs, however, impose unique challenges due to the high fluctuation in available spectrum as well as the diverse quality of service (QoS) requirements [3]. Specifically, in CRANs, the distributed multihop architecture, the dynamic network topology, and the time- and location-dependent spectrum availability are some of the key distinguishing factors. These challenges necessitate novel design techniques that simultaneously address a wide range of communication problems spanning several layers of the network protocol stack.

Although the basic idea of CR is simple, the efficient design of CRNs imposes new challenges that are not present in traditional wireless networks [18]. The heterogeneous spectrum environment and the importance of protecting the transmission of the licensed users of the spectrum mainly differentiate CRANs from classical ad hoc networks [9]. Specifically, identifying the time-varying channel availability imposes a number of nontrivial design problems to the medium access control (MAC) layer. One of the most difficult (but important) design problems is how the SUs decide when and which channel they should tune to in order to transmit or receive their packets without interference with the PUs. This problem becomes even more challenging in wireless ad hoc networks where there are no centralized controllers, such as base stations (BSs) or access points (APs).

9.2 CR Characteristics and Functions

The increase of service quality and channel capacity in wireless networks is severely limited by the scarcity of energy and bandwidth, which are the two fundamental resources for communications. Therefore, researchers are currently focusing on new communications and networking paradigms that can intelligently and efficiently utilize these scarce resources [1,12,19]. CR is one critical enabling technology for future communications and networking that can utilize the limited network resources in a more efficient and flexible way. It differs from traditional communication paradigms in that the radios or devices can adapt their operating parameters, such as transmission power, frequency, modulation type, and so on, to the variations of the surrounding radio environment [20]. Before CRs adjust their operating mode to environment variations, they must first gain necessary information from the radio environment. This kind of characteristic is referred to as *cognitive capability* [2], and it enables CR devices to be aware of the transmitted waveform, radio frequency spectrum, communication network type and protocol, geographical information, locally available resources and services, user needs, security policy, and so on. After CR devices gather their needed information from the radio environment, they can dynamically change their transmission parameters according to the sensed environment variations and achieve optimal performance, which is referred to as *configurability* [2].

CR technology is the key technology that enables CRNs to use the spectrum in a dynamic manner. Because most of the allocable spectrum is already assigned, the most important challenge in CRNs is to share the licensed spectrum without interfering with the transmissions of other licensed users. CR enables the usage of temporarily unused spectrum, which is referred to as a *spectrum hole* or *white space* [2].

If this frequency spectrum (band) is further reclaimed by a licensed user, the CR moves to another spectrum hole or stays in the same band, altering its transmission power level or modulation scheme to avoid interference with PUs. Figure 9.1 illustrates opportunistic access of the spectrum white space and switching of the frequency bands by a CR user at the incidence of use by a PU.

A typical duty cycle of CR, as illustrated in Figure 9.2, includes detecting spectrum white space, selecting the best frequency bands, coordinating spectrum access with other users, and vacating the frequency when a PU appears. Such a cognitive cycle is supported by the following functions:

- Spectrum sensing and analysis
- Spectrum management and handoff
- Spectrum allocation and sharing

Through spectrum sensing and analysis, CR can detect the spectrum white space (see Figure 9.1)—that is, a portion of frequency band that is not being used by the PUs—and utilize the spectrum. By contrast, when PUs start using the licensed spectrum again, CR can detect their activity through sensing, so that no severe interference is generated due to SU transmissions.

After recognizing the spectrum white space by sensing, the spectrum management and handoff functions of CR enable SUs to choose the best frequency band and hop among multiple bands according to the time-varying channel characteristics to meet various QoS requirements. For instance, when a PU reclaims its frequency band, the SU that is using the licensed band can direct its transmission to other available frequencies, according to the channel capacity determined by the noise and interference levels, path loss, channel error rate, holding time, and so on.

In dynamic spectrum access (DSA), an SU may share the spectrum resources with PUs, other SUs, or both. Hence, a good spectrum allocation and sharing

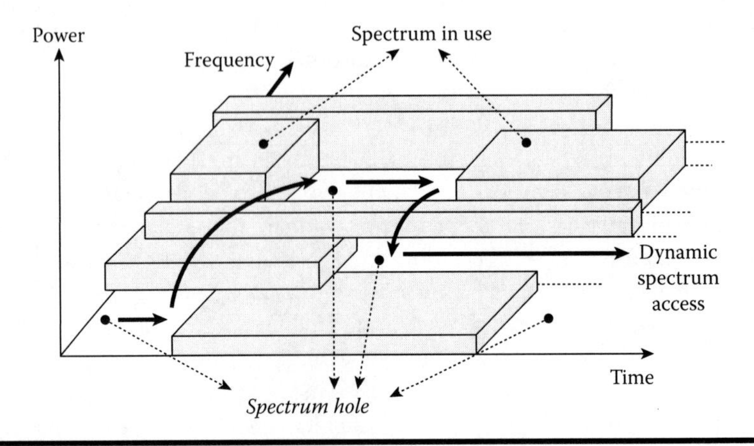

Figure 9.1 Concept of spectrum hole or white space.

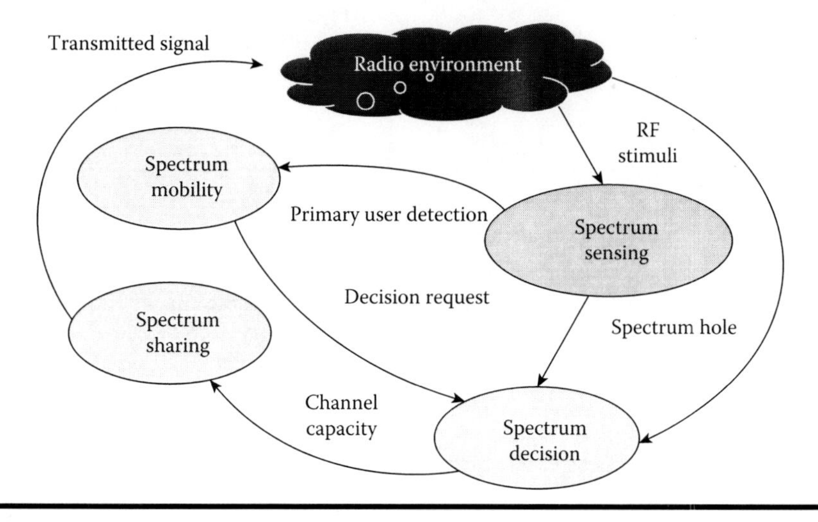

Figure 9.2 Cognitive cycle. RF, radio frequency.

mechanism is critical to achieve high spectrum efficiency. Because PUs own the spectrum rights, when SUs coexist in a licensed band with PUs, the interference level due to secondary spectrum usage should be limited by a certain threshold. When multiple SUs share a frequency band, their access should be coordinated to alleviate collisions and interference.

The capabilities of CRs as nodes of a CRN can be classified according to their functionalities based on the definition of *CR*. A CR senses the environment (cognitive capability), analyzes and learns sensed information (self-organized capability), and adapts to the environment (reconfigurable capabilities).

9.2.1 Cognitive Capability

Cognitive capability refers to the ability of the radio technology to capture or sense the information from its radio environment. This capability cannot simply be realized by monitoring the power in some frequency bands of interest. Rather, more sophisticated techniques, such as autonomous learning and action decision, are required in order to capture the temporal and spatial variations in the radio environment and avoid interference with other users. Through this capability, the portions of the spectrum that are unused at a specific time or location can be identified. Consequently, the best spectrum and appropriate operating parameters can be selected. We can summarize cognitive capability as follows:

- **Spectrum sensing:** A CR can sense spectrum and detect spectrum holes, which are those frequency bands not used by the licensed users or having limited interference with them. A CR could incorporate a mechanism that would enable sharing of the spectrum under the terms of an agreement between a licensee and a third party. Parties may eventually be able to negotiate for spectrum use on an ad hoc or real-time basis, without the need for prior agreements between all parties.

- **Location identification:** Location identification is the ability to determine its location and the location of other transmitters and then select the appropriate operating parameters, such as the power and frequency allowed at its location. In bands such as those used for satellite downlinks that are receive-only and do not transmit a signal, location technology may be an appropriate method of avoiding interference because sensing technology would not be able to identify the locations of nearby receivers. However, such location identification should be based on relative information for more system flexibility and overall spectrum utilization, instead of fixed universal location to create system or network control overheads.
- **Network/system discovery:** For a CR terminal to determine the best way to communicate, it should first discover the available networks around it. These networks are reachable either via directed one-hop communication or multihop relay nodes. For example, when a CR terminal has to make a phone call, it discovers if there is Global System for Mobile communication (GSM) Base Transceiver Station (BTS)s or Wi-Fi APs nearby. If there is no directed communication link between the terminal and the BTSs/APs but through other CR terminals some access networks are reachable, it can still make a call in this circumstance. The ability to discover access networks one or multiple hops away is important.
- **Service discovery:** Service discovery is almost similar to network/system discovery. Network or system operators provide their services through their access networks. A CR terminal tries to find appropriate services to fulfill its demands throughout the network. CR terminals discover the service by identifying nearby Bluetooth and Wi-Fi devices based on the time-binning approach, the bit-comparison approach, and so on.

9.2.2 Reconfigurable Capability

The cognitive capability provides spectrum awareness whereas reconfigurability enables the radio to be dynamically programmed according to the radio environment. More specifically, the CR can be programmed to transmit and receive on a variety of frequencies and to use different transmission access technologies supported by its hardware design. CR supports various reconfigurability, which are summarized as follows:

- **Frequency agility:** This is the ability of a radio to change its operating frequency. This ability usually combines with a method to dynamically select the appropriate operating frequency based on the sensing of signals from other transmitters or on some other method.

- **Dynamic frequency selection:** This is defined in the rules as a mechanism that dynamically detects signals from other radio frequency systems and avoids co-channel operation with those systems. The method that a device could use to decide when to change frequency or polarization could include spectrum sensing, geographic location monitoring, or an instruction from a network or another device. Such capability can be generalized as dynamic selection of logic channels and physical channels in wireless communications.
- **Adaptive modulation/coding:** This has been developed to approach channel capacity in fading channels. Adaptive modulation/coding can modify transmission characteristics and waveforms to provide opportunities for improved spectrum access and more intensive use of spectrum while "working around" other signals that are present. A CR could select the appropriate modulation type for use with a particular transmission system to permit interoperability between systems.
- **Transmit power control:** This is a feature that enables a device to switch dynamically between several transmission power levels in the data transmission process. It allows transmission at the allowable limits when necessary but reduces the transmitter power to a lower level to allow greater sharing of spectrum when higher power operation is not necessary.
- **Dynamic system/network access:** For a CR terminal to access multiple communication systems or networks that run different protocols, the ability to reconfigure itself to be compatible with these systems is necessary. It is therefore useful in coexisting multiradio environments to exploit heterogeneous wireless networking fully.

9.2.3 Self-Organizing Capability

The self-organizing capability provides the ability to derive optimized parameterization based on the gathered information and the end-to-end goals. With more intelligence than communication terminal devices, CRs are able to self-organize their communication based on sensing and reconfigurable functions as discussed before. Like wireless sensor networks (WSNs), CRNs have a limited energy supply, so the CR nodes need to cooperate and self-organize to provide smooth network operation by either putting off the communication of nodes that are not required or giving permission to use the radio of nodes that are required. We can summarize the self-organized capability of CR as follows:

- **Spectrum/radio resource management:** An improved spectrum management scheme is necessary to efficiently manage and organize information on spectrum holes between CRs. Using these spectrum holes, the CR terminals can meet their demands.
- **Mobility and connection management:** Routing and topology information is becoming increasingly complex day by day due to the heterogeneity

of CRNs. Thus, mobility and connection management techniques can help neighborhood discovery and provide related information. By using this information, CRNs can detect the available Internet access and support vertical handoffs, which help CRs to select routes and networks.

- **Security management:** The various heterogeneous nature of CRNs (e.g., wireless access technologies, system/network operators) introduce several security issues [21]. Security is a very challenging issue in CRNs, as different types of attacks are very common to CR technology compared with the general wireless network. Several techniques to ensure CRNs security have been explored, such as public key infrastructure [21], the trust-based approach [22], and so on.

9.3 CRN Architectures

Traditional wireless network architectures employ heterogeneity in terms of both spectrum policies and communication technologies [23]. Some portion of the wireless spectrum is already licensed for different purposes, whereas some bands remain unlicensed. Thus, for the development of communication protocols for CRNs, a clear description of CRN architecture is essential. In CRNs, the CR users must learn about the state of the network and physical environment through local coordination with other users. This coordination becomes a challenge when the available spectrum varies with time, and further if the node has mobility.

CRNs are composed of various kinds of coexisting multiradio communication systems, including CR systems. CRNs can be viewed as a sort of heterogeneous network composed of various communication systems. The heterogeneity exists in the wireless access technologies, networks, user terminals, applications, service providers, and so on. The design of the CRN architecture aims towards the objective of improving network utilization. From the users' perspective, the network utilization means that they can fulfill their demands anytime and anywhere by accessing CRNs. From the operators' perspective, they can not only provide better services to mobile users but also allocate radio and network resources in a more efficient way. A CRN architecture includes components corresponding to both the SUs (secondary network) and the PUs (primary network) as shown in Figure 9.3.

A *secondary network* refers to a network composed of a set of SUs with or without a secondary BS. SUs can only access the licensed spectrum when it is not occupied by a PU. The opportunistic spectrum access of SUs is usually coordinated by a secondary BS, which is a fixed infrastructure component serving as a hub of the secondary network. Both SUs and secondary BSs are equipped with CR functions. If several secondary networks share one common spectrum band, their spectrum usage may be coordinated by a central network entity, called a *spectrum broker* [24]. The spectrum broker collects operation information from each secondary network and allocates the network resources to achieve efficient and fair spectrum sharing.

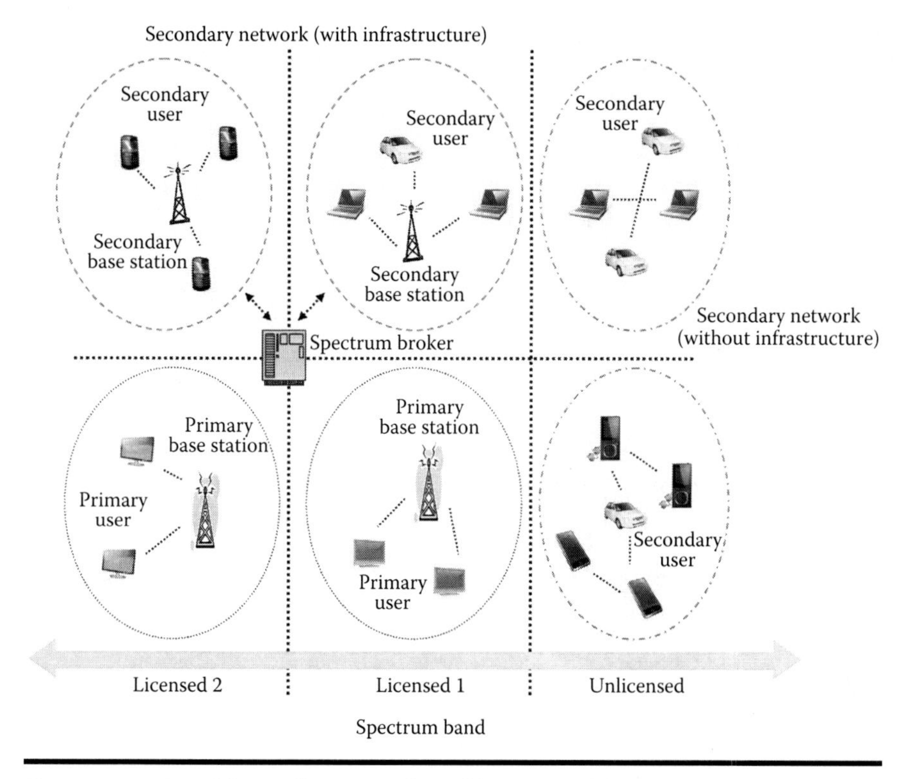

Figure 9.3 Cognitive radio network architecture with primary and secondary user networks. (From Wang, B. and Ray Liu, K. J., *IEEE J. Select. Topics Sig. Process.*, 5(1), 5–23, 2011.)

A primary network comprises PUs and one or more primary BSs, all of which are in general not equipped with CR functions. Hence, if a secondary network shares a licensed spectrum band with a primary network, the secondary network is required to be able to detect the presence of a PU and direct the secondary transmission to another available band that will not interfere with the primary transmission. Figure 9.3 illustrates CRN architecture with both the PU network and the SU network (with and without infrastructure—BS support).

The CRN can be deployed in network-centric, distributed, ad hoc, and mesh architectures and serve the needs of both licensed and unlicensed applications. The basic components of CRNs are the mobile stations (CR users), BSs/APs, and backbone/core networks. These three basic components compose three kinds of network architectures in CRNs: infrastructure, ad hoc, and mesh architectures.

9.3.1 Infrastructure-Based Architecture

The components of the infrastructure-based (or centralized) CRN architecture, as shown in Figure 9.4 [25], can be classified into two groups: the primary network

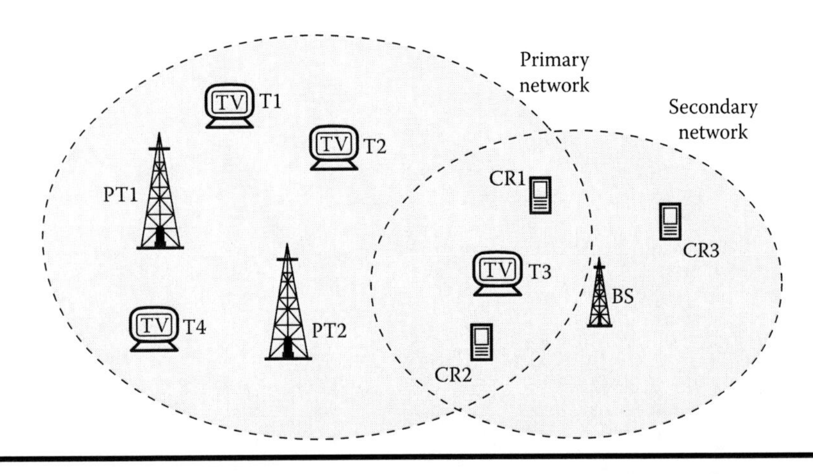

Figure 9.4 Infrastructure-based architecture of a cognitive radio network. (From Amaral de Souza, R. A., et al., *Vehicular Technologies—Deployment and Applications*, pp. 171–198, Intech Publishers, 2013.)

and the CRN. The primary network is referred to as the legacy network that has an exclusive right to a certain spectrum band. Examples include the common cellular and TV broadcast networks. In contrast, the CRN does not have a license to operate in the desired band. Hence, spectrum access is allowed only in an opportunistic manner [26].

A secondary network with a BS is referred to as the *infrastructure-based CRN*; the BS acts as a hub collecting the observations and results of spectrum analysis performed by each CR user and deciding on how to avoid interference with the primary networks. As per this decision, each CR user reconfigures its communication parameters. In the infrastructure-based architecture shown in Figure 9.4, a CR user can only access a BS/AP in a one-hop manner. CR users within transmission range of the same BS/AP will communicate with each other through the BS/AP. Communications between different cells are routed through backbone/core networks. The BS/AP may be able to run one or multiple communication standards or protocols to fulfill different demands from CR users. A CR user can also access various kinds of communication systems through their BS/AP.

In this figure, the coverage areas of the primary transmitters PT1 and PT2 are reaching the CRs, CR1 and CR2. Thus, both CRs can check if these primary transmitters are active in a given band (channel) and, if not, the secondary network can opportunistically use that spectrum band. At regular time intervals, the CRs must interrupt their transmissions and verify that the channel is still unoccupied by the primary network. If the channel becomes occupied, the secondary network must stop transmission and start the search for another vacant channel.

9.3.2 Ad Hoc Architecture

There is no infrastructure support in ad hoc architecture. The network is set up on the fly. A secondary network without a BS is referred to as an infrastructureless CRAN. In a CRAN, the CR users employ cooperation schemes to exchange locally observed information among the devices to broaden their knowledge on the entire network. They decide on their actions based on this perceived global knowledge. If a MS (CR user) recognizes that there are some other MSs nearby and they are connectable through certain communication standards/protocols, then they can set up a link and thus form an ad hoc network. Note that these links between nodes may be set up by different communication technologies. In addition, two CR terminals can either communicate with each other by using existing communication protocols (e.g., Wi-Fi, Bluetooth) or by dynamically using spectrum holes.

As shown in Figure 9.5, the primary network and the CRAN coexist in the primary bands. The primary network is composed of PUs that may be stationary or mobile. Television broadcast stations and receiver sets form the static topology PU network, whereas devices in the public service band may be mobile. For CR users, different ad hoc architectures are possible and each of these may entail a specialized protocol development. CR users can communicate with each other in a multihop manner on both licensed and unlicensed spectrum bands with the help of a single transceiver [27].

In ad hoc architecture, each CR node has all CR capabilities and is responsible for determining its next events based on the local information that it observes. Because the CR user cannot predict the influence of its actions on the entire network based on its local observations, cooperation schemes are essential, where the observed information can be exchanged between devices to broaden the knowledge on the network [10,22].

9.3.3 Mesh Architecture

Ad Hoc networks also include specialized architectures such as wireless mesh networks (WMNs) and WSNs. A typical WMN consists of mesh routers (MRs) forming the backbone of the network, interconnected in an ad hoc fashion [28] as shown in Figure 9.6 [29]. Each MR can be considered an AP serving a number of possibly mobile users or mesh clients (MCs). The MCs direct their traffic to their respective MRs, which then forward it over the backbone, in a multihop manner, to reach the gateway that links to the Internet. WSNs are composed of simple, resource-constrained nodes reporting to a BS and are being increasingly used for military, environmental monitoring, and data-gathering applications [30].

The mesh architecture is a combination of the infrastructure and ad hoc architectures plus enabling the wireless connections between BSs and APs. This network architecture is similar to hybrid wireless mesh networks. In this architecture, the BSs and APs work as wireless routers and form wireless backbones. MSs can either

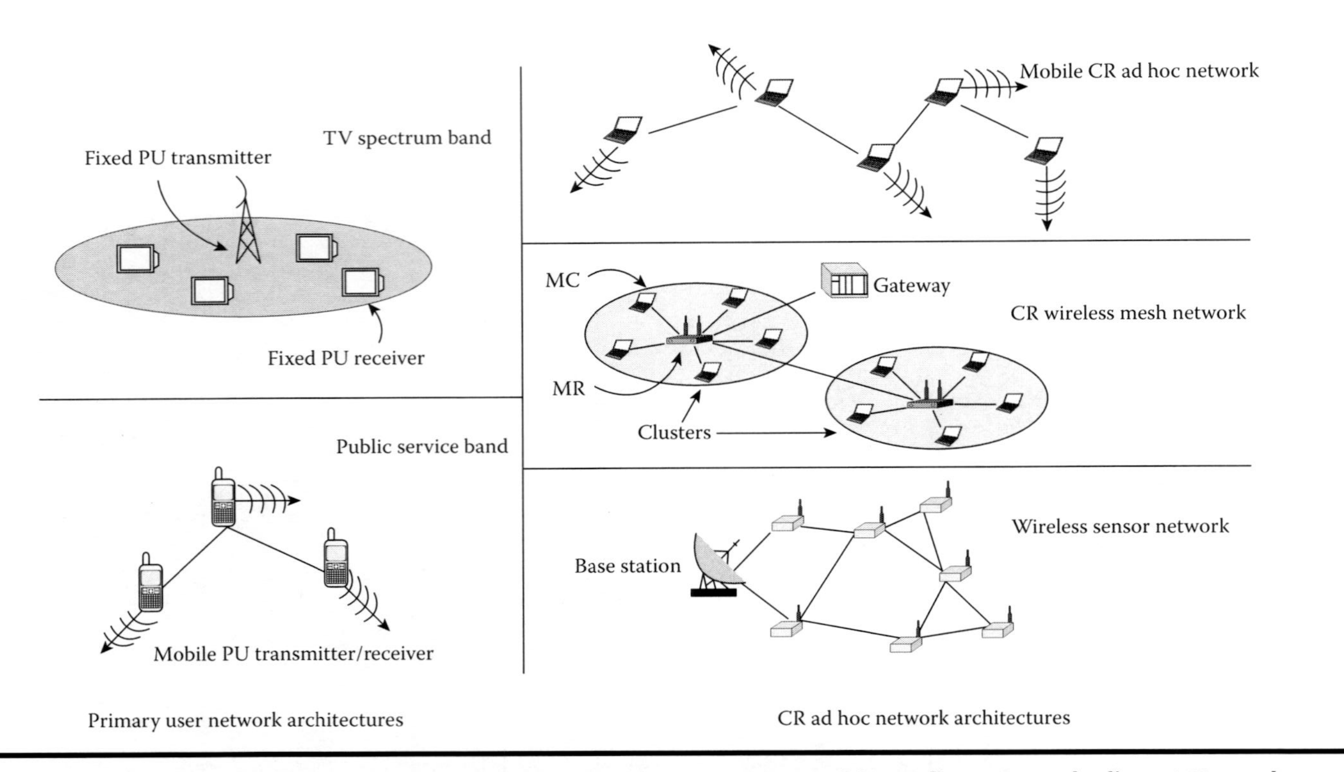

Figure 9.5 The cognitive radio ad hoc network (CRAN) architecture. CR, cognitive radio; MC, mesh client; MR, mesh router; PU, primary user. (From Chowdhury, K. R., Communication protocols for wireless cognitive radio ad hoc networks, *Ph.D. Thesis*, School of Electrical and Computer Engineering, Georgia Institute of Technology, 2009).

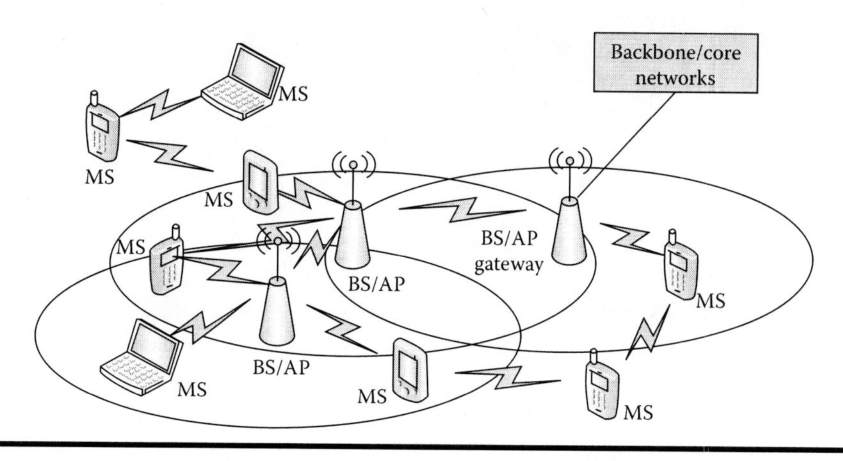

Figure 9.6 Mesh architecture of a cognitive radio network. (From Chen, K.-C. and Prasad, R., *Cognitive radio networks*, Wiley, 2009). MS, mobile station; BS, base station; AP, access point.

access the BSs and APs directly or use other MSs as multihop relay nodes. Some BSs/APs may connect to the wired backbone/core networks and function as gateways. Because BSs and APs can be deployed without necessarily connecting to the wired backbone/core networks, there is more flexibility and less cost in planning the locations of BSs and APs. If the BS or AP has CR capabilities, it may use spectrum holes to communicate with each other. Due to the inefficiency of current spectrum utilization, there may be lots of spectrum holes detected. So the capacity of wireless communication links between CR BSs/APs may be large and it makes it feasible for the wireless backbone to serve more traffic.

9.4 CR and DSA

Due to the rapid advance of wireless communications, a tremendous number of different communication systems exist in licensed and unlicensed bands, suitable for different demands and applications such as GSM/GPRS, IEEE 802.11, Bluetooth, UWB, ZigBee, 3G (CDMA series), HSPA, 3G LTE, IEEE 802.16, and so on. In contrast, radio propagation favors the use of spectrums under 3 GHz due to non-line-of-sight propagation. Consequently, many more devices, up to 1 trillion wireless devices by 2020, require radio spectrum allocation in order to respond to the challenge for further advances in wireless communications [29].

The tremendous growth of mobile devices and applications has triggered a need for additional radio frequency spectrum to satisfy this demand. Because most of the frequency spectrum has been licensed for different purposes (satellite, TV, radio, radar, cellular, etc.), the recent concept of DSA is being explored to provide additional spectrum with little disruption to existing licensed users and their devices.

A recent manifestation of DSA is in the TV spectrum. In 2008, the Federal Communications Commission (FCC) passed a historic ruling that allowed unlicensed devices (similar to Wi-Fi) to operate in the locally unoccupied TV spectrum (also called the *TV white spaces* or simply *white spaces*). Devices similar to the Wi-Fi devices of today are required to detect the available spectrum before using it for communication. According to the 2010 FCC Second Report and Order, the white space devices can detect available spectrum using either spectrum sensing or by querying a geolocation web service over the Internet [7].

In the past, spectrum allocation was based on the specific band assignments designated for a particular service. A number of reports from various parts of the world showed that there is a significantly unbalanced usage of the wireless spectrum [5–7], with a small portion of spectrum (e.g., cellular band, unlicensed band) increasingly crowded; most of the remaining spectrum is underutilized. The FCC Spectrum Policy Task Force [31] reported vast temporal and geographic variations in the usage of the allocated spectrum with use ranging from 15% to 85% in the bands below 3 GHz that are favored in non-line-of-sight radio propagation. The spectrum utilization is even tighter in the range above 3 GHz. In other words, a large portion of the assigned spectrum is used sporadically, leading to underutilization of a significant amount of spectrum.

Although the fixed spectrum assignment policy generally worked well in the past, there has been a dramatic increase in the access to limited spectrum for mobile services and applications in recent years. This increase is straining the effectiveness of the traditional spectrum polices. The limited available spectrum due to the nature of radio propagation and the need for more efficiency in the spectrum usage necessitates a new communication paradigm to exploit the existing spectrum opportunistically. Inspired by the successful global use of multiradio coexisting at 2.4 GHz, unlicensed Industrial, Scientific, and Medical (ISM) bands, and others, DSA has been proposed as a solution to these problems of current inefficient spectrum usage. The inefficient usage of the existing spectrum can be improved through opportunistic access to bands licensed by existing users (PUs).

The key enabling technology of DSA is CR technology, which provides the capacity to share the wireless channel with the licensed users in an opportunistic way. CRs are envisioned to be able to provide the high bandwidth to mobile users via heterogeneous wireless architectures and DSA techniques. The networked CRs also impose several challenges due to the broad range of available spectrum as well as diverse QoS requirements of applications.

In order to share the spectrum with licensed users without disturbing them and meet the diverse QoS requirement of applications, each CR user in a CRN must:

- Determine the portion of spectrum that is available, which is known as *spectrum sensing*
- Select the best available channel, which is called *spectrum decision*

- Coordinate access to this channel with other users, which is known as *spectrum sharing*
- Vacate the channel when a licensed user is detected, which is referred to as *spectrum mobility*

As mentioned in Ref. [32], each CR has the capability of being cognitive, reconfigurable, and self-organized to fulfill the functions of spectrum sensing, spectrum decision, spectrum sharing, and spectrum mobility that each CR must require.

An overview of different spectrum sharing models, namely open sharing, hierarchical access, and dynamic exclusive usage models, was proposed by Zhao and Sadler [33]. Spectrum management is an important functionality in CRNs, which involves DSA/sharing and pricing, and it aims to satisfy the requirements of both PUs and SUs. It is also mentioned in Ref. [34] that actual spectrum usage measurements obtained by the FCC's Spectrum Policy Task Force tell a different story: at any given time and location, much of the prized spectrum lies idle. This paradox indicates that spectrum shortage results from the spectrum management policy rather than the physical scarcity of usable frequencies. The underutilization of spectrum has stimulated a flurry of exciting activities in engineering, economics, and regulation communities in searching for better spectrum management policies and techniques.

The availability and quality of a spectrum band may change rapidly with time due to PU activity and competition from other SUs. In order to utilize the spectrum resources efficiently, SUs need to be able to address issues such as when and how to use a spectrum band, how to coexist with PUs and other SUs, and which spectrum band they should sense and access if the current one in use is not available. The current spectrum allocation and sharing schemes are classified according to three criteria: (1) spectrum bands in use by a CR user, (2) network architecture, and (3) access behavior of CR users. These are described below.

Spectrum bands used by the CR user: Based on the spectrum bands in use by a SU, the spectrum sharing scheme could be classified as an open spectrum sharing and hierarchical spectrum access model. In the open spectrum sharing model, the SUs access the unlicensed spectrum band and no user owns any spectrum license; hence, all users have the same access rights in using the unlicensed spectrum. In the hierarchical spectrum access model [33], the SUs share the licensed spectrum bands with the PUs. Because PUs need not be equipped with CR, they have all the priority to use the spectrum band. Hence, when a PU reclaims a spectrum band for use, the SUs currently using the spectrum band and the nearby bands will have to adjust their operating parameters (such as power, frequency, and bandwidth) to avoid interrupting the PUs. The hierarchical spectrum access model can be further divided into two categories, depending on the access restrictions on the SUs:

- *Spectrum underlay*: With this model, the secondary CR users coexist along with the PUs and use the licensed spectrum band without exceeding the interference temperature limit/threshold. If PUs transmit data all the time in a constant mode, there is no need for the secondary CR users to detect available spectrum band; instead, they can just continue to use the spectrum (of course, only for short-range communication).

- *Spectrum overlay*: With this model, the secondary CR users can only use the licensed spectrum when the PUs are not transmitting. So, there is no need for the CR users to operate under an interference temperature limit; however, the trade-off is that the CR users need to repeatedly sense the licensed frequency band and detect the spectrum white space, to avoid interfering with the PUs. If a PU is detected, the CR users have to change to another spectrum.

Network architecture: Based on the network architecture, the spectrum sharing model can be divided into centralized and distributed architectures. Under the centralized model, a central entity controls and coordinates the spectrum allocation and access of SUs. With the distributed spectrum sharing model, the users make their own decisions regarding spectrum access based on their local observations of the spectrum dynamics. The centralized controller model is expensive and also not suitable for ad hoc emergency or military use. The distributed spectrum sharing model is relatively less expensive and can be used in infrastructureless mode.

Access behavior of secondary CR users: Based on the access behavior of SUs, the spectrum sharing model can be categorized as either cooperative or noncooperative. Under the cooperative model, the SUs often belong to the same service provider and coordinate between themselves to collectively maximize the benefit to the entire group. On the other hand, under the noncooperative model, SUs access the open spectrum band and aim to maximize their own benefit from using the spectrum resources.

CR can efficiently improve spectrum utilization at link level. It also demonstrates that cooperative relay among CRs and nodes in the PS can greatly enhance the network capacity by constructing a general sense CRN. This suggests that a CR will sense available networks and communication systems around it, to complete networking functions beyond utilizing the spectrum hole at link level. Thus, CRNs are not just another network with interconnecting CRs. They are composed of various kinds of coexisting multiradio communication systems, including CR systems. CRNs can be viewed as a sort of heterogeneous network composed of various communication systems. The heterogeneity exists in wireless access technologies, networks, user terminals, applications, service providers, and so on. The design of the CRN architecture has the objective of improving network utilization. From the users' perspective, the network utilization means that they can always fulfill their demands anytime and anywhere by accessing CRNs. From the operators' perspective, they can not only provide better services to mobile users but also allocate radio and network resources in a more efficient way.

9.5 Spectrum Sensing in CRNs

A CR is designed to be aware of and sensitive to the changes in its surroundings, which makes spectrum sensing an important requirement for the realization of CRNs. Spectrum sensing enables CR users to exploit the unused spectrum portion adaptively to the radio environment. This capability is required in the following cases: (1) CR users find available spectrum holes over a wide frequency range for their transmission (out-of-band sensing) and (2) CR users monitor the spectrum band during transmission and detect the presence of primary networks so as to avoid interference (in-band sensing). As shown in Figure 9.7, the CRAN necessitates the following functionalities for spectrum sensing:

- *PU detection*: The CR user observes and analyzes its local radio environment. Based on these location observations of itself and its neighbors, CR users determine the presence of PU transmissions and accordingly identify the current spectrum availability.
- *Cooperation*: The observed information in each CR user is exchanged with its neighbors so as to improve sensing accuracy.
- *Sensing control*: This function enables each CR user to perform its sensing operations adaptively to the dynamic radio environment. In addition, it coordinates the sensing operations of the CR users and its neighbors in a distributed manner, which prevents false alarms in cooperative sensing.

In order to achieve high spectrum utilization while avoiding interference, spectrum sensing needs to provide high detection accuracy. However, due to the lack of a central network entity, CR ad hoc users perform sensing operations independently of each other, leading to an adverse influence on sensing performance.

A wealth of literature on spectrum sensing focuses on primary transmitter detection based on the local measurements of SUs, since detecting the PUs that are receiving data is in general very difficult. According to the *a priori* information they require and the resulting complexity and accuracy, spectrum sensing techniques can be categorized into the following types, which are summarized in Figure 9.8.

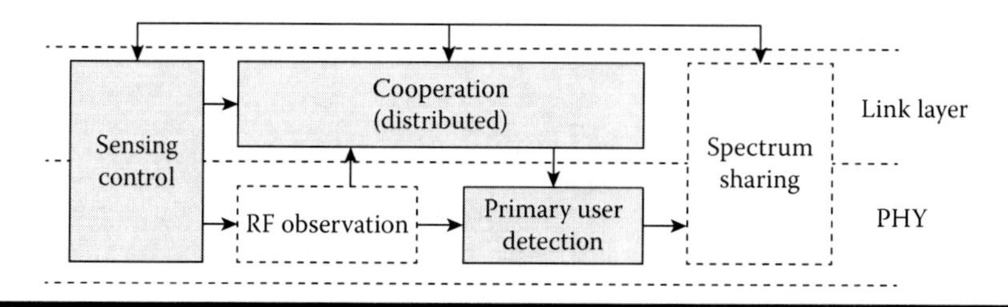

Figure 9.7 Spectrum sensing structure for CRANs. (From Akyildiz, I. F., et al., *Ad Hoc Netw.*, 7(5), 810–836, 2009.)

Type	Test statistics	Advantages	Disadvantages
Energy detector	Energy of the received signal samples	• Easy to implement • Not require prior knowledge about primary signals	• High false alarm due to noise uncertainty • Very unreliable in low SNR regimes • Cannot differentiate a primary user from other signal sources
Feature detector	Cyclic spectrum density function of the received signal, or by matching general features of the received signal to the already known primary signal characteristics	• More robust against noise uncertainty and better detection in low SNR regimes than energy detection • Can distinguish among different types of transmissions and primary systems	• Specific features, e.g., cyclostationary features, must be associated with primary signals • Particular features may need to be introduced, e.g., to OFDM-based communications
Matched filtering and coherent detection	Projected received signal in the direction of the already known primary signal or a certain waveform pattern	• More robust to noise uncertainty and better detection in low SNR regimes than feature detector • Require less signal samples to achieve good detection	• Require precise prior information about certain waveform patterns of primary signals • High complexity

Figure 9.8 Summary of main spectrum sensing techniques SNR, signal-to-noise ratio. (From Wang, B. and Ray Liu, K. J., *IEEE J. Select. Topics Sig. Process.*, 5(1), 5–23, 2011.)

Energy detection: Energy detection is the most common type of spectrum sensing because it is easy to implement and requires no prior knowledge about the primary signal [9]. The energy detector is optimal to detect the unknown signal if the noise power is known. In energy detection, CR users sense the presence/absence of the PUs based on the energy of the received signals.

While energy detection is easy to implement, it has several shortcomings. Energy detection suffers from longer detection time compared with matched filter detection. Furthermore, because energy detection depends only on the signal-to-noise ratio (SNR) of the received signal, its performance is susceptible to uncertainty in noise power. If the noise power is uncertain, the energy detector will not be able to detect the signal reliably, as the SNR is less than a certain threshold. In addition, the energy detector can only determine the presence of the signal but cannot differentiate signal types. Thus, energy detection often results in false detection triggered by unintended CR signals.

Feature detection: Feature detection determines the presence of PU signals by extracting their specific features such as pilot signals, cyclic prefixes, symbol rate, spreading codes, or modulation types from its local observation. These features introduce built-in periodicity in the modulated signals, which can be detected by analyzing a spectral correlation function. The feature detection leveraging this periodicity is also called *cyclostationary detection*.

The main advantage of feature detection is its robustness to the uncertainty in noise power. Furthermore, it can distinguish the signals from different networks. This method allows the CR user to perform sensing operations independently of those of its neighbors without synchronization. Although feature detection is most effective for the nature of CRANs, it is computationally complex and requires significant sensing time [35]. Different from an energy detector, which uses time-domain signal energy as test statistics, a cyclostationary feature detector performs a transformation from the time domain into the frequency feature domain and then conducts a hypothesis test in the new domain.

Matched filtering and coherent detection: The matched filter is the linear optimal filter used for coherent signal detection to maximize the SNR in the presence of additive stochastic noise. If SUs know information about a PU's signal *a priori*, then the optimal detection method is matched filtering, since a matched filter can correlate the already known primary signal with the received signal to detect the presence of the PU and thus maximize the SNR in the presence of additive stochastic noise. The merit of matched filtering is the short time it requires to achieve a certain detection performance such as a low probability of missed detection and false alarm [36], since a matched filter needs fewer received signal samples.

However, the matched filter necessitates not only *a priori* knowledge of the characteristics of the PU signal but also the synchronization between the PU transmitter and the CR user. If this information is not accurate, then the matched filter performs poorly. Furthermore, CR users need to have different multiple matched filters dedicated to each type of PU signal, which increases the implementation cost and complexity.

9.6 MAC in CRNs

MAC in CRNs refers to the policy that controls how an SU should access a spectrum band licensed by a PU [14]. Because the control and coordination of communication over wireless channels happen mainly at the MAC layer, designing a smart and efficient MAC protocol is the most important obligation for successful deployment of any CRN. In CRNs, due to the second priority on accessing spectrum resource of CR users, MAC protocols have to perform DSA functions, including spectrum sensing, spectrum access, spectrum allocation, spectrum sharing, and spectrum mobility, in addition to conventional control procedures. As a result, designing MAC protocols for CRNs requires more complicated consideration than that needed for conventional wireless networks.

Various MAC protocols have been proposed in wireless networking such as carrier sense multiple access/collision avoidance (CSMA/CA) and slotted ALOHA. Due to the new features of CRNs, such as PU collision avoidance and dynamics of spectrum availability, new MAC protocols need to be designed to address the new challenges in CRNs. Research has shown that MAC solutions designed for traditional distributed wireless systems cannot meet the requirements for CRANs efficiently.

A number of MAC protocols that have also been proposed for CRNs are available in the literature [14]. MAC protocols for CRNs can be broadly divided into two classes, CR infrastructure-based MAC and CR ad hoc MAC, depending on the architecture of the CRN under consideration. In an infrastructure-based CRN, the BS typically controls the spectrum access of SUs. By contrast, CRANs usually do not have a central controller or BS to assist SUs in spectrum sensing or spectrum access functions.

The MAC protocols for both categories of CRNs can be either time-slotted, random access, or both. The time-slotted MAC protocols require network-wide synchronization and operate by dividing time into discrete slots for both the control channel and data transmission. By contrast, the random access protocols do not require time synchronization and are based on the CSMA/CA principle wherein a CR user monitors the spectrum band to detect the presence of any transmission from peer CR users and if detected, transmits after backing off for a random duration, to reduce collisions due to simultaneous transmissions.

It is anticipated that the CRAN may be practical as a future generation CRN due to its easier and faster deployment and low cost of implementation. CRAN also offers more challenging research issues due to lack of a central control unit in such networks. The MAC protocol and resource allocation approach should be able to smartly adapt to the unique features of CRANs and maintain robust performance in the presence of a highly dynamic environment. In CRANs, radio spectrums are opportunistically available to SUs that may be relinquished by PUs anytime. Such resource uncertainty makes QoS provisioning to SUs an important research issue. Moreover, the design of an energy-efficient MAC protocol is crucial for CRANs to enhance the network lifetime as mobile CR nodes are battery operated. Furthermore, efficient resource sharing is crucial to enhance the utilization of resources in CRNs, as resources are limited and available only opportunistically to SUs.

9.7 Routing in CRNs

The problem of routing in multihop CRNs refers to the creation and maintenance of wireless multihop paths among the CR users (also called *SUs*) by deciding the relay nodes and the spectrum to be used on each of the links in the path [37]. Even though the above problem definition exhibits similarities with routing in multichannel, multihop ad hoc networks, and mesh networks, the challenge in the form of dynamic changes in the available spectrum bands due to simultaneous transmissions involving PUs needs to be handled. Any routing solution for multihop CRNs needs to be tightly coupled with spectrum management functionalities [38] so that the routing modules can make more accurate decisions based on dynamic changes in the surrounding physical environment. As the topology of multihop CRNs is highly influenced by the behavior of the PUs, the route metrics should be embedded with measures on path stability, spectrum availability, PU presence, and so on. For instance, if PU activity is low to moderate, then the topology of the SUs is almost static, and classical routing metrics adopted for WMNs could be employed. By contrast, if PUs become active very frequently, then the routing techniques employed for ad hoc networks could be more applicable [3]. In addition, the routing protocols should be able to repair broken paths (in terms of nodes or used channels) due to the sudden reappearance of a PU [39].

With respect to the issue of spectrum awareness, the routing solutions for CRNs could be classified as those based on full spectrum knowledge and local spectrum knowledge. In the former case, the spectrum availability between any two nodes in the network is known to all the nodes (or to a central control entity). This is often facilitated through a centrally maintained spectrum database to indicate channel availabilities over time and space. The routing solutions built on top of the availability of full spectrum knowledge are mostly based on a graph abstraction of the CRN and, though not often practically feasible for implementation, are used to derive benchmarks for routing performance. The routing module is not tightly coupled with the spectrum management functionalities for centralized full spectrum knowledge-based solutions. On the other hand, for local spectrum knowledge-based solutions, information about spectrum availability is exchanged among the network nodes along with traditional network state information (such as the routing metrics, node mobility, traffic, and so on). On these lines, the local spectrum knowledge-based routing protocols could be further classified as those that aim to minimize the end-to-end delay, maximize the throughput, and maximize path stability. In addition to the above, we have also come across probabilistic approaches for routing [40,41] in which CR users opportunistically transmit over any spectrum band available during the short idle periods of the surrounding PUs.

In distributed CRNs, routing algorithms are important to find a route between a source SU and a destination SU. Because SUs must be adaptive to the dynamic changes in spectrum utilization by PUs, routing in CRNs has been challenging and it differs from the routing in traditional wireless networks, particularly multichannel networks in which static sets of channels are available for the nodes. In CRNs, SUs may have different sets of channels that impose more routing challenges, especially in multihop communication. For this reason, routing must be spectrum-aware in CRNs, and thus routing schemes for traditional wireless networks cannot be readily applied to CRNs [42].

Routing is a challenging issue when dealing with the multihop CRNs. The stability of a route is highly influenced by the behavior of the PU. The more frequently the PU changes its activity, the less stable spectrum availability the cognitive user has. This causes tremendous route stability fluctuation in CRNs.

9.8 Transport Layer Issues in CRNs

Research on transport layer protocols for CRNs is very much in the promising stages [10]. Packet losses in a CRN involving mobile wireless nodes may occur due to one of the following factors: (i) traditional network congestion that could be further aggravated due to reduced link capacity and loss of connection, (ii) link error, (iii) collision due to simultaneous transmissions, (iv) mobility of a node from one BS to another, (v) mobility of the intermediate forwarding nodes, (vi) the intermittent spectrum sensing undertaken by the CR users, (vii) the switching of a CR node between transmitting and spectrum sensing states, (viii) the activity of the primary licensed users of the spectrum, and (ix) large-scale bandwidth variation due to spectrum availability. Factors vi through ix are characteristic of CRNs, and these factors have not been considered in the design of the transport layer protocols for mobile ad hoc networks or sensor networks.

As the transport protocol usually runs at the end nodes (source and destination), it has limited knowledge of the conditions of the intermediate nodes. Unknown to the source, the route may be disconnected due to node mobility. In addition, packet losses may be wrongly attributed to network congestion rather than bad channel conditions at the link layer. Classical TCP suffers from some of the above issues and efforts have been made to address them for wireless scenarios in Refs. [43,44].

However, these protocols for classical wireless ad hoc networks do not consider the cases that may arise in CRANs. As an example, in a classical wireless ad hoc network, packets may incur a longer round trip time (RTT) owing to network congestion or due to a temporary route outage. In CRANs, a similar effect on the packet RTT may be caused if an intermediate node on the route is engaged in spectrum sensing and hence is unable to forward packets. In addition, the sudden appearance of a PU may force the CR nodes in its vicinity to limit their transmissions leading to an increase in the RTT. In such cases, the network is partitioned until a new channel is identified and coordinated with the nodes on the path. The duration of the periodic spectrum sensing decides, in part, the end-to-end performance—a shorter sensing time may result in higher throughput but may affect the transport layer severely if a PU is misdetected. While several works have focused on spectrum sensing algorithms in the last few years [3], the integration of the channel information collected at the nodes and the study of the performance of these approaches from the viewpoint of an end-to-end protocol remains an open challenge.

9.9 Applications of CRNs

There are many emerging CRN applications [45]. In the following, we illustrate some of these applications.

Public safety and homeland security system: Wireless communications are extensively used for emergency services like police and medical services to respond to emergency situations. A CRN can also be implemented to enhance public safety and homeland security. A natural disaster or terrorist attack can destroy existing communication infrastructure, so an emergency network becomes indispensable to aid the search and rescue. Because a CR can recognize spectrum availability and reconfigure itself for much more efficient communication, this provides public safety personnel with dynamic spectrum selectivity and reliable broadband communication to minimize information delay. These safety workers are equipped with various modules like mobiles, videotelephony, and so on to improve their efficiency and to be always be in touch with their co-workers and central authorities. The CRNs are extensively proposed for these kinds of services, as they promise high spectrum coverage in such situations where large bandwidth might be required. Moreover, CR can facilitate interoperability between various communication systems. Through adapting to the requirements and conditions of another network, the CR devices can support multiple service types, such as voice, data, video, and so on.

Military communications: The capacity of military communications is limited by radio spectrum scarcity because static frequency assignments freeze bandwidth into unproductive applications, where a large amount of spectrum is idle. CR using DSA can alleviate the spectrum congestion through efficient allocation of bandwidth and flexible spectrum access. Therefore, CR can provide the military with adaptive, seamless, and secure communications.

Smart grid networks: A smart grid is a network that attempts to intelligently guess the behavior of all electric power users and smartly respond to their actions in order to efficiently deliver reliable, economic, and sustainable electricity service. This increases the reliability and efficiency of electricity transmission and reduces cost for consumers and producers. These grids make use of Wi-Fi, ZigBee, TV white spaces, and cellular networks in rural areas as the primary spectrum for their cognitive nodes for easy communication.

CR sensor networks: Sensor networks generate bursty data. The transmission of such bursty data at any given time requires a high bandwidth for a small period of time. Otherwise low bandwidth or a small number of channels may cause high collisions and make the system inefficient. Realizing CRSNs involves a few challenges such as the design of power-efficient and low-cost CR sensor nodes, as well as opportunistic multihop routing over licensed and unlicensed spectrum bands. These could be used for real-time applications, multimedia applications, or indoor sending applications.

9.10 Wireless Medical Networks: A Case Study

In recent years, there has been increasing interest in implementing ubiquitous monitoring of patients in hospitals for vital signs such as temperature, pressure, blood oxygen, and electrocardiogram. Normally these vitals are monitored by on-body sensors that are then connected by wires to a bedside monitor. Medical body area networks (MBANs) [45] are a promising solution for eliminating these wires, thus allowing sensors to reliably and inexpensively collect multiple parameters simultaneously and relay the monitoring information wirelessly so that clinicians can respond rapidly [46]. Introduction of MBANs for wireless patient monitoring is an essential component to improving patient outcomes and lowering health care costs. Through low-cost, wireless devices, universal patient monitoring can be extended to most if not all patients in many hospitals. With such ubiquitous monitoring, changes in a patient's condition can be recognized at an early stage and appropriate action taken. By getting rid of wires and their management, the associated risks of infection are reduced using MBANs. Additionally, MBANs would increase patient comfort and mobility, improve the effectiveness of caregivers, and improve the quality of medical decision-making. Patient mobility is an important factor in speeding up patient recovery.

QoS is a key requirement for MBANs, and hence the importance of having a relatively clean and less crowded spectrum band. Today, MedRadio and Wireless Medical Telemetry Service (WMTS) band are used in many medical applications but the bandwidth is limited and cannot meet the growing need [46,47]. The 2.4 GHz ISM band is not suitable for life-critical medical applications due to the interference and congestion from IT wireless networks in hospitals. By having the 2360–2400 MHz band allocated for MBAN on a secondary basis, QoS for these life-critical monitoring applications can be better ensured. Moreover, the 2360–2400 MHz frequency band is immediately adjacent to the 2400 MHz band, for which many devices exist today that could be easily reused for MBANs, such as IEEE 802.15.4 radios. This would lead to low-cost implementations due to economies of scale and ultimately wider deployment of MBANs and hence improvement in patient care.

MBAN communication will be limited to transmission of data (voice is excluded) used for monitoring, diagnosing, or treating patients. MBAN operation is permitted either by health care professionals or authorized personnel under license by rule. It is proposed that the 2360–2400 MHz frequency band be classified into two bands: 2360–2390 MHz (Band I) and 2390–2400 MHz (Band II). In the 2360–2390 MHz band, MBAN operation is limited for indoor use only to those health care facilities that are outside exclusion zones of Aeronautical Mobile Telemetry (AMT) services. In the 2390–2400 MHz band, MBAN operation is permitted everywhere—all hospitals, in homes, mobile ambulances. There are a number of mechanisms for MBAN devices to access spectrum on secondary basis while protecting incumbents and providing a safe medical implementation. An unrestricted contention-based protocol such as Listen-Before-Transmit (LBT) is proposed for channel access. The maximum emission bandwidth of MBAN devices is proposed to be 5 MHz. The maximum transmit power is not to exceed the lower of 1 mW and 10logB dBm (where B is the 20 dB bandwidth in MHz) in the 2360–2390 MHz band and 20 mW in the 2390–2400 MHz band. The maximum aggregated duty cycle of an MBAN is not to exceed 25%. A geographical protection zone along with an electronic key (e-key) MBAN device control mechanism is further used to limit MBAN transmissions. E-key device control is used to ensure that MBAN devices can access the 2360–2390 MHz frequency band only when they are within the confines of a hospital facility that is outside the protection zone of AMT sites.

Figure 9.9 illustrates both an in-hospital and an out-of-hospital solution for using the 2360–2390 MHz band. Any hospital that plans to use the AMT spectrum for

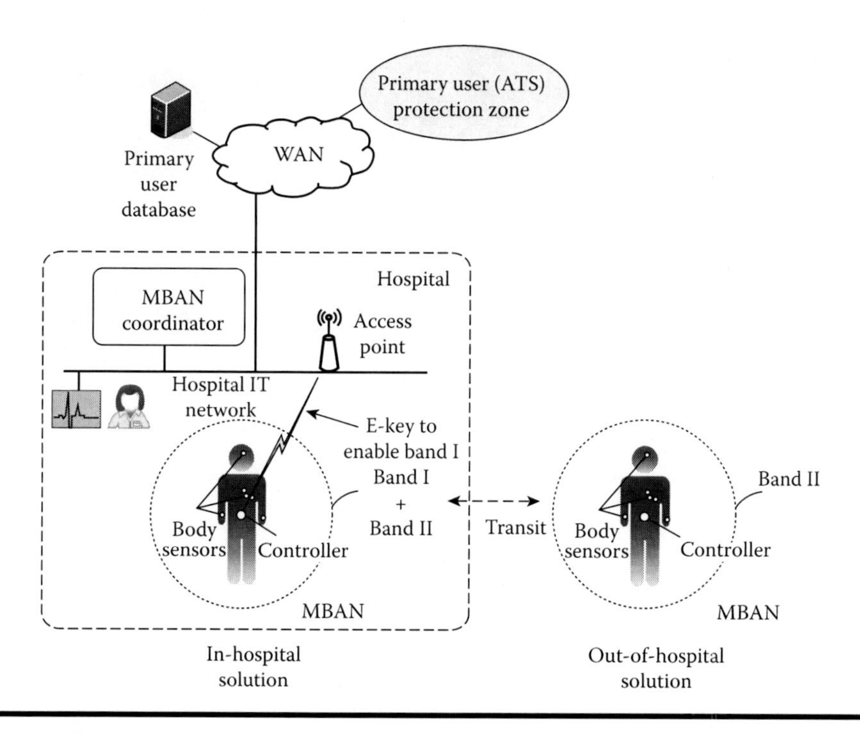

Figure 9.9 Medical body area networks (MBANs). (From Jianfeng, W., et al., *IEEE Commun. Mag.*, 49(3), 74–81, 2011.)

MBAN has to register with an MBAN coordinator. The MBAN coordinator determines whether a registered hospital is within the protection zones of AMT sites (with possible coordination with PUs). If a hospital is outside protection zones, then the MBAN coordinator will issue an e-key specifically for that hospital to enable MBAN devices within that hospital to access AMT spectrum. Without a valid e-key, by default MBAN devices can only use the 2390–2400 MHz band. The distribution of e-keys to MBAN devices that are connected to the hospital IT network can be automatically done either through wired or wireless links. MBAN devices must have a means to automatically prevent transmissions in the 2360–2390 MHz AMT band when devices go outdoors. Once a sensor in an MBAN loses its connection to its hub device, it stops transmission within the 2360–2390 MHz AMT spectrum or transitions to the 2390–2400 MHz band. The 2390–2400 MHz band can be used anywhere without restriction and hence without an e-key. Simulations have shown that these technologies would work well to protect AMT from interference while also maintaining the QoS required for the MBAN applications.

The IEEE has been working on MBAN standardization. In addition to ongoing activities in IEEE 802.15.6 on body area networks, a new 802.15 Task Group 4j was started in December 2010 to specifically develop standards for MBANs in the 2360–2400 MHz band, by leveraging the existing IEEE 802.15.4 standard.

9.11 Conclusion

CRNs have been proposed in recent years as a revolutionary solution towards more efficient utilization of scarce spectrum resources in an adaptive and intelligent way through opportunistic use of frequencies reserved for licensed users of the bands. In this chapter, we provided a unified view of CRNs through step-by-step introduction of different associated concepts and techniques. We also highlighted different applications of CRNs and discussed a case study of wireless medical networks. The study of CRNs is relatively new and many challenges are yet to be addressed. Many researchers are currently engaged in developing the communication technologies and protocols required for CRNs. However, challenges still remain because CR-enabled networks have to coexist with PUs as well as SUs and need to mitigate interference in such a way that they can provide better end-to-end support to ensure efficient spectrum-aware communication.

References

1. B. Wang, and K. J. Ray Liu, Advances in cognitive radio networks: A survey, *IEEE Journal of Selected Topics in Signal Processing*, vol. 5, no. 1, pp. 5–23, 2011.
2. S. Haykin, Cognitive radio: Brain-empowered wireless communications, *IEEE Journal of Selected Areas in Communications*, vol. 23, no. 2, pp. 201–220, 2005.
3. I. F. Akyildiz, W.-Y. Lee, M. C. Vuran, and S. Mohanty, Next generation/dynamic spectrum access/cognitive radio wireless networks: A survey, *Computer Networks*, vol. 50, pp. 2127–2159, 2006.
4. H. Kim, Efficient identification and utilization of spectrum opportunities in cognitive radio networks, Ph.D. Thesis, The University of Michigan, 2010.
5. M. Mchenry, Spectrum white space measurements, in *New America Foundation Broadband Forum*, vol. 1, 2003.
6. S. Yin, D. Chen, Q. Zhang, M. Liu, and S. Li, Mining spectrum usage data: A large-scale spectrum measurement study, *IEEE Transactions on Mobile Computing*, vol. 11, no. 6, pp. 1033–1046, 2012.
7. X. Ying, J. Zhang, L. Yan, G. Zhang, M. Chen, and R. Chandra, Exploring indoor white spaces in metropolises, in *Proceedings of the ACM 19th Annual International Conference on Mobile Computing and Networking* (*MobiCom*), pp. 255–266, September 2013.
8. Q. Zhang, J. Jia, and J. Zhang, Cooperative relay to improve diversity in cognitive radio networks, *IEEE Communications Magazine*, vol. 47, no. 2, pp. 111–117, 2009.
9. X. Hong, C.-X. Wang, H.-H. Chen, and Y. Zhang, Secondary spectrum access networks, *IEEE Vehicular Technology Magazine*, vol. 4, no. 2, pp. 36–43, 2009.
10. I. F. Akyildiz, W. Y. Lee, and K. R. Chowdhury, CRAHNs: Cognitive radio ad hoc networks, *Ad Hoc Networks*, vol. 7, no. 5, pp. 810–836, 2009.
11. Q. Zhao, L. Tong, A. Swami, and Y. Chen, Decentralized cognitive MAC for opportunistic spectrum access in ad hoc networks: A POMDP framework, *IEEE Journal of Selected Areas in Communications*, vol. 25, no. 3, pp. 589–600, 2007.
12. P. Ren, Y. Wang, and Q. Du, CAD-MAC: A channel-aggregation diversity based MAC protocol for spectrum and energy efficient cognitive ad hoc networks, *IEEE Journal on Selected Areas in Communications*, vol. 32, no. 2, pp. 237–250, 2014.

13. S. M. Kamruzzaman, An energy efficient multichannel MAC protocol for cognitive radio ad hoc networks, *International Journal of Communication Networks and Information Security* (*IJCNIS*), vol. 2, no. 2, pp. 112–119, 2010.
14. A. D. Domenico, E. C. Strinati, and M.-G. Di Benedetto, A survey on MAC strategies for cognitive radio networks, *IEEE Communications Surveys & Tutorials*, vol. 14, no. 1, pp. 21–44, 2012.
15. Federal Communications Commission, *Notice of proposed rule making and order*, ET Docket No. 03-222, FCC, Washington, D.C., 2003.
16. R. W. Thomas, L. A. DaSilva, and A. B. MacKenzie, Cognitive networks, in *Proceedings of IEEE DySPAN*, pp. 352–360, 2005.
17. S. M. Kamruzzaman, E. Kim, and D. G. Jeong, An energy efficient QoS routing protocol for cognitive radio ad hoc networks, in *Proceedings of the 13th IEEE International Conference on Advanced Communication Technology* (*ICACT-2011*), Phoenix Park, Korea, pp. 344–349, 13–16 February 2011.
18. M. Devroye, P. Mitran, and V. Tarohk, Limits on communications in a cognitive radio channel, *IEEE Communications Magazine*, pp. 44–49, 2006.
19. T. M. Salem, S. M. Abd El-kader, S. M. Ramadan, and M. Z. Abdel-Mageed, Opportunistic spectrum access in cognitive radio ad hoc networks, *International Journal of Computer Science Issues* (*IJCSI*), vol. 11, no. 2, pp. 41–50, 2014.
20. Federal Communications Commission, *Spectrum policy task force report*, ET Docket No. 02-135, FCC, Washington, D.C., 2002.
21. A. G. Fragkiadakis, E. Z. Tragos, and I. G. Askoxylakis, A survey on security threats and detection techniques in cognitive radio networks, *IEEE Communications Surveys & Tutorials*, vol. 15, no. 1, pp. 428–445, 2013.
22. S. Parvin, F. K. Hussain, O. K. Hussain, S. Han, B. Tian, and E. Chang, Cognitive radio network security: A survey, *Journal of Network and Computer Applications*, vol. 35, no. 6, pp. 1691–1708, 2012.
23. I. F. Akyildiz, Y. Altunbasak, F. Fekri, and R. Sivakumar, AdaptNet: Adaptive protocol suite for next generation wireless internet, *IEEE Communications Magazine*, vol. 42, no. 3, pp. 128–138, 2004.
24. R. W. Brodersen, A. Wolisz, D. Cabric, S. M. Mishra, and D. Willkomm, *Corvus: A cognitive radio approach for usage of virtual unlicensed spectrum*, Berkeley Wireless Research Center (BWRC) White paper, University of California, Berkely, CA, 2004.
25. R. A. Amaral de Souza, D. A. Guimarães, and A. Antônio dos Anjos, Simulation platform for performance analysis of cooperative eigenvalue spectrum sensing with a realistic receiver model under impulsive noise, in *Vehicular technologies—Deployment and applications*, edited by Lorenzo Galati Giordano and Luca Reggiani. pp. 171–198, Intech Publishers, Croatia: European Union, 2013.
26. W.-Y. Lee, K. R. Chowdhury, and M. C. Vuran, Spectrum sensing algorithms for cognitive radio networks, in *Cognitive Radio Networks, edited by Yang Xiao, Fei Hu*. pp. 1–35, CRC Press, Boca Raton, FL, 2009.
27. K. R. Chowdhury, Communication protocols for wireless cognitive radio ad-hoc networks, Ph.D. Thesis, School of Electrical and Computer Engineering, Georgia Institute of Technology, West Sussex, UK,2009.
28. I. F. Akyildiz, X. Wang, and W. Wang, Wireless mesh networks: A survey, *Computer Networks*, vol. 47, pp. 445–487, 2005.
29. K.-C. Chen, and R. Prasad, *Cognitive radio networks*, Wiley, West Sussex, UK, 2009.

30. I. F. Akyildiz, W. Su, Y. Sankarasubramaniam, and E. Cayirci, Wireless sensor networks: A survey, *Computer Networks*, vol. 38, pp. 393–422, 2002.
31. Federal Communications Commission, *First report and order*, FCC 02-48, FCC, Washington, D.C., 2002.
32. K.-C. Chen, Y.-J. Peng, N. Prasad, Y.-C. Liang, and S. Sun, Cognitive radio network architecture: Part I—General structure, in *Proceedings of ACM ICUMIC*, Seoul, 2008.
33. Q. Zhao, and B. M. Sadler, A survey of dynamic spectrum access: Signal processing, networking, and regulatory policy, *IEEE Signal Processing Magazine*, vol. 24, no. 3, pp. 79–89, 2007.
34. K.-C. Chen, Y.-J. Peng, N. Prasad, Y.-C. Liang, and S. Sun, Cognitive radio network architecture: Part II—Trusted network layer structure, in *Proceedings of ACM ICUMIC*, Seoul, 2008.
35. Y. Hur, J. Park, W. Woo, J. S. Lee, K. Lim, C.-H. Lee, H. S. Kim, and J. Laskar, A cognitive radio (CR) system employing a dual-stage spectrum sensing technique: A multi-resolution spectrum sensing (MRSS) and a temporal signature detection (TSD) technique, in *Proceedings of the IEEE Globecom*, 2006.
36. A. Sahai and D. Cabric, A tutorial on spectrum sensing: Fundamental limits and practical challenges, in *Proceedings of IEEE Symposium of New Frontiers Dynamic Spectrum Access Networks (DySPAN)*, Baltimore, MD, November 2005.
37. N. Meghanathan, A survey on the communication protocols and security in cognitive radio networks, *International Journal of Communication Networks and Information Security (IJCNIS)*, vol. 5, no. 1, pp. 19–38, 2013.
38. J. Zhao, H. Zheng, and G.-H. Yang, Spectrum sharing through distributed coordination in dynamic spectrum access networks, *Wireless Communications and Mobile Computing*, vol. 7, no. 9, pp. 1061–1075, 2007.
39. S. M. Kamruzzaman, E Kim, D. G. Jeong, and W. S. Jeon, Energy-aware routing protocol for cognitive radio ad hoc networks, *IET Communications*, vol. 6, no. 14, pp. 2159–2168, 2012.
40. H. Khalife, N. Malouch, and S. Fdida, Multihop cognitive radio networks: To route or not to route, *IEEE Network Magazine*, vol. 23, no. 4, pp. 20–25, 2009.
41. A. C. Talay and D. T. Altilar, ROPCORN: Routing protocol for cognitive radio Ad Hoc networks, in *Proceedings of the International Conference on Ultra Modern Telecommunications & Workshops*, pp. 1–6, October 2009.
42. S. M. Kamruzzaman, E. Kim, and D. G. Jeong, Spectrum and energy aware routing protocol for cognitive radio ad hoc networks, in *Proceedings of IEEE International Conference on Communications (ICC)*, pp. 1–5, Kyoto, Japan, 5–9 June 2011.
43. J. Liu and S. Singh, ATCP: TCP for mobile Ad Hoc networks, *IEEE Journal on Selected Areas in Communications*, vol. 19, no. 7, pp. 1300–1315, 2001.
44. K. Sundaresan, V. Anantharaman, H.-Y. Hsieh, and R. Sivakumar, ATP: A reliable transport protocol for ad hoc networks, *IEEE Transactions on Mobile Computing*, vol. 4, no. 6, pp. 588–603, 2005.
45. W. Jianfeng, M. Ghosh, and K. Challapali, Emerging cognitive radio applications: A survey, *IEEE Communications Magazine*, vol. 49, no. 3, pp. 74–81, 2011.
46. M. Patel and J. Wang, Applications, challenges, and prospective in emerging body area networking technologies, *IEEE Wireless Communications*, vol. 17, no. 1, pp. 80–88, 2010.
47. B. Zhen, H. Li, K. Takizawa, K. Yazdandoost, and R. Kohno, Frequency band consideration of SG-MBAN, *IEEE 802.15-MBAN-07-0640-00*, March 2007.

Chapter 10

Never Die Networks*

Norio Shiratori and Yoshitaka Shibata

Contents

10.1 Introduction

Facing the challenges posed by nature such as global warming and natural disasters have become important concerns for twenty-first century science and technology. The Great East Japan Earthquake on March 11, 2011, that left 20,000 people dead or missing brought the strong emotional impact of an earthquake and tsunami to

* Invited Chapter.

the whole world. In addition, recently the Japanese government reported that an M9 Nankai Trough earthquake would trigger a big tsunami that may result in as many as 323,000 deaths and devastate a wide area of the coastline in Japan [1]. Figures 10.1 through 10.3 show the serious damages caused by the Great East Japan Earthquake and Tsunami. Casualties such as death or people washing away, and infrastructural collapse became serious because of the disruption in information communication systems such as landline telephone and broadband network, local area network, cellular wireless network, and so on, as shown in Table 10.1. This led researchers and engineers to start considering the development of an information communication system infrastructure that can withstand disasters. In this chapter, we discuss Never Die Networks (NDN), which are a type of disaster-resilient information communication system related to natural disasters such as big earthquakes and tsunamis. We provide examples based on our painful experiences in the Great East Japan Earthquake [26].

Before going into a detailed discussion of NDN, we briefly introduce an overview of the great earthquakes and tsunamis in Japan as well as worldwide [28]. In Japan and outside Japan, we have had very great earthquakes over the years. Many people either died or went missing as a result of these earthquakes and the corresponding tsunamis. To decrease the number of deaths and missing persons, it is strongly expected that disaster-resistant information communication systems will be

Figure 10.1 A prefectural road recovered by removing tsunami rubble in Miyagi Prefecture.

Figure 10.2 A boat swept by the tsunami about 2 km away from the sea in Miyagi Prefecture.

Figure 10.3 A gas station damaged by the tsunami, located about 1 km away from the coast in Miyagi Prefecture.

Table 10.1 Network Conditions of Various Communication Systems

System	*Condition*	*Details*
Radio	○	Small area service such as local FM radio was useful for evacuators.
TV	×	It did not work because of blackout in a wide area.
Fixed phone	×	Devices were damaged; service blackout.
Cellular phone (audio)	△	Highly congested.
Internet (cellular phone)	△	Highly congested.
Iwate Information Highway (government's information network system)	×	Broken down network and devices.
LAN in City Hall	×	Broken down network and devices.
Disaster Government Radio System	△	Unable to hear inside of house or car.
Amateur radio	○	Worked but only a few devices and licensed users.
Wireless LAN	○	Worked but electricity was needed.
Satellite system (Internet)	○	Worked.

(○ - worked well, △ - worked partially, × - did not work)

developed worldwide. The authors had painful experiences in the Great East Japan Earthquake on March 11, 2011. Based on the lessons learnt through our experiences, we have been promoting research activities toward the realization of NDN.

The contents of this chapter are a modified version of our paper [2, 27].

10.2 History of Disaster-Resilient Information Communication Systems

Before the March 11 disasters, no systematic study on the impact of big natural disasters on information communication systems had been conducted. Rather, the matter of network failure had been addressed either by providing dedicated "robust" infrastructure or with packet-based networks, developing mechanisms to overcome some failure in the network.

The military and police forces historically own closest network, which is used during disasters. However, no equivalent infrastructure exists for the general public for use in such situations.

By contrast, there are mainly two approaches to resolving network failure in packet-based networks. These are (1) the Network Fault Tolerance System [3] and (2) the Resilient Overlay Network [4]. We can then summarize the research on disaster-resilient information communication systems, including our proposed NDN, as follows.

1. *Network Fault Tolerance System* [3]: In the Network Fault Tolerance System, the network failure caused by a physical component failure of the Local Area Network (LAN) or interconnected LAN environment, such as cable, network interface card, switch, and router, is considered. In order to recover from network component failure in this way, redundancy is introduced. For example, more than one pair of network sets including a cable and router is installed in parallel. If one of the network sets fails, the failure is detected by the network management protocol (e.g., Simple Network Management Protocol (SNMP)) and the other network is selected to recover from the situation. Thus, the Network Fault Tolerance System deals with hardware-based recovery by dual network component sets.
2. *Resilient Overlay Network* [4]: By contrast, the Resilient Overlay network is an application-layer overlay on top of the existing Internet routing substrate. The Resilient Overlay Network nodes monitor the functioning and quality of the Internet paths among themselves. This information is used to decide whether to route packets directly over the Internet or through other Resilient Overlay Network nodes, optimizing application-specific routing metrics even if some of the network nodes become nonfunctional.

   As common characteristics of those networks, both networks consider only stable wired configurations, not wireless or ad hoc networks. Neither includes autonomous network reconfiguration ability through multiple layers, multiple links, or multiple channel environments. Furthermore, there is no guarantee of connection between any network nodes. Therefore, those may not always be used as the disaster information network.
3. *Never Die Networks* [5,6]: The author of this chapter has experienced the following two big earthquakes:
   a. June 12, 1978: Miyagi-oki Earthquake
      (M7.4, Maximum Seismic Intensity 5)
   b. March 11, 2011: Great East Japan Earthquake
      (M9.0, Maximum Seismic Intensity 7)

When the Miyagi-oki Earthquake mentioned above in (A) occurred, as Professor Shiratori, the first author of this chapter recalls in first person, "I was in my laboratory at Tohoku University, located at the center of Sendai City. At that time, files, documents, books, and other items flew off shelves and scattered throughout the room. The hard strike by the earthquake caused blackouts and then people could not establish telephone communications. On the roads, traffic lights were broken down and therefore just about everywhere, roads were jammed with people and cars. For a while after the earthquake, I could not confirm the safety of my wife who supposed to stay at home 12 km away from my laboratory at Tohoku University. I realized that if there was a way to hear her voice [for] only a few seconds, then I could know about her safety. Generally, when people are rescued within the first 72 hours after the earthquake, their chances of survival is high: the sooner … the better to increase [survival] rates of victims."

Such a dreadful personal experience led Professor Shiratori to conceive of the need for a "Never Die Network"—a network that remains functional even after big natural disasters and ensures communication for general people. As he recalls, "From the frustrating experience of [the] Miyagi-oki Earthquake, I have keenly realized the necessity to develop [a type of] network named Never Die Networks which would never break down during the time of disaster. Hence as early as June 1978, I was thinking of the idea of Never Die Network. In order [to] attain its objectives, the Never Die Network has to be capable of addressing the following points."

Original Concept of Never Die Networks
1. Real-time services 2. Ability to support massively large number of simultaneous users 3. Service offered for a short duration

The basic concept of NDN [5,6] was first made public through a paper by the authors in 2003 based on the above three features 25 years after the original conception of the idea. The main technical issue to realize the above requirements of NDN is the development of an autonomous network control method (communication method and protocol) and management method to assist the NDN system to continue to work even in case of emergency, such as a disaster. This is achieved by the use of technologies to collect and analyze network information (communication connectivity, traffic congestion, etc.) for expressing the network status in real time. Thus, NDN aspires to provision of uninterrupted communication even in case of an emergency such as a disaster. The authors' group has succeeded in standardizing one such technology in Internet Engineering Task Force (IETF) as a basic technology [7].

10.3 Characteristics of NDN

Based on the original concepts of NDN, we have been conducting research on different aspects of the topic for the last several years [8–11, 30, 31]. We have revised different models and applications of NDN in a step-by-step manner. The concept of NDN can be characterized by the following five points.

1. Even if a part of the system is damaged due to external factors, such as a disaster, NDN will continue to work without halting the whole system by autonomously finding an alternative route or a node and establishing a communication link. Thus, NDN aspires to provision of uninterrupted communication.
2. In order to improve the "never die" characteristic, NDN tries to control divisions and multiplexing of the available bandwidth.
3. Moreover, by introducing a layered structure such as wired and wireless, NDN aspires to provision of uninterrupted communication services even if the degree of system damage spreads.
4. Comparing to the Network Fault Tolerance System and Resilient Overlay network mentioned above, the concept of NDN includes not only hardware but also software aspects, communication methods such as wired or wireless media, satellite, balloon, mobile, and so on and the corresponding protocols. Hence, the concept of NDN includes the concepts of both the Network Fault Tolerance System and Resilient Overlay Network.
5. Performance characteristics such as connectivity, throughput, and so on of NDN are as shown in Figure 10.4.

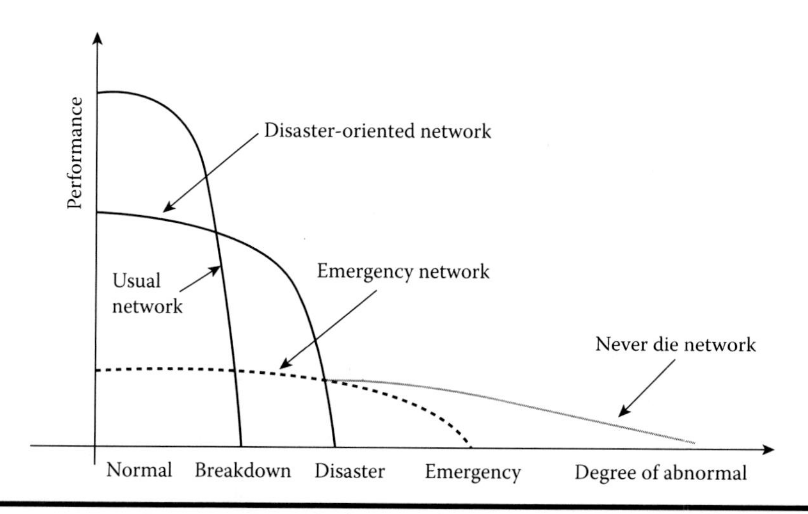

Figure 10.4 Performance characteristics (connectivity, throughput, etc.) of Never Die Networks.

Research Activities in Japan after March 11, 2011

The Great East Japan Earthquake had a great impact not only on the ways of our lives but also future of science and technology. Just after the earthquake, the government, companies, and universities in Japan started various types of research and development for disaster reconstruction. For example, the Japanese government launched the Reconstruction Agency.

Regarding information and communication system, the Information Processing Society Japan solicited and published special journal issues on information and communication technologies for disaster reconstruction [12]. The Institute of Electronics, Information, and Communication Engineers also prepared special issues of their journals on disaster communication [13]. Furthermore, government agencies, companies, and universities started research projects. For example, the National Institute of Information and Communication Technology started research projects on disaster information networks with groups of domestic communication enterprises such as NTT, NTT Docomo, KDDI, NEC, NHK, and so on in collaboration with Tohoku University and supported by the government [14]. Kyushu University initiated research on a national disaster information system particularly for the disaster caused by heavy rain in Kyushu, in collaboration with the Ministry of Land, Infrastructure, Transport, and Tourism [15]. Hokkaido University, in collaboration with the telecom operator SoftBank, started experimentation on balloon-mounted wireless base stations for wireless communication networks for potential serious disaster scenarios when the communication infrastructure on the ground is completely damaged [16].

Regarding NDN, Iwate Prefectural University is developing a new research stream, which is organized as multilayered and multilinked architecture with heterogeneous wireless networks including the small mobile satellite IP networks, millimeter-wave wideband networks, and balloon-mounted wireless ad hoc networks in addition to conventional Wi-Fi and Wi-Max connectivity. All of those network components are operated and controlled by the software-defined network framework to perform the following functions:

1. Self-charging power supply when necessary
2. Self-cognition and diagonal for network quality and failure detection function
3. Dynamic reconstruction from failure to always work even in the worst case where all network components on the ground are completely damaged and destroyed by a huge disaster on a scale larger than that of the Great East Japan Earthquake

Currently, the project team, led by Prof. Yoshitaka Shibata, is constructing a prototype and evaluating the functionality and performance of a new NDN that covers several areas on the Sanriku coast seriously damaged by Great East Japan Earthquake.

10.4 Serious Problems Caused by the Great East Japan Earthquake

10.4.1 Problems That Became Obstacles to Rescue Activities

The Great East Japan Earthquake on March 11, 2011, caused severe and tremendous damage over a wide area of northern Japan. A massive 9.0 earthquake destroyed many buildings and equipment. Moreover, devastating waves of tsunami swept over cities and coastal residential areas. The earthquake and tsunami left 15,868 dead; 2,848 missing; and 6,109 injured [17]. Among the major earthquakes recorded in world history, it was the fourth largest earthquake, next to the Great Chilean earthquake in 1960 (M9.5), the Great Alaskan Earthquake in 1964 (M9.2), and the Indian Ocean earthquake and tsunami in 2004 (M9.1) [10]. However, this tragedy shocked the world.

10.4.2 Problems of Information and Communications Technology Caused by Limited Network Conditions

In addition to the casualties, the earthquake caused many secondary disasters such as blackouts, lack of food, and lack of fuel along a wide area of Japan (Figure 10.5). These problems were compounded by the disconnection of communication networks, which

Figure 10.5 Lack of food in a convenience store in Iwate Prefecture.

had severe impacts. Disconnection of mobile phones and the Internet created great confusion in today's highly developed IT society. It also caused the delay of various rescue and support activities. Shibata et al. reported on the problems from communication network disconnections by the earthquake and their network reactivating efforts on the coastal side of Iwate Prefecture in their papers [18,19].

Their papers reported various problems that emerged during the activities after the disaster. First of all, the breakdown of transportation systems caused a scarcity of fuel and food in the stores. The lack of fuel and food spread in a vast area of Japan. It led to a delay in rescue and support activities. Most food such as rice and bread disappeared from the cities, and many people could not go to the coastal areas for support because of the lack of gasoline. Second, lack of information about topics such as damage and evacuation affected rescue activities. Because the damage was spread over a vast area, it was very hard to obtain local disaster information.

The following are additional major problems they noted during their activities:

1. Fuel for cars was difficult to get.
2. Electricity supply and batteries for information network systems were damaged.
3. Network devices and servers were damaged.
4. Wired networks were disconnected.
5. Cellular phone systems were damaged and congested.
6. The Government Disaster Radio System had broken down.
7. TV could not be watched.
8. In many evacuation shelters, handwritten papers were used to obtain disaster information and to search for missing persons.

The papers also reported damage to various communication systems. Cellular phones and Internet connectivity were dysfunctional in particular because of severe congestion and damage to devices. It had a great effect on various aspects. According to the Ministry of Internal Affairs and Communication, just after the earthquake, the number of telephone calls over cellular phones had increased about 10 times compared with its usual numbers. Furthermore, up to 95% of audio calls (the maximum) were blocked [20].

Table 10.1 summarizes the network conditions in Iwate Prefecture.

The authors pointed out that only the wireless network and satellite system were useful for network recovery activities. It was also identified that network connectivity was very important in the early stages of the disaster, because the early information mainly consisted of life-related information such as rescue and evacuation, and the information is preferred to be text contents such as Web service. Therefore, they mentioned that in the early stages of the disaster, disaster-resilient information communication systems should be focused on the connectivity more than throughput or delay of data transmission. Hence the authors introduced these concepts to extend the design of NDN.

10.5 Required Information for Disaster Response

In designing NDN, Shibata et al. observed from the history of previous large natural disasters [21,22] that the required information changes with time, before and after the disaster. The transition of the required information over time is shown in Table 10.2. This information and knowledge were gained from the experiences of past disasters, especially from the Great East Japan Earthquake on March 11, 2011.

According to Table 10.2, the forecast and evacuation information are required during normal time, t_1. Evacuation information, however, is not so important at this stage. In the event of an indication of disaster such as news or rumor during t_2, evacuation and disaster prevention become more important than usual. When a disaster occurs, very significant activities relating to human lives such as rescue, evacuation, and safety status information become essential. Because the information is strongly related to human lives during the time t_3, NDN should be focused on network connectivity more than throughput or latency. In addition, text contents are mainly used for disaster evacuation, resident safety, and disaster status information. In later phases, delay or throughput become important while the connectivity is less focused after recovery stages.

10.6 Examples of NDN

Based on our painful experiences of past disasters as discussed in Serious Problems Caused by the Great East Japan Earthquake, we give two examples of NDN. One is an example of NDN system behavior, and the other is an example of NDN system construction.

NDN is a robust network method proposed by Shiratori et al. [5,6], and it is aimed at ensuring an effective network control method for a disaster-resistant information communication system. Refs. [5,6] mention that NDN is defined as a robust network that will be unaffected by any change in the environment such as a sudden degradation or fluctuation of the quality of network capability. To realize NDN, they proposed a method of network control by both infrastructure-dependent mode and autonomous mode. Once a disaster occurs, the network switches to autonomous mode. In autonomous mode, the network condition is grasped, and a new administration node is selected for route reconstruction. Shibata et al. extended their proposal and discussed NDN more fastidiously by comparing it with the current wired and wireless networks [9].

For their proposed NDN in the paper [18], data connection is robustly kept as shown in Figure 10.6.

As shown in Figure 10.6, wired networks are easily affected by disaster. The connectivity of a wireless network or cognitive wireless network (CWN) is more robust against disasters than a wired network. However, their connectivity may

Table 10.2 Required Information over Time

Subject	*Necessary Information/Time*	t_1	t_2	t_x	t_3	t_4	t_5	t_6
Victims (inside of the area)	Disaster prevention	△	○					
	Evacuation		○		◎			
	Safety				◎	◎	○	△
	Stricken area				◎	○		
	Traffic				○	◎	◎	
	Relief supplies					◎	○	
	Public service					○	◎	
	Lifeline					○	◎	
	Local government					○	◎	
Relatives volunteers (outside of the area)	Safety				◎	◎	○	
	Stricken area				◎	○	△	
	Relief supplies					○	◎	

	Condition	*Activity*	*Period*
t_1	Normal		
t_2	Indication of disaster	Indication, rumor, etc.	2 weeks before – disaster
t_x	Occurrence of disaster		During disaster
t_3	Just after disaster	Rescue, evacuation, safety information	Disaster – 2 days
t_4	Calm down from disaster	Relief materials, safety information	3 days – 2 weeks
t_5	Restoration from disaster	Restore lifeline, residences, and so on	3 weeks – several months
t_6	Revival		

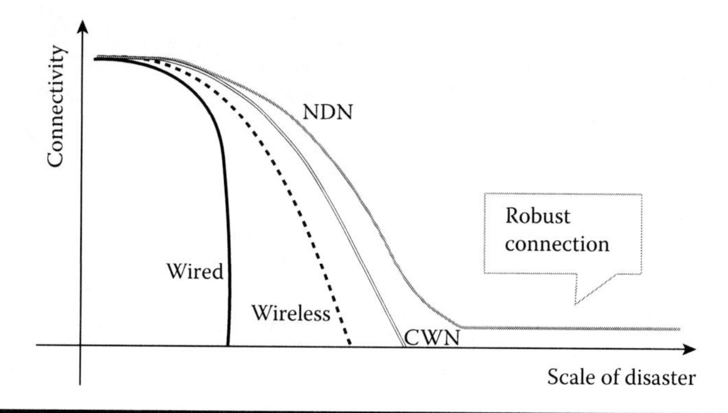

Figure 10.6 Schematic system failure for different types of networks by scale of disaster.

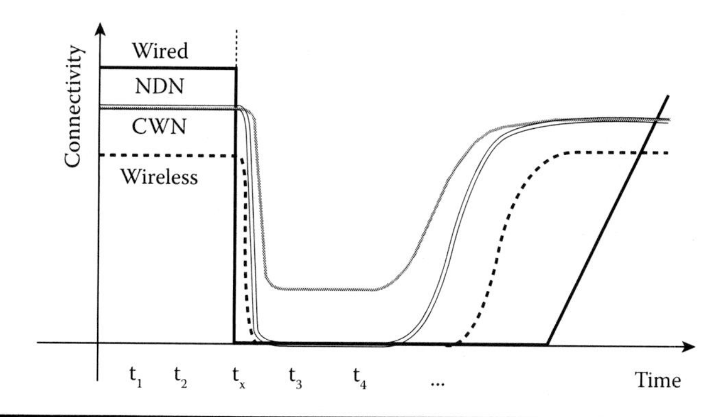

Figure 10.7 Schematic system failure for different types of networks through elapsed time.

be disrupted by strong disasters. On the contrary, NDN is capable of maintaining data connection even if the transmission quality, like bandwidth, decreases as shown in Figure 10.7.

Figure 10.7 shows the data connectivity of various networks through elapsed time. NDN guarantees minimal data connection even in case of a very strong natural disaster. At t_x, when a severe disaster occurs, network conditions rapidly degrade. However, unlike other traditional information communication systems, it is possible to provide minimal data transmission support with NDN.

10.6.1 An Example of System Behavior of NDN

There are some approaches trying to realize NDN today. First of all, Shibata et al. proposed the optimal network control methods for CWN and satellite system [23,24].

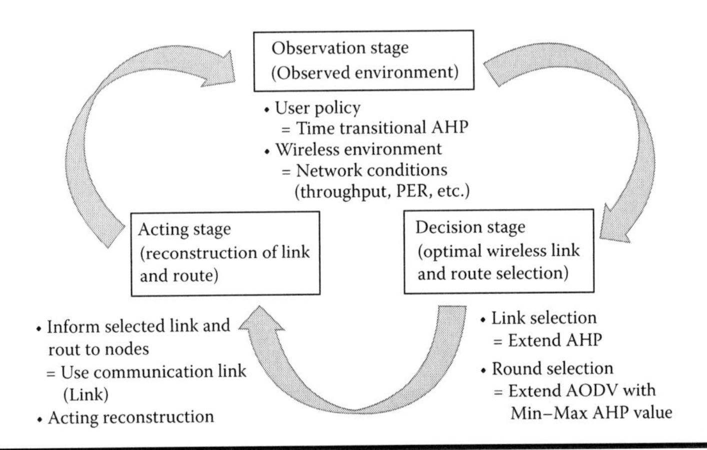

Figure 10.8 Stages in the proposed cognition cycle. AHP, Analytic Hierarchy Process; PER, packet error rate; AODV, Ad Hoc On-Demand Distance Vector.

They proposed the new cognition cycle, which consists of three stages: the observation stage, the decision stage, and the acting stage, as shown in Figure 10.8. Each stage is continuously cycled in order to perform link or route configuration.

At the observation stage, users and wireless environments are observed by the change in required information that is mentioned in Serious Problems Caused by the Great East Japan Earthquake and by the observation values such as throughput or packet error rate. Then, these parameters are used for the calculation of link and route selection at the decision stage. In this stage, they proposed to extend the Analytic Hierarchy Process (AHP) method for link selection in order to consider the user policies. The extended AHP results are deployed to extend Ad Hoc On-Demand Distance Vector routing with Min–Max AHP values. Finally, these computational results are used for network reconstruction.

The authors of Refs. [23,24] evaluated the robustness and quick recovery capabilities of their system through simulations as shown in Figure 10.9. As the results of the simulation show (Figure 10.9), the proposed method realizes the requirements of NDN as shown in Figures 10.6 and 10.7.

10.6.2 An Example of NDN System Construction

In order to realize the never die property, we designed an NDN consisting of a three-layered architecture by using wired and wireless networks [10]. Figure 10.10 shows an overview of the system construction. The first layer is constructed by the Wi-Fi full-mesh public wireless network in accordance with the IEEE 802.11s standard. The second layer is a core network that connects the central areas of the distributed branches. This layer is constructed based on the 4.9 GHz long-distance wireless network. However, when the two branches are close, they will be

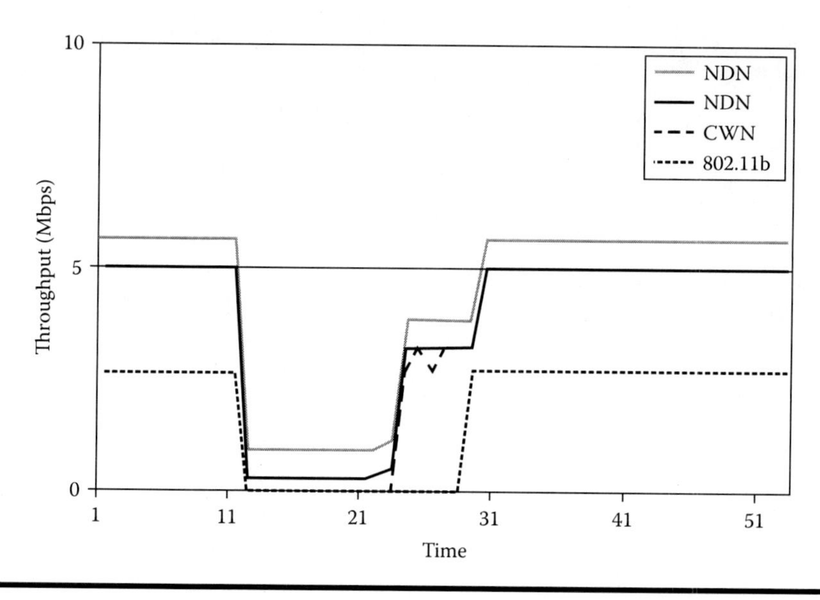

Figure 10.9 Proposed cognition cycle.

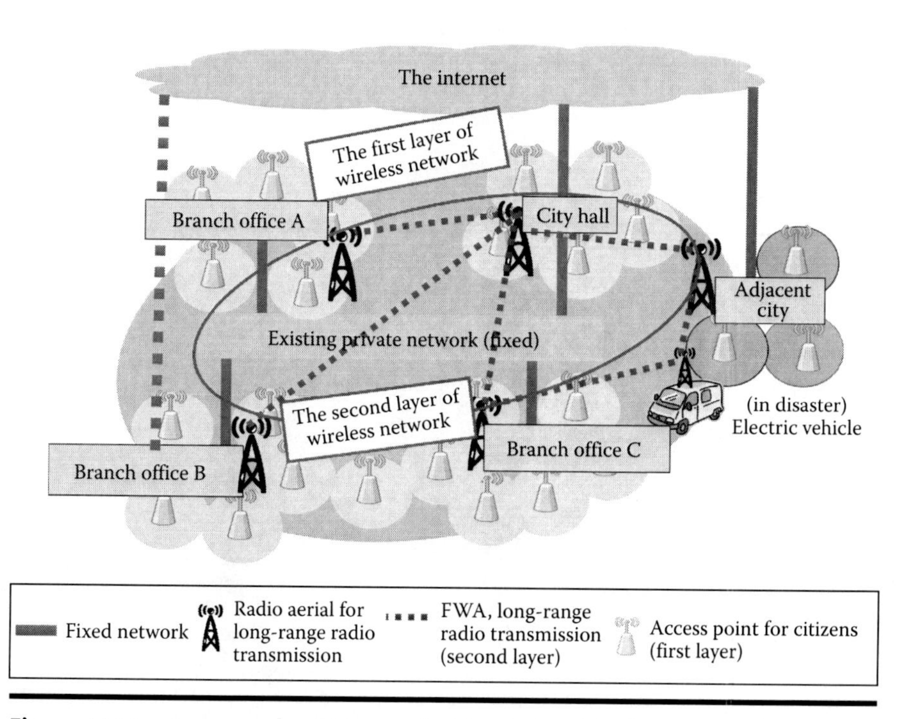

Figure 10.10 An example of Never Die Networks with three-layered hierarchical network structure.

connected by the full-mesh Wi-Fi network. By introducing this architecture, we can keep connectivity to the Internet in the event of a disaster and ensure the never die property of the information communication systems.

Moreover, the Never Die Messaging System, as shown in Figure 10.11, performs bulk delivery of important information to a variety of media to protect the safety of people in the initial phase after the disaster. This system uses public information such as the J-Alert warning system, public information commons, information from the Japan Meteorological Agency, and so on, as input to the system. Then, this system outputs the information to users by using multiple media. After the Great East Japan Earthquake, accurate information could not be delivered to the residents due to the damage to the public disaster-prevention announcement system. Our proposed system aims to deliver serious lifesaving information tenaciously, via several devices that have already been retained by residents.

Next, to realize fault-tolerant file systems at the time of disaster, the single point of failure should be eliminated. Therefore, we need a disaster-resilient file system that cannot be affected totally by damage even if a part of the file system in the file server is lost. We investigate the Never Die File System as shown in Figure 10.12 as an infrastructure of the platform to achieve centralized management of information on safety, delivery of aid, and so on. We install storage in the buildings of several branch offices based on the distributed storage architecture for disaster recovery. If one of the branch offices is affected by the disaster, the rest of the system provides function of the total file system.

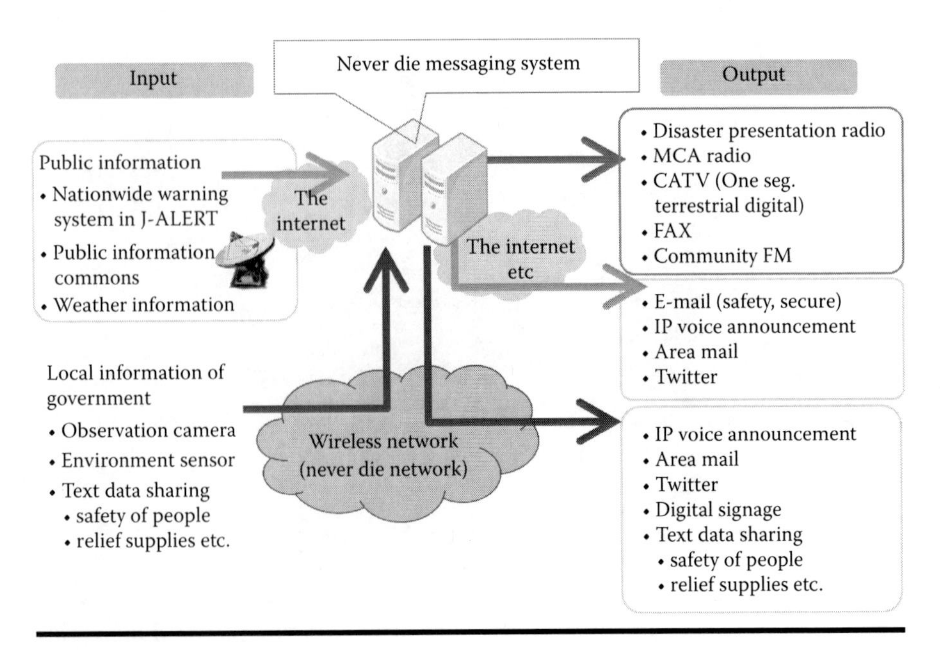

Figure 10.11 An example of the Never Die Messaging System.

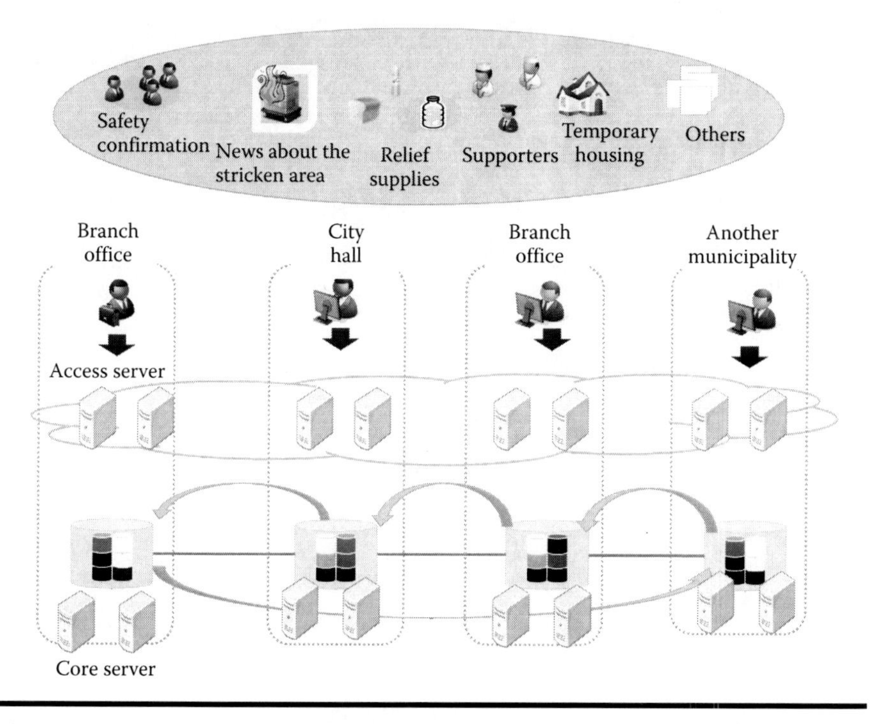

Figure 10.12 An example of the Never Die File System.

10.7 Delay-Tolerant Network: Toward Interplanetary Internet

In 2003, NDN was proposed as a network that is resilient to disaster. In the same year, delay-tolerant network (DTN) was also introduced toward the realization of an interplanetary Internet. Table 10.3 shows a comparison of NDN and DTN.

The basic principle of DTN is to temporarily store messages that cannot be transmitted at the moment due to errors in or unavailability of communication channels. These messages are retransmitted after the recovery of the communication channels. So as per the Japanese saying, it takes a Ieyasu Tokugawa-type approach (Table 10.3): *If the little cuckoo does not sing, I will wait until it does.*

On the contrary, NDN aims to provide uninterrupted real-time services through routing, line allocation controls, and switching and multiplexing of transmission media (both wired and wireless). So as per the Japanese saying, it takes a Hideyoshi Toyotomi-type approach (Table 10.3): *If the little cuckoo does not sing, I will find or create a way so that the bird sings.* Japanese researchers are leading research on application of DTN in disaster situations [25, 29].

It can be envisioned that both NDN and DTN will lead to the creation of a framework for an interplanetary network in the future.

Table 10.3 Comparison of NDN and DTN Approaches

Proposal of NDN in 2003
1. Purpose: Realization of disaster communications 2. Original concept: Active-type protocol 3. Hideyoshi Toyotomi (Japanese shogun) approach: "If the little cuckoo does not sing, I will find or create a way so that the bird sings." 4. Public attention: NDN attracted much attention in the wake of the Great East Japan Earthquake of March 2011.
Proposal of DTN in 2003
1. Purpose: Realization of interplanetary communications 2. Original concept: Passive-type protocol 3. Ieyasu Tokugawa(Japanese shogun) approach: "If the little cuckoo does not sing, I will wait until it does." 4. Public attention: DTN gained attention when the National Aeronautics and Space Administration successfully completed communication experiments with the EPOXI satellite, in November 2008. 5. Others: Toward application of DTN to disaster communication, its research is promoted and led by Japanese researchers [25].

NDN, Never Die Networks; DTN, Delay-tolerant networks.

10.8 Conclusion

In this chapter, we have discussed the concept of NDN, which is one sort of disaster-resilient information communication system. The Great East Japan Earthquake brought strong impacts of natural disasters all over the world, and many problems still remain to be resolved in North and East Japan. The information and communication system has been identified as one of the significant enablers that can reduce the damage caused by natural disasters. According to the authors' painful experiences of the earthquake, uninterrupted communication was very important no matter how bad the throughput or delay in the network. Hence, achieving disaster-resistant information communication systems such as NDN has become one of the major challenges for researchers and engineers.

We expect to promote research on fundamental technologies toward realization of NDN to reduce the damage caused by disasters such as major earthquakes and tsunamis in the coming years.

Acknowledgments

We have had generous support from the staff of the Student Support Room at Iwate Prefectural University, Dr. Akihiro Yuze at Shizuoka Prefectural University, Dr. Kaoru Sugita at the Fukuoka Institute of Technology, Mr. Yuji Ohashi at GFJ, Mr. Toshihiro Tamura at NetBridge, Saitama Institute of Technology, Japan Science and Technology Agency, KDDI, and Coretec for our disaster activities. The authors are grateful to Dr. Tsutomu Inaba, NTT East Miyagi, for his numerous discussions on the design of system construction of NDN. This research is partially supported by JSPS Kakenhi-Kiban (A)-26240012 (2014–2016).

References

1. The Asahi Shinbun, http://ajw.asahi.com/article/0311disaster/analysis/AJ201208300060.
2. N. Shiratori, N. Uchida, Y. Shibata, and S. Izumi, Never Die Network towards Disaster-Resistant Information Communication Systems, *ASEAN Engineering Journal Part D*, Vol. 1, No. 2, pp. 1–22, 2013.
3. J. Sullivan, Network Fault Tolerance System, A Thesis Submitted to the Faculty of the Worcester Polytechnic Institute in Partial Fulfillment of the Requirements for the Degree of Master of Science in Electrical and Computer Engineering, May 2000.
4. D. Andersen, H. Balakrishnan, F. Kaashoek, and R. Morris, Resilient Overlay Network, *SOSP '01 Proceedings of the Eighteenth ACM Symposium on Operating Systems Principles*, pp. 131–145, Dec 2001.
5. T. Suganuma, G. Kitagata, T. Katoh, and N. Shiratori, Configuration of Never Die Network Service Function for Wireless Network, *Proceedings of the IEICE General Conference, The Institute of Electronics, Information and Communication Engineers 2003_Communication* (2), 376, Mar 2003.
6. N. Shiratori, Never Die Network: Towards Network Operating under Worsening Environment. *Database of Grants-in-Aid for Scientific Research.* http://kaken.nii.ac.jp/d/p/19650007.en.html.
7. G. Keeni, RFC4498: The Managed Object Aggregation MIB, 2006.
8. W. Kuji, K. Koide, and N. Shiratori, *Development of Network Management Technology towards Never Die Network*, Information Processing Society of Japan, Tohoku-branch, 07-5-A-23, 2008.
9. W. Kuji, G. Satou, K. Koide, Y. Shibata, and N. Shiratori, Never-Die Network and Disaster-Control System, *IPSJ SIG DPS*, Vol. 2008, No. 54, pp. 131–135, June 2008.
10. T. Inaba, T. Ogasawara, N. Kita, N. Nakamura, T. Suganuma, and N. Shiratori, Greenoriented Never Die Network Management: The Concept and Design, in *Proceedings of 2012 International Conference on Systems and Informatics* (*ICSAI 2012*), pp. 529–535, May 2012.
11. N. Shiratori, T. Inaba, N. Nakamura, and T. Suganuma, Disaster-Resistant Greenoriented Never Die Network, *IPSJ Journal*, Vol. 53, No. 7, pp. 1821–1831, 2012.

12. N. Shiratori, and I. Noda, Special Issue on ICT which Encourages Society, *IPSJ Journal*, Vol. 53, No.7, pp. 1663–1664, 2012.
13. IEEE Transactions on Communications, Special Issue. http://www.ieice.org/cs/jpn/cs-edit/CFP/cfp_JB_2013.6.pdf.
14. National Institute of Information and Communications Technology. http://www.nict.go.jp/en/index.html.
15. http://www.kyushu-u.ac.jp/pressrelease/2012/2012_01_23.pdf.
16. Softbank Group. http://www.softbankmobile.co.jp/ja/news/press/2012/20120510_01/.
17. Japan Police Department, The Great East Japan Disaster, http://www.npa.go.jp/archive/keibi/biki/index.htm (Aug 29, 2012).
18. N. Uchida, K. Takahata, and Y. Shibata, Disaster Information System from Communication Traffic Analysis and Connectivity (Quick Report from Japan Earthquake and Tsunami on March 11th, 2011), in *The 14th International Conference on Network-Based Information Systems* (*NBIS2011*), pp. 279–285, Sep 2011.
19. Y. Shibata, N. Uchida, and Y. Ohashi, Problem Analysis and Solutions of Information Network Systems on East Japan Great Earthquake, in *Fourth International Workshop on Disaster and Emergency Information Network Systems* (*IWDENS2012*), pp. 1054–1059, Mar 2012.
20. Ministry of Internal Affairs and Communication, About How to Secure Communication Systems on Emergent Affair such as A Large Scale Disaster, http://www.soumu.go.jp/main_sosiki/kenkyu/saigai/index.html.
21. T. Watanabe, T. Oishi, K. Watanabe, H. Oishi, T. Hashimoto, S. Oishi, S. Watanabe, and H. Sanbonmatsu, Research and Development of Disaster People/Local Government Support Information System, in *The Second Convention of Japan Society for Disaster Information Studies*, pp. 163–172, 2000.
22. D. Nakamura, N. Uchida, H. Asahi, K. Takahata, K. Hashimoto, and Y. Shibata, Wide Area Disaster Information Network and Its Resource Management System, in *AINA, 03*, Mar 2003.
23. N. Uchida, K. Takahata, Y. Shibata, and N. Shiratori, Proposal of Never Die Network with the Combination of Cognitive Wireless Network and Satellite System, in *13th International Conference on Network-Based Information Systems* (*NBIS2010*), pp. 365–370, Sep 2010.
24. N. Uchida, K. Takahata, Y. Shibata, and N. Shiratori, A Large Scale Robust Disaster Information System based on Never Die Network, in *The 26th IEEE International Conference on Advanced Information Networking and Applications* (*AINA2012*), pp. 89–96, Mar 2012.
25. Masato Tsuru, Masato Uchida, Tetsuya Takine, Akira Nagata, Takahiro Matsuda, Hiroyoshi Miwa, and Shinya Yamamura, Delay Tolerant Networking Technology—The Latest Trends and Prospects, *IEICE Communications Society Magazine*, Vol. 2011, No. 16, pp. 16.57–16.68, 2011.
26. Guest Editorial, Lessons of the Great East Japan Earthquake, *IEEE Communications Magazine*, March 2014, pp. 21–22.
27. Y. Shibata, N. Uchida, and N. Shiratori, Analysis and Proposal of Disaster Information Network from Experience of the Great East Japan Earthquake, *IEEE Communications Magazine*, March 2014, pp. 44–48.
28. Japan Meteorological Agency, Past Reports of Earthquake and Tsunami, http://www.seisvol.kishou.go.jp/eq/higai/higai-1995.html.

29. K. Fall, A Delay-Tolerant Network Architecture for Challenged Internets, in *Proceedings of the 2003 Conference on Applications, Technologies, Architectures, and Protocols for Computer Communications* (*ACM SIGCOMM2003*), pp. 27–34, 2003.
30. Japan Society for the Promotion of Sciences (JSPS) "Kakenhi", Grants-in-Aid for Scientific Research on Challenging Exploratory Research, in *Never Die Networks: Towards Networks Resilient to Worsened Network Environment*, 2007–2009.
31. Japan Society for the Promotion of Sciences (JSPS) "Kakenhi", Grants-in-Aid for Scientific Research(A), in *Green-Oriented Never Die Networks which Adapt to Disaster Situation and Maximizes Satisfaction Levels of Connectivity by Multiusers*, 2014–2016.

Index